EUROPE'S GREEN RING

Europe's Green Ring

Edited by
LEO GRANBERG, IMRE KOVÁCH AND HILARY TOVEY

Ashgate
Aldershot • Burlington USA • Singapore • Sydney

Published by
Ashgate Publishing Limited
Gower House
Croft Road
Aldershot
Hampshire GU11 3HR
England

Ashgate Publishing Company
131 Main Street
Burlington, VT 05401-5600 USA

Ashgate website: http://www.ashgate.com

British Library Cataloguing in Publication Data
Europe's green ring. - (Perspectives on rural policy and planning)
1.Peasantry - Europe - History 2.Agriculture - Europe - History 3.Europe - Rural conditions
I.Granberg, Leo II.Kovách, Imre III.Tovey, Hilary
338.1'094

Library of Congress Control Number: 2001090081

ISBN 0 7546 1754 8

Printed in Great Britain by
Antony Rowe Ltd, Chippenham, Wiltshire.

Contents

List of Figures and Maps

List of Tables

List of Contributors

Alanen, Ilkka
Professor, Department of Social Sciences and Philosophy, University of Jyväskylä, Finland.

Begg, Bob
Professor, Department of Geography and Regional Planning, Indiana University of Pennsylvania, USA.

Bruckmeier, Karl
Associate Professor, Department for Interdisciplinary Studies of the Human Condition, Human Ecology Section, University of Göteborg, Sweden.

Camarero, Luis A.
Professor, Universidad Nacional de Educación a Distancia, Spain.

Fonte, Maria
Associate professor of agricultural economics, Dipartimento di Teoria e Storia dell' Economia Publica, Università di Napoli "Federico II", Italy.

Gorlach, Krzysztof
Associate Professor of Sociology, Institute of Sociology, Jagiellonian University, Poland.

Granberg, Leo
Docent, Aleksanteri Institute, University of Helsinki, Finland.

Jonsson, Ivar
Professor, Social Implication of Technical Change, University of Sussex/ University of Bifrost, Iceland.

Kasimis, Charalambos
Professor, Department of Economics, University of Patras, Greece.

Kopytina, Marina Olegowna
Senior Researcher, Russian Academy of Sciences, Institute for International Economic and Political Studies, Russia.

Kovách, Imre
Head of Department, holder of István Széchenyi professorship,
Institute for Political Sciences, Hungarian Academy of Sciences, Hungary.

Majerová, Věra
Head of Department of Humanities/Rural Sociology Centre, Faculty of Economics and Management Czech University of Agriculture Prague, the Czech Republic.

Meurs, Mieke
Associate Professor, Department of Economics, American University, Washington, DC, USA.

Moreira, Manuel Belo
Professor, Departamento de Economia Agrária e Sociologia Rural – ISA/UTL, Portugal.

Oliva, Jesús
Professor, Department of Sociology, Universidad Pública de Navarra, Spain.

Papadopoulos, Apostolos G.
Lecturer, Department of Farm Organisation and Management, University of Ioannina, Greece.

Peltonen, Matti
Professor, Department of Social Science History/Economic and Social History, University of Helsinki.

Rodrigo, Isabel
Professor, Departamento de Economia Agrária e Sociologia Rural – ISA/UTL, Portugal.

Starosta, Pawel
Associate Professor of Sociology, Institute of Sociology, University of Łódź, Poland.

Tovey, Hilary
Senior Lecturer, Department of Sociology, Trinity College Dublin, Ireland.

Varis, Eira
Director for International Relations, City of Joensuu, Finland.

Introduction

LEO GRANBERG, IMRE KOVÁCH AND HILARY TOVEY

This anthology looks at peasantisation and de-peasantisation – the making, and particularly the unmaking, of peasant agriculture – in fourteen European countries: Finland (and the Scandinavian region), Iceland, Ireland, Portugal, Spain, Italy, Greece, Hungary, the Czech Republic, Bulgaria, the Republic of Karelia in Russia, Poland, East Germany, and Estonia. These countries constitute what we are here calling a 'Green Ring' around Europe: they have followed pathways into modernity which are different from those of the early industrialising European core, and have been characterised, until recent decades, by a strong agrarian presence in their politics, economy and culture. We use this concept of a 'Green Ring' as a way of emphasising the similarities, and also addressing the divergences, between the experiences and historical roles of the agrarian populations of these countries – on one side, those subject to European integration and agricultural modernisation processes and, on the other side, to post-communist changes. Do both lead to the same outcomes for peasant populations; and if so, how are these outcomes best understood?

The main concept used to explore trends in the rural parts of these societies is de-peasantisation. In its broadest sense, de-peasantisation refers to the decreasing economic, social, cultural and political importance of farmers and food production workers in European society in general, and specifically within the 'Green Ring'. The book chapters discuss different ways of theorising de-peasantisation, recreate the history of de-peasantisation in different European societies, and try to trace the outlines of the new societies which are coming into being and the place which 'the rural' and rural inhabitants may have within them in future. De-peasantisation is explored in the book in two ways: as a very long-drawn-out trend, covering the late 19th and the 20th centuries, and as a much more recent phase in rural development which is connected particularly to intensifying trends towards global and European integration.

Enormous changes have occurred in European agriculture in the second half of the 20th century. Within the EU, agriculture has gone from employing 23 percent of the workforce in the 1960s, to less than 10 percent in the 1990s (in some countries it is down to less than 2 percent), and similar dramatic trends have occurred in the 1990s alone in Eastern and Central Europe. Both core industrial regions of Europe, and regions where industrial modernisation occurred later in time, have experienced extensive rural transformation; but in the regions on the boundaries of Europe, rural transformation has been a particularly dominant element in the overall experience of social change, reflecting the dominant role which agriculture and the peasantry earlier played in their economies, societies and national cultures.

Many aspects of the transformation of agriculture have been well researched and documented elsewhere. We argue, however, that adequate attention has not been given to the long term, historical aspect of social transformation, nor to the question of the comparability of processes taking place across geographically and politically distinct areas of Europe. Across rural Europe some trends – such as population decline, counter-urbanisation, and social and especially community restructuring – seem almost universal. These shifts tend to have been addressed by social science researchers primarily as quantitative modifications: decrease or expansion of structural trends. Discussions of the qualitative impacts of the changes are less common. In this book, we want to assess whether European peasantries no longer exist, and if so, what qualitative changes – social, cultural and political – follow for European society. Franklin (1969), Mendras (1970) and others proclaimed the death of the European peasantry over a third of a century ago but their work reflected the situation of the peasantry in core regions of Europe. It is not so long ago that it still seemed meaningful to speak of rural groups as peasants in other European regions from Alentejo to Lapland, and from Connemara to the Eastern steppes. Participants in the European Conference for Rural Sociology in 1986 visited a weekly free market in Barcelos (in Minho, Northern Portugal), where hundreds of retailers and consumers wore traditional costume, and arrived and departed by horse-drawn carts. Nowadays modern dress has replaced traditional clothes, and lorries the teams of horses.

This volume is the outcome of a workshop in Veszprém, Hungary, where a group of researchers from various countries met in 1998 to discuss de-peasantisation in the late industrialising regions of Europe. An earlier conference in Erdőtarcsa, Hungary (see Granberg and Kovách eds. 1998) had discussed the social map of the late modern European countryside. The Veszprém workshop went on to catalogue empirical evidences of de-peasantisation in 'Green Ring' regions in the context of increasing pressures for European integration, and of the recent fall of communist regimes in Central and Eastern Europe. We feel it is very important to theorise the transformations of Nordic, Mediterranean, Celtic, Central and Eastern European rural regions in ways which locate these as comparable phenomena, and to address in an integrated way the experiences of rural and agricultural groups around the European periphery. This gives an opportunity to identify similarities, as well as contrasts, not just in the processes and trends of change, but also in the ways in which these countries have tried to 'manage' their rural areas in the face of the challenges set by processes of industrialisation and modernisation, political integration, and economic and cultural globalisation. A key goal of the book is to problematise standard models of 'rural development' which have dominated western thinking for most of the past half century. By covering both existing EU member states and post-socialist aspirants to membership, we wish to open up discussion of the problems which implementation of this standard 'modernisation' model may cause for societies around the core of Europe.

Understanding 'Peasant'

The book, then, is about de-peasantisation in the European rim, but how should we understand the term 'de-peasantisation'? Much of the rest of this introduction addresses that issue, and attempts in the process to generate a theoretically as well as empirically rich perspective on European social change. However we start by considering what the concept of 'peasant' itself might mean – what is it that we are suggesting has been or is in process of being lost from European countrysides and cultures? Is there a shared understanding of 'peasant' across even the countries we include in the 'green ring'? Has there ever been a coherent European peasantry, or are we in fact trying to compare social groups and characteristics which are very divergent?

One way of trying to answer this question is to look at how terms such as 'peasant' are actually used in the everyday language of the different societies studied here. Gorlach and Starosta report that in Polish the concept of peasant traditionally refers to a social category characterised by a specific way of life based on involvement in small-scale agriculture. Peasants nowadays are mainly thought of as the opposite of urban dwellers, but in the past, particularly under feudal relations, they were thought of as opposed to lords (gentry, aristocracy, large land owners): the peasant was rural and not an urban inhabitant but, first of all, he was not a lord, a member of the gentry. In other words 'peasant' was a kind of synonym for somebody who occupied a low position in the social hierarchy.

Thus the term had some negative emotional connotations. The Polish word for peasant (*chlop*) first appeared in the fifteenth century, a period when the position of peasants significantly worsened because of the emergence of the system of serfdom. It came from old Russian in which the word *cholop* was used simply as 'slave'. By the seventeenth century, when serfdom was entrenched, *chlop* had come to mean all people in rural areas who were subordinated by and forced to work for landowners. In the ideology of the gentry, 'peasants' were a different type of people to themselves: they were created by God to work on the fields and to follow the gentry's orders. This etymological history has had a great impact on the concept of 'peasant' even in twentieth century Poland. The sociologist Jozef Chalasinski argues that in modern Poland, peasants were still treated by the intelligentsia (because of its post-gentry cultural heritage and attitudes) as a type of second class citizen. In communist Poland, during rapid and massive processes of industrialisation and urbanisation, negative attitudes towards peasants as people who live in villages and work in their small farms were evident. The peasants themselves preferred to call themselves 'farmers' or 'agrarian producers', even if they only possessed small and quite unmodernised farms. Similarly, rural politicians often preferred to call their parties or unions 'people' (*ludowy*) or 'farmers' (*rolniczy*) – only a small minority claimed to represent 'peasants' (*chlopski*) in order to stress a particular political tradition and cultural roots. In Hungary also 'peasant' has negative connotations; 'peasant', 'villager' and 'countrysider' are often used as overlapping categories both in political or scientific discourses and in everyday language.

The Polish and Hungarian (and we might include here also the Irish) attitudes towards 'peasants' are very different from those of the Nordic countries, where a tradition of a 'free peasantry' has been highly valued. In Sweden a peasant army was one major element in the making of the Swedish Empire from the 16th Century. In Finland peasants (*talonpoika* =

'son of the house') are associated with the history of national independence and are appreciated as guarantors of national freedom, not least during the period of the Second World War. They have also been seen as builders of the country during periods of peace. Given that there are such differences in the cultural understandings of the peasantry around the European rim, is this a concept which is useful at all for comparative social analysis?

Western rural sociology, in the late 1970s and 1980s, became for a time preoccupied with discussions of the meaning, social position, and historical fate of the peasantry and with attempts to provide a definitive meaning for the concept of 'peasant'. Two contrasting ways of using the concept fought for supremacy. One, associated particularly with Marxist theorising, insisted that its use be limited to analyses of the feudal mode of production. In this view, 'peasant' represents a specific set of relations between the direct cultivators of the land and other groups – to do with land ownership and land use rights, and with the appropriation of surplus from land use – which is only found in feudal social systems. Feudal peasants are not pure subsistence producers, since they always have to produce some surplus for appropriation, which may be greater or smaller depending on economic and political circumstances. Outside feudalism, what are loosely referred to as peasants are better called family farmers, or simply farmers. For those putting forward this perspective, 'de-peasantisation' essentially meant de-feudalisation: the replacement of a feudal by a capitalist mode of production. In the transition from feudalism to capitalism, peasants, it was assumed, necessarily become small-scale capitalists.

The other perspective on 'peasant' emphasised its distinctiveness as a form of production, and as the basis for a distinctive (sub)society or set of social arrangements. This derived largely from Chayanov (1966), whose analysis of processes of production and reproduction within individual peasant farms clearly assumed the presence within peasant localities of societal institutions (communes, co-operatives) which legitimated and policed the cultural orientations shaping the peasant form of production. Redfield's view of peasants as part of a 'peasant society' rather than just as a relation within a larger mode of production was also influential. In this model, peasants live within a set of (communal) institutions and collective practices which add up less to a class relationship than to a distinctive social and cultural world. Both however took for granted that peasant societies are always only 'part societies': they do not possess their own political and state forms of organisation, they are dominated by a state which is always owned or controlled by other groups in the total society. This could have led to emphasising the relatedness of peasants to other social groups, but in practice what was often stressed was rather the capacity of peasants to insulate themselves from such relations, or to protect themselves against their transforming effects. Thus they were represented as capable of surviving under almost any set of larger economic and political circumstances – feudalism, capitalism, even socialism – because their most significant characteristic is, as Chayanov suggested, in a sense an internal one: their focus on social reproduction rather than economic accumulation.

Chayanov argued that all peasant practices, whether in relation to markets, to co-operatives or to communal groups, are explicable only as part of a strategy for reproduction. He believed that trying to explain peasant practices as the outcome of a search for profit makes them unintelligible. This tended to blur the distinction, so important to the more orthodox Marxist position, between 'peasant' and 'family farmer'. Differences in land tenure relations between feudal and capitalist systems, for example, began to look far less significant as a way of defining the boundaries of 'peasantness' than changes in orientation towards agricultural production and the use of the farm holding. From a Chayanovian perspective, 'de-peasantisation', came to be mean the destruction of family-oriented farming. That could not simply be inferred from changes in agrarian structure; it involved changes in the subjective attitudes or cultural orientations of farmers.

Attempts to conceptualise peasants on these lines engendered many and complex theoretical debates. One important debate was about how far peasants or family farmers can accommodate to different sorts of commoditisation pressures without ceasing to be peasant in orientation to their farm. Friedmann (1986) for example, suggested that peasant farming can survive the commoditisation of all the relations in which it is involved except that of the farm labour process itself, because as soon as hired labour is used on the farm, it ceases to be a family enterprise concerned only or primarily with farm-family reproduction. Over time this led to the more general question: at what point, or under what conditions, do peasant farmers living in a highly commoditised society (that is, under capitalism) begin to adopt an economically rational and instrumental orientation to their farming which displaces their definitive concern with social reproduction, and signals the end of the peasantry? A second debate focussed on the status of this supposed orientation to social reproduction (or use value) as the defining characteristic of a peasantry. Should this be seen as some sort of invariant 'structural imperative' of the peasant form of farming? Or should it be seen rather as a cultural norm, therefore variable, a creation of specific social conditions and historical periods? These debates raised issues which are still of significance for any attempt to understand contemporary 'de-peasantisation' processes.

The interest in peasants and peasant production which developed in the 1970s and 1980s ebbed away in the 1990s. This was only partly because, as became clear, the differences between participants in the debate were ultimately irreconcilable. It was partly also a response to changing fashions within rural sociology and sociology in general – which were not trivial but, with hindsight, can be seen as part of the wider shift from modern to 'postmodern' concerns. The 'family' side of family farming – gender and generational relations and conflicts, and shifts in these over time as farming became for example more 'feminised' in some localities, more 'masculinised' in others and so on – began to attract more attention than the 'farm' side – the organisation and meaning of agricultural production. Rural sociology in general began to follow the lead of western society, and withdraw from an interest in the countryside as a location of production, in favour of seeing it as a site for consumption and a site of environmental hazards. It is our belief that this shift away from studying the long-term fate of agrarian – in the sense of food-producing or natural-resource-using – groups in Europe is ready to be reversed. There are already signs of a revival of interest in agrarian studies, largely driven by the experiences of Eastern European countries in the post-socialist era and their search for ways of conceptualising the rather specific agrarian formations and politics which they currently exhibit. But if 'the agrarian question' (Kautsky 1899) is being reborn in Eastern and Central Europe, answers to it must be sought in a wider social, political and geographical framework.

Nordic, Central European, and Mediterranean scholars have argued that the debates of the 1970s and 1980s around the meaning and the historical fate of the peasantry are too narrow to address many of their own concerns. They point out, for example, that the Marxist and Chayanovian theoretical traditions were developed on the basis of two extremes of European rural social development: core, early industrialised countries, and the situation of late 19th century Russia, often considered a unique development case. They argue for a broadening of the notion of 'peasant', and consequently of 'de-peasantisation', in a number of ways.

The Nordic understanding of 'peasantry' refers historically to early modern or even later phases of society in which a 'free peasantry' struggled and collaborated with other social groups in contributing to the making of national identity and political democracy. Serfdom had been in use in Denmark but for external and internal reasons a radical shift took place to what has been called a 'model country of peasantry'. From the end of the 18th century peasantisation became a state-led process, and radical land reforms became instruments to solve economic and social problems. The first reforms were implemented long before any pressures from urbanisation or the modernisation of agrarian production. The free peasant thereby became the ideal type of the ordinary person, even if the reality was far more complicated.

In Ireland a semi-feudal agrarian structure of large landlords and varied types of tenant producer developed around the commercial export of commodities (primarily beef and butter) to global markets, as early as the 17th century (Tovey, in this volume), which gave way only in the late 19th century to 'peasant proprietorship' as the agricultural norm. In Mediterranean areas such as southern Italy and southern France, and in many regions of the Iberian peninsula, the spatial relations under which the peasantry existed were very different from those in Northern core areas. In some cases agricultural producers (peasants or farmers) lived in rural towns or agro-towns (for example *chiantishire* in Central Italy), where agriculture was the major economic sector before mass tourism began to displace it, while some other areas have been fashioned historically as urbanised countryside (Fonte, in this volume). In the Great Hungarian Plain there were also hundreds of thousands of city dwellers who were landowners and agricultural producers and were called 'peasants' – *paraszt* or a more special term for 'city dweller peasant', *civis* (from Latin *civitas*). Such examples remind us to avoid conceptualising 'peasants' in narrowly ethnocentric ways.

First, we suggest that peasants should be understood as small-scale producers (always remembering, as Friedmann (1978) argued, that smallness of scale is relative to the level of development of the technology used to work the holding in any given place or time) who, without necessarily being owners of the means of production, exercise a substantial degree of control over decisions as to their allocation and use.

Second, 'peasants' or 'family farmers' do not exist on their own, but always in significant, even definitive, relations with other groups in society. We may not wish to tie the term 'peasant' to specific relations in feudalism – peasants are more characteristically found in societies which are going through the stages of early modernity – but the argument that 'peasant' is a relational term remains. Among other things it implies that when we investigate 'de-peasantisation' we are interested as much in how other groups have changed over time, as in changes among agricultural producers themselves – in particular, changes in the power relations or political structures within society.

Third, peasants are not passive adaptors nor an unreflexive mass. They calculate and pursue their own interests, and contribute to the construction of society in many ways in doing so. Peasants have been participants in revolutions and in fascist movements, agents of resistance in collectivisation campaigns, and also peaceful state makers, educators and implementers of Enlightenment ideas. De-peasantisation means also the loss of capacity of this kind of social group to have an impact on social change.

Fourth, alongside the relations of farmers with other social groups, we also need to study their own practices, and to treat these as not just economic (instrumentally rational) but also cultural (meaningful). Peasant farming is a culturally meaningful activity which requires for its reproduction a specific social and cultural world. We do not put forward an evolutionary modernisationist approach, which identifies 'peasant' with 'backward' and treats 'de-peasantisation' as a process of change from 'traditional culture' to 'modern rationality' in regard to farming practices. But we do argue that evidence of cultural and civilisational transformation, and of the transformation of 'community', need to be included in the analysis of de-peasantisation.

Fifth and last, definitions of 'peasant' should include the idea that the farm, and its productive use, are central to the self-conceptions and identity of the farm family and its individual members. Peasants may obtain incomes from other sources as well, in fact historically this has probably always been the norm, but they remain peasants as long as their off-farm work is seen as contributing to the survival of the functioning farm. Once the farm becomes just a place of residence, the site of reproduction of a labour force which gets its primary livelihood elsewhere, then it is questionable whether this is still usefully thought of as peasant farming (rather than hobby farming, for example). The term 'part-time farming' is thus often extremely ambiguous, and needs to be analytically unpacked with great care.

The Making and Unmaking of Peasants

Our focus in this volume, as already noted, is primarily on the unmaking of peasants, or de-peasantisation. But we have argued for an approach which takes a long historical perspective, and this should include some brief account of how we understand also the making of the European peasantry. As noted earlier, we do not wish to treat European peasants as some sort of 'traditional' social group, if this is taken to imply that they existed in some sense 'outside history', or as some sort of 'natural economy' whose presence requires no social explanation. In our view, while historical timings and circumstances vary considerably between the different countries of Europe, they all share in common the experience, at some point, of a socio-political project to constitute the peasantry as a significant social stratum within their society. What marks out the 'green ring' countries in particular is how recently this occurred, and how closely it was bound up with broader projects for social and political governance and for economic modernisation.

A number of green ring countries experienced a successful political project to develop a peasantry, and to let it develop into a class with important economic and cultural influences. During the rise of capitalist industrialisation, peasants often grew in importance compared to the nobility on one hand and the landless population on the other. In many countries they succeeded in acquiring some political power, and in some cases even dominance, for a period, in political and cultural life (see Jonsson, Granberg and Peltonen, Tovey, in this volume). Nordic countries experienced a radical turn from a part-feudal, part-subsistence economy in agriculture (combined with forestry and fishing) to a liberated peasantry which became a strong participant in state-making processes from the establishment of political democracy to the birth of modern welfare states. A key peasantisation process in this context was the early land enclosure movement in both Denmark and Sweden which started at the end of the 18th century as a state-led process. During the first half of the 20th century, further land reforms indicated the political victory of peasant movements. Nordic welfare states later became the outcome of political action by a wide variety of coalitions between peasants' and workers' political representatives. The cultural components of Nordic peasant transformation are also important: the prestige of the peasantry and of rurality has been so high in Nordic countries, and elements of peasant culture from decorative arts to leisure activities so organically integrated into national and everyday culture, that the peasantry has been said to be the basis, or at least one significant cornerstone, of modern Nordic nationhoods. Similar points can be made about other cases discussed here.

If the creation of peasants in given European societies was part of a broader socio-political project, then their de-constitution should be seen in a similar light. De-peasantisation is not an outcome of 'natural' or 'inevitable' modernisation trends, any more than 'peasantisation' was. The variety of possible meanings for 'peasant' discussed above suggest that de-peasantisation needs to be addressed on a number of different levels. In this introduction, we try to include in our analysis at least three different levels:

- 'cultural' de-peasantisation as transformation of the meanings given to farming practices by agricultural producers and their families (often included under the concepts of 'modernisation' or 'farmerisation'),
- 'structural' de-peasantisation as the differentiation and disappearance of a social class and its interest organisations, and
- 'societal' de-peasantisation as the broad consequences of these two processes in the entire society, including in particular the reconstruction of national identity, and the transformation of rural community life.

Analysing de-peasantisation as a multidimensional process of societal transformation is supported by a number of theories of social change (Giddens 1984, Sztompka 1993). For Sztompka, "depending on what is seen as changing – what aspect, fragments, dimensions of the system are involved in change – various kinds of change may be distinguished. This is because the overall state of the system is not simply one-dimensional, but rather emerges as the combined, aggregated results of the state of various components..." (1993, 5). Recent sociology questions the validity of organic models of society and the idea of a dichotomy between social statics and social dynamics. The methodological consequence of this is a rejection of the validity of purely synchronic studies, giving a preference to diachronic (historical) perspectives.

Sztompka sees changes as rarely isolated or singular; they are interlinked with others, therefore the study of any change should recognise complexity in addressing the typical forms of such linkages. The study of de-peasantisation should examine the clusters of changes which result in the disappearance of the 'peasant state'. Sztompka's ideas on concrete forms of social change can help to concretise the dimensions of de-peasantisation:

- Changes in the composition of the peasantry (through outmigration, social mobility, depopulation of rural regions, dispersion of agricultural producers, recruitment to new forms of wage employment and to entrepreneurial farmers).
- Changes in agrarian social structure (the appearance of new types of inequalities, structural impacts of globalisation, crystallisation of new power relations as consequences of EU rural development policies and redefinitions of nation states, regionalisation and decentralisation, establishing co-operative or competitive relationships, emergence of new friendship-ties and new networks).
- Changes of practices (specialisation and differentiation of jobs as a consequence of, first, the industrialisation, then the globalisation of food production, distribution and consumption; the emergence of new actors in the rural economy, re-interpretions of the role of families in farming and in pluriactivity, indoctrinating effects of cultural institutions or the mass-media).
- Changes of boundaries (merging of formerly different groups, conquest and incorporation of peasant groups by other rural and urban classes, changing notions of urban and rural, urban tourism in rural places, the development of the 'culture economy').
- Changes in the relations of subsystems or institutional complexes (ascendancy of political regime over economic organisation, the emerging redistributive power of the EU, EU control of economic actors rather than market relations in LEADER, new organised controls on agricultural producers' families and on civil society or the private sphere in general).
- Changes in the natural environment and in relations to nature (sustainability becoming an aim and criterion of political control, organic or ecological production becoming an element of innovations in the countryside).

These distinctions make the task of analysing any given case of de-peasantisation immensely complex. To increase complexity further, social scientists trying to capture the social changes surrounding the collapse of the peasantry around Europe have used a range of concepts, of which 'de-peasantisation' is only one. Particularly significant debates have taken place around proletarianisation, embourgeoisement, and de-traditionalisation or the movement to a 'post-traditional rurality'. How does 'de-peasantisation' relate to these terms?

From Proletarianisation to Embourgeoisement

Within the Marxist approach to agrarian social change, de-peasantisation was widely taken to mean simply proletarianisation: as capitalist relations of production become institutionalised in society, the small-holding peasantry find themselves stripped of their control over the means of food production and, becoming landless, would be forced into joining the army of those who sell their labour power to survive. Marx's own writings on the history of agrarian groups in England encouraged this interpretation (Newby 1978, Goodman and Redclift 1981). As the agrarian writings of Lenin and Kautsky became more widely known, differentiation, rather than universal proletarianisation, began to be claimed as the future for European peasantries. Within particular regions, patterns of land accumulation and dispossession could be traced which suggested that emerging out of the peasantry were two class groups, large farmers employing wage labour, and landless rural labourers. Between regions – particularly regions which formed for some purposes a single labour market, such as rural regions of Ireland and expanding industrial cities in Britain during much of the 19th and early 20th centuries – differentiation could lead to the consolidation of a geographically separated urban proletariat and rural family-farming 'petit bourgeoisie' (Crotty 1986). Lenin's interpretation of Russian agrarian change suggested that proletarianisation need not necessarily take the form of landlessness, but of holding too little land to support the family, or lacking the means (draught animals, tools) with which to work their land – over time, the 'lower peasantry' everywhere were increasingly likely to find themselves renting out their land to large farming neighbours (if they did not simply let it lie fallow) as well as selling their labour off the farm to survive. Kautsky, and later 'articulation' theory, argued that small production could survive in agriculture in symbiosis with activities of industrial capitalism (Meillassoux 1981). Less attention was paid to transformations among the middle peasantry, such as the movement of large segments of the old (rural) middle class into the new (urban) middle class as a result of educational certification and migration to towns.

Similar ideas and debates developed in the work of social scientists in the socialist countries of Europe. Here, transformations of the peasantry were a consequence of politically enforced collectivisation, and of re-privatisation/liberalisation after the fall of socialist system. These took different forms in different countries: in Poland and in the former Yugoslavia, collectivisation was narrowly implemented. In other countries, private ownership of land and of the means of production was succeeded by common (co-operative) or state ownership; both land-owning peasants and landless agricultural workers become members of co-operatives. During the period of industrialisation and collectivisation of agriculture in the state socialist societies, the 'historical peasantry' disappeared from much of Central and Eastern Europe. Sociological theories and empirical studies took distinctive perspectives towards this remarkable social change. The peasantry had been a dominant social stratum in many communist states before collectivisation, and its transformation under communism focused analytical attention on the re-stratification of rural society. In Hungary for example, which developed a productive collectivised agricultural sector, 'the agricultural miracle' (symbioses of co-operatives, state farms and part-time family farms) created opportunities for new forms of private economic activities, and these were widely interpreted as mechanisms by which rural people tried to resist the pressures of collectivisation and industrialisation (Szelényi 1988, Gorlach 1993, Kovách 1994, Csité 1998).

Three theories in particular were developed to describe the complex rural changes occurring (Szelényi 1988). Proletarianisation theorists interpreted the rural working class – members of co-operatives, and commuters who run part-time farms outside their working hours – as different from the urban proletariat, but nevertheless as a stratum within the working class. Marxist social scientists, in both East and West, agreed that East European part-time farmers were members of the working class, and that under the logic of socialist industrialisation rural pluriactivity would soon cease, nor would any form of traditional peasant life survive socialist collectivisation and industrialisation. The peasant-worker theory, thirdly, adopted a concept of 'post-peasantry' from Mendras (Márkus 1973) to analyse the new cultural and economic pattern of behaviour exhibited by that social class which lives between the urban and the rural world.

István Kemény (1972) suggested that half the industrial working class were first-generation workers who had rural/agricultural/peasant origins. This 'new working class' was seen by Kemény, and by some Polish sociologists, as having dualistic social characteristics: they lived in the countryside and continued part-time farming, and they worked in urban centres. This worker-peasant status was still seen as a transitory period on the road to proletarianisation, but now agreed to be a long-drawn-out one – workers who had country homes could complement their 'urban' incomes from their small-scale, intensive, commodity-oriented family farming for a long period, even the lifetime of several generations. Part-time family agricultural production was a successful and innovative strategy of rural people who were better off than the 'classical' urban proletariat, but the main reason for continuing farming was not the survival of a peasant tradition but a new strategy by workers (Szelényi 1988), and de-peasantisation was thus seen as ultimately irreversible.

The transformation or disappearance of a historical peasantry is also a focus of embourgeoisement theory. This theory developed first in the inter-war period, in the work of Ferenc Erdei, using themes from Weberian theories of embourgeoisement and of the Protestant ethic. Erdei argued that the Hungarian peasantry had been on a trajectory of embourgeoisement for centuries. The 'peasant-burgers' of the Hungarian Great Plain where he was born became the heroes of his many studies. Erdei argued for a non-class interpretation of 'peasant', saying that peasants live within a community (mainly in market towns) which includes artisans, craftsmen and landless agricultural workers, and the identity of the rural way of life, and exclusion from the capitalist class structure, are stronger constituents of peasant existence than occupational position. Thus in his approach, all rural occupations (even landlords, intellectuals, officials, lower state officials and tradesmen) are classified as within the peasant world. This implies a double (both capitalist and feudal) structure in inter-war Hungarian society, in which the peasantry are a social stratum or estate excluded from the class structure of capitalist society. Embourgeoisement is that process of social change which removes certain peasant groups from this excluded condition so that they become integral parts of capitalist society, as agricultural entrepreneurs/farmers and wage-labourers in agriculture and in industry. The historical problem of the peasantry – and at the same time, of the entire Hungarian society – is that embourgeoisement has been interrupted and could not develop further.

This 'interrupted embourgeoisement' theory was addressed again by a number of sociologists during the socialist period. In his impressive reinterpretation of it Szelényi (1988, see also Csité 1998) saw the entrepreneurial strata emerging in the non-collectivised (or so-called 'second') economy as crucial phenomena of Hungarian society in the 1970s and 1980s. In trying to explain the crises of the late socialist system, he studied the emergence of this entrepreneurial stratum particularly in part-time family farming and in the structure of rural society.

Szelényi tries to test empirically the hypothesis that embourgeoisement is primarily found in the life-paths of those coming from the middle or rich peasantry. This allows him to develop a structural model of 'socialist embourgeoisement'. The 1980s in Hungary brought about a 'second (market) hierarchy' and a new entrepreneurial class. According to Szelényi, the middle peasantry is the social stratum that was able to continue embourgeoisement in socialist areas, through part-time farming. His understanding of embourgeoisement, then, points in particular towards continuity of social position: middle and rich peasants' descendants, essentially, were able to preserve or regain their social structural position. He considered this permanent coexistence of a dual economy and a dual reproduction order as a most characteristic feature of Hungarian social structure.

Critics of Szelényi have pointed to the existence of historical continuity also in small producers' habits, in the typology of farms, and in the lasting presence of pluriactivity itself. They stressed that while on one hand the socialist attempt at the transformation of the society did not radically change certain long-term trends in the development of Hungarian society, on the other hand embourgeoisement remained a weak social process (Kovách 1988, 1994; Juhász 1979, 1982, 1983, 1985; Márkus 1972, 1973, 1979, 1980). After the fall of socialism, embourgeoisement became a stable element of political rhetoric but the reclamation, privatisation, and marketisation of land interrupted massive embourgeoisement (Harcsa, Kovách and Szelényi 1998). Kovách (1998) argued that throughout the 20th century, attempts at adaptation and modernisation, involving aggressive state interventions into the rural economy and society, blocked rural embourgeoisement.

Embourgeoisement, understood as the adjustment of local groups to the macro environment, to the conditions of the market economy, can be traced in more than one dimension of social existence (Benda 1991, Juhász 1991, Kovách 1991, Laki 1991, Harcsa 1991, Vajda 1991). Nearly all of those who use or debate the concept see it as working through four significant dimensions: transformation of the farm enterprise (including transformations of economic mentality), individualisation and autonomy (including transformation of community – see Gorlach and Starosta in this volume), becoming a citizen, and civilisation, or the spread of 'bourgeois' forms of way of life and thinking.

Embourgeoisement has positive connotations, in the sense that the term indicates a direction and an ultimate aim for social transition, while de-peasantisation is a transformation from a concrete social status (peasantry) to somewhere that we cannot forecast. However, the theoretical conclusion of the embourgeoisement debates is that analyses of such complex change as is involved in the disappearance of a social stratum should not solely concentrate on structural shifts: there are different layers of transformation of peasantry. In exploring de-peasantisation in Green Ring countries, it might be useful to start from the assumption that shifts can occur independently in the different dimensions of de-peasantisation.

De-ruralisation and Post-traditional Rurality

Peasant culture is understood to be basic to national culture and identity in many countries. If this is so, we must ask what is the cultural impact of de-peasantisation in these national societies. Can de-peasantisation be analysed just as isolated structural change within rural society, if in Green Ring countries the peasantry was such a dominant social class only decades ago? What else has vanished, with the vanishing of this historical peasantry?

Economic and political changes among the peasantry are often linked to the advent of modernity. Can the cultural and societal implications of de-peasantisation be linked to late modernity via the concept of 'de-traditionalisation'? Discussions of de-traditionalisation may help to expand the analysis of de-peasantisation to include its cultural and societal dimensions which, as we noted above, are often overlooked in debates about proletarianisation or embourgeoisement as social structural processes.

We need to stress again, here, that we are not identifying peasants with 'traditional social groups', nor accepting evolutionary modernisation theorists' account of rural transformation as a process of change from backward to modern. De-traditionalisation can be understood theoretically as either a structural or a cultural and societal process. Sztompka (1993) and Shils (1981), for example, interpret tradition as the continued existence of the past in the present. Continued existence may take two forms: material and ideal, or objective and subjective. In Sztompka's perspective, tradition refers to "the totality of objects and ideas deriving from the past but which are actually to be found in the present, those which have not been destroyed, damaged, abandoned, or forgotten" (1993, 60). A second, narrower sense restricts tradition to those fragments of past heritage which are not simply survivals into the present but retain a strong, intimate link with the present. This can be related to a third use of the category of tradition, in for example Hobsbawm and Ranger (1983), as 'contrived' and 'invented' tradition. Hobsbawm suggests that 'invented tradition' may have a range of uses – for example, symbolising and expressing social cohesion; legitimising status, institutions, or authority; or inculcating values and norms in the socialisation process.

Other theorists of de-traditionalisation (see Heelas, Scott and Morris eds, 1996) represent this as a relative concept. Heelas offers a working definition: "It entails the decline of belief in the pre-given or natural orders of things" which is not tied to any particular historical period (or to a lower and more primitive level of social integration), and change in which indicates the occurrence of complex cultural change. Heelas distinguishes two versions of the concept of de-traditionalisation – a radical and a coexistence thesis. In the radical version, traditional societies are informed by belief in established, timeless orders which are rooted in the past and legitimated by authorities. The self is formed in a collectivist or communal order, which is systematically organised. De-traditionalisation leads to a world where the person ceases to be able to act as a coherent or determinate moral subject. The coexistence thesis stresses that traditions are always open to human agency; they are continuously reinterpreted. This draws attention to so-called 'quasi-traditions' – the routines, rules, procedures, regulations and so on which continue to control people's practices, and which reveal a strong coexistence of new and past traditions. In this thesis, between the tradition-oriented past and the post-traditional present or future there are many continuities. Any period of the cultural history of the West could then be analysed to discover what in it is traditional and what is relatively de-traditionalised.

Calhoun (1992) in an innovative development of the concept, suggests that 'traditional' is most useful as a way of conceptualising differences in the forms of transmission of knowledge, in societies which have become increasingly dependent on and organised around the communication of knowledge. The 'traditional' way in which knowledge is passed on is through predominantly face-to-face, informal or localised channels of communication, whereas increasingly, in contemporary society, knowledge is transmitted by formal authorities and institutions and transferred at a distance through mediating channels such as text. This use of 'traditional' has the advantage of not making any presuppositions about how receivers of the knowledge evaluate its status, as 'natural' and 'authoritative', or otherwise. Thus, our discussion suggests that 'de-traditionalisation' can include a range of very different ways in which sociologists have thought about broad cultural changes within societies. Its use to summarise the cultural transformations experienced by societies undergoing de-peasantisation may lead to vagueness or confusion, unless we are very careful to elaborate precisely how we wish the term to be understood.

Cultural de-peasantisation may be thought of as the broad consequences of structural de-peasantisation and the 'modernisation' or 'farmerisation' of a peasant economy. Cultural changes may also be thought of as having structural and economic consequences. On one hand, forms of social and cultural reproduction which connect the present or future of farming life to the past may be interrupted; on the other, elements of the historical past may re-emerge as culturally significant in radically new contexts to re-shape social reproduction. A number of recent theoretical works (Ray 1998, Jollivet 1997, Bruckmeier in this volume) emphasise the cultural component of 'modern' rural development. The emergence of notions like 'culture economy' and 'consumption countryside' imply that definitive changes are occurring in cultural understandings of rurality, and in an age of reflexive modernity, these can restructure agricultural or peasant ways of work and life. Cultural re-shaping of the late modern rural may suggest that the links between past and present/future structures must be completely re-interpreted – in the form, perhaps, of a 'heritage industry' or 'museumisation' of rural life. That might suggest that de-peasantisation leads not so much to de-traditionalisation as to re-traditionalisation, or to Hobsbawm's 'invented tradition'. It then becomes critical to ask what such contrived traditions may play in the process of change. Do EU and national state rural development policies, for example, shape the process of de-peasantisation by introducing 'invented' traditions?

Bruckmeier (this volume) uses the concept of a 'colonisation' of agriculture or of the rural to describe processes through which peasant society comes under the influence of other social structures and cultural values, although the peasantry keeps a role and a tradition as a part of the changing rural society. The question is whether such colonisation of agriculture/the rural, which may be linked to modernisation or globalisation processes of change and restructuring in industry as well, brings about a definitive break with earlier modes of rural cultural reproduction. Could any form of peasant cultural resistance survive the policy and knowledge interventions of late modernity? Is there evidence of difference between Green Ring and core European countries in the contemporary position of rural or peasant culture within each?

Re-peasantisation?

Some chapters in this collection touch on the question of whether any existing or developing trends within European society may be leading to reproduction or recreation of a peasantry, particularly in 'green ring' areas. The possibility of a meaningful re-peasantisation is something which should be addressed briefly here too.

Within the EU, it might be argued that although mainline budgetary supports for agriculture through FEOGA have tended to promote processes of concentration and commercialisation within farming which have accelerated de-peasantisation, nevertheless other aspects of EU rural development policy go some way towards countering these processes. LEADER supports for local community development may create new opportunities for farm families to diversify farm income (for example through promoting rural tourism), which in turn helps to maintain owners of very small holdings to survive in the locality which otherwise they would have had to leave. Income subsidies paid on a basis of land holding, rather than output, can support extensive methods in farming instead of a market-oriented intensification of production, and similar effects follow from the increasing variety of environmental subsidies available in EU member states. In some countries (for example Ireland) the way in which EU agri-environmental measures have been implemented has led to agri-environmental financial supports being targetted particularly on smaller producers – although these producers may also find themselves increasingly constrained in what they can do with their agricultural means of production, and may decide to leave farming as a result. The emergence of new trends in food consumption – such as the increasing consumer interest in organic products, and/or in products which carry a clear local provenance and significance – may also help to sustain and reproduce small farmers who can take advantages of the new opportunities opened up, or may even be bringing new entrants into farming on a small scale. The growth of a 'culture economy' (Ray 1998) can give local groups of small producers chances to re-skill and re-valorise their work, through the development of exclusionary controls over the labelling of their produce in the context of an increased interest among affluent consumers in differentiated and origin-distinctive forms of food.

More research needs to be done to identify whether and where trends towards the re-constitution of a class of viable small farmers is taking place in locations around the rim of Europe. However it would be important, in the light of our preceding discussion of peasants and de-peasantisation, not to assume that an increase, or a halt in the decline, in numbers of smallholders alone can be taken as evidence for 're-peasantisation' in such societies: that requires much more multi-dimensional analytic approaches. One question which could be addressed in this context is whether the increased concern about environmental problems, and interest in maintaining 'nature' in West European societies, which often identifies green ring countries as particularly important locations for experiencing and relating to nature, is helping to reverse the derogatory attitudes to 'peasants' and 'peasant production' which we earlier noted to be a widespread if not universal feature of modern Europe, and hence to reconstitute the possibilities for a culturally valued and self-confident peasantry. Particularly critical in this context perhaps will be the relations which develop between small farmers and environmental managers and scientific experts: the latter have considerable power over whether the general public comes to perceive peasant farmers as ignorant and backward destroyers of nature, or as exceptional within modern farming in their concern for sustainability of food production and stewardship of natural resources.

Some of the same processes which might be expected to recreate a peasantry or at least to halt its further decline in Western Europe – such as the growth of organic production, and the commoditisation of local cultural distinctiveness to facilitate marketing of food products – can also be identified within Central and Eastern European rural areas. However here a critical factor affecting possibilities of re-peasantisation has to be the programs of land restoration implemented in the different post-socialist contexts. It is often assumed that these programs, in attempting to restore land to those who owned it in the pre-war or pre-Communist period, are leading towards the re-constitution of the historical peasantry in these societies. However post-socialist rural societies, as was also the case before socialism, are far from a single or 'uniform' model (Szelényi 1998). In Poland, for example, collectivisation was never fully realised, and there what has been called 'the last reservation of European peasants' survived communism. The question is whether they will survive capitalism. In Hungary, in Eastern Germany and in the territory of post-Czechoslovakian states, collectivisation and re-privatisation re-stratified rural society. Modernisation, commercialisation and concentration of agricultural production, on one side, and proletarianisation on the other, seem more significant trends here than re-peasantisation. In the post-Soviet states and in Bulgaria and Romania, some features of peasant economy such as pluriactive survival strategies, private land ownership, and the household as the social framework of farming have revived, but in stringent conditions of competition. This is due to low producer prices, material damage caused during the privatisation process, the lack of state intervention or public innovations to safeguard the first years of new holdings (no trade barriers, no investment funds, etc.), hostility towards collaboration and the absence of a third or voluntary sector. Such factors made a significant contribution to peasantisation processes about hundred years earlier. Clearly what researchers today identify as re-peasantisation looks a very different process from early peasantisation (see Varis, Alanen in this volume).

In the post-socialist parts of Europe, the dramatic changes of the last decade have reduced the agricultural sector and agricultural employment by between 25 and 60 percent, depending on the country. At the same time, some hundred thousands of new family farms have come into being, and many of these seem to adopt elements of peasant survival strategies, but even here it would be difficult to argue that we are witnessing the rebirth of the 'historical peasant'. As we see it, the thesis of a complex renaissance of the peasantry across Europe's Green Ring still calls for more investigation.

This book aims to build towards *the development of a comparative framework* for studying European peasantries – east and west, north and south of the European rim. This also underpins how we have organised the chapters. We start with the experiences of most dramatic change in the 20th century, in Central and Eastern European Countries – Poland, Hungary, the Czech Republic, the Baltic states, the Republic of Karelia in north-west Russia, and rural development in East Germany in the transition to integration with the west. Mediterranean countries are analysed next in 4 chapters (Greece, Italy, Spain and Portugal). The country analysis ends with three chapters from Northern Europe, describing the long historical path of the Finnish and Irish country people to peasantry and beyond, and the route followed by the fishery-state Iceland into a modern welfare state. The final chapter will summarise some lessons suggested by the case by case analyses.

Acknowledgements

The project to produce this book was long and it is impossible to list all the helpful contributors, collaborators and commentators – we do not even attempt it here. With this book we finish the work which started from two seminars in Hungary and a series of smaller meetings between the writers and editors. The small but necessary seed money to this process came from Finland's Ministry of Agriculture and Forestry, Rural Development Fund, and from Hungarian Academy of Sciences to the project 'Actors in the European Countryside'. It was, in fact, not long after these seminars that the idea of this book took shape and Hilary Tovey joined the team as an editor. Furthermore, she took the heavy responsibility for final language checking. We are grateful to András Csite, who made much of organising work at this early phase, as we also are to all other participants at these innovative meetings. Royal Irish Academy sponsored a short term scholarship and a meeting of two editors. As many institutes and organisations as there are writers have supported the project. The list of these institutes is presented in the list of contributors. Minna Lehtola has worked as the network coordinator and the technical editor of the book. This work took place in Swedish School of Social Science, University of Helsinki. Many other scientists of this institute have contributed to the work, Erland Eklund most of all.

References

Benda, G. (1991), 'A polgárosodás fogalmának történeti értelmezhetősége' (The possibility of historically interpreting the concept of embourgeoisement), *Századvég*, No. 2-3, pp. 169-176.

Chayanov, A.V. (1966), *The Theory of Peasant Economy*, translated and edited by D. Thorner, R.E.F. Smith, B. Kerblay, Irwin, London (original publication in Russian, 1925).

Crotty, R. (1986), *Ireland in Crisis – a Study in Capitalist Colonial Underdevelopment*, Brandon Press, Dingle.

Csité, A. (1998), 'Embourgeoisement theories and debates', *Review of Sociology*, special issue.

Franklin, S.H. (1969), *The European Peasantry – the Final Phase*, Methuen, London.

Friedmann, H. (1978), 'World market, state and family farm – social bases of household production in the era of wage labour', *Comparative Studies in Society and History* 20, 4.

Friedmann, H. (1986), 'Family enterprises in agriculture – structural limits and political possibilities', in G. Cox, P. Lowe, M. Winter (eds), *Agriculture: People and Policies*, Allen and Unwin, London.

Giddens, A. (1984), *The Constitution of Society – Outline of a Theory of Structuration*, Polity, Cambridge.

Goodman, D. and M. Redclift (1981), *From Peasant to Proletarian – Capitalist Development and Agrarian Transitions*, Blackwell, Oxford.

Gorlach, K. (1993), 'The embourgeoisement trajectory: how it works in Polish society', in J. Szmatka, Z. Mach and J. Mucha (eds), *Eastern European Societies on the Threshold of Change*, The Edwin Mellon Press, pp. 175-192.

Granberg, L and I. Kovách (eds, 1998), *Actors on the Changing European Countryside*, Institute for Political Science of the Hungarian Academy of Sciences, Budapest.

Harcsa, I. (1991),'Polgárosodás' (Embourgeoisement), *Századvég*, 2-3.

Harcsa, I., Kovách, I. and Szelényi, I. (1998), 'The Crisis of Post-Communist Transformation in the Hungarian Countryside and Agriculture', in I. Szelényi (ed), *Privatising the Land: Rural Political Economy in Post-Socialist Societies*, Routledge, London, pp. 214-245.

Heelas, P., S. Lash and P. Morris (eds, 1996), *De-traditionalisation – Critical Reflections on Authority and Identity*, Blackwell, Oxford.

Hobsbawm, E.J. and T.O. Ranger (1983), *The Invention of Tradition*, Cambridge University Press, Cambridge.

Jollivet, Marcel (ed, 1997), *Vers un rural postindustriel*, l'Harmattan, Paris.

Juhász, P. (1979), *A'második' gazdaság mezögazdasági kistermelés köré szervezödö alrendszeréröl* (About the 'second' economy, as a subsystem of agricultural family production), Unpublished manuscript.

Juhász, P. (1982), 'Agrárpiac, kisüzem, nagyüzem' (Agrarian market, small producers and latifundia), *Medvetánc*, No.1, pp. 117-139.

Juhász, P. (1983), 'Az agrárértelmiség szerepe és a mezögazdasági szövetkezetek' (The role of agricultural intellectuals and the agricultural cooperatives), *Medvetánc*, No. 1, pp. 191-213.

Juhász, P. (1985), 'A mezögazdaság fejlödésében megjelenö tehetetlenségröl' (On the continuity in agrarian development), Közgazdasági és Jogi Könyvkiadó. Szövetkezeti Kutató Intézet, *Közlemények*, No. 93, Budapest.

Juhász, P. (1991), 'Polgárosodás' (Embourgeoisement), *Századvég*, 2/3.

Kautsky, K. (1988), *The Agrarian Question* (Translated from the German by Peter Burgess), Zwan (Original publication 1899), London.

Kemény, I. (1972), 'A magyar munkásosztály rétegződése' (Social stratification of Hungarian working class), *Szociológia*, 1(1), pp. 36-48.

Kovách, I. (1988), *Termelök és vállalkozók* (Producers and entrepreneurs), Társadalomtudományi Intézet, Budapest.

Kovách, I. (1991), 'A polgárosodás fogalom értelmezéséhez' (Contribution to the interpretation of the concept of embourgeoisement), *Századvég*, 2/3.

Kovách, I. (1994), *Embourgeoisement Theories and Processes*, Politikatudományi Intézet, Budapest.

Kovách, I. (1998), 'Postsocialism and embourgeoisement', *Sociological Review*, special issue, pp 20-40.

Laki, L. (1991), 'Polgárosodás Magyarországon az 1970-es 1980-as években' (Embourgeoisement in Hungary in the 1970s and 1980s), *Századvég*, 2/3.

Márkus, I. (1972), *Kifelé a feudalizmusból* (Moving out from feudalism), Szépirodalmi Könyvkiadó, Budapest.

Márkus, I. (1973), 'Az utóparasztság arcképéhez' (Portrait of post-peasantry), *Szociológia*, 2(1), 56-67.

Márkus, I. (1979), *Nagykőrös* (Ethnography of city of Nagykőrös), Szépirodalmi Könyvkiadó, Budapest.

Márkus, I. (1980), 'Az ismeretlen föszereplö – a szegényparasztság' (The unknown actor of history - the poor peasants), *Valóság*, No. 4, pp. 13-39.

Meillassoux, Claude (1981), *Maidens, Meal and Money – Capitalism and the Domestic Economy*, Cambridge University Press, Cambridge.

Mendras, H. (1970), *The Vanishing Peasant: Innovation and Change in French Agriculture*, MIT Press, London.

Newby, H. (1978), 'The rural sociology of advanced capitalist society', in H. Newby (ed), *International Perspectives in Rural Sociology*, Wiley, Chichester.

Ray, C. (1998), 'Culture, intellectual property and territorial rural development', *Sociologia Ruralis*, 38, 1, pp. 3-20.

Shils, E. (1981), *Tradition*, Faber, London.

Szelenyi, I. (1988), *Socialist Entrepreneurs*, University of Wisconsin Press, Madison.

Sztompka, P. (1993) *The Sociology of Social Change*, Blackwell, Oxford.

Vajda, Á. (1991), *A nem vállalt polgárosodás* (Unaccepted embourgeoisement), *Századvég* 2-3.

PART I:

CENTRAL AND EASTERN EUROPEAN COUNTRIES IN THE GREEN RING

1 De-peasantisation or Re-peasantisation?: Changing Rural Social Structures in Poland after World War II

KRZYSZTOF GORLACH AND PAWEL STAROSTA

Introduction

One may argue generally that processes of modernisation in rural society mean simply its de-peasantisation. Changes from an agrarian to an industrial economy, from traditional structures to modern institutions, from particular to universal values, reflect processes of modernisation occuring on various levels of social organisation. The most concise formulation of such a theory may be found in the well-known theory of modernisation developed more than thirty years ago by Smelser (1964).

The changes taken into consideration above are described in general terms by various well-known sociological theories founded on the contrast between traditional (agrarian, premodern) and contemporary (industrial, modern) types of social order. One may point out in this context theories by Durkheim (mechanical versus organic solidarity), Tönnies (*Gemeinschaft* versus *Gesellschaft*), Spencer (military versus industrial), Becker (sacred versus secular), Maine (status versus contract), Weber (traditional versus rational) and Redfield (folk versus urban).

Changes in the image and logic of farm operation are also part of this broad process of modernisation. Typical changes reflect the transition from a self-subsistence peasant farm, focused on preserving its assets and the survival of the farm family, run in a traditional way using enduring techniques based on the experience of past generations, to modern farms using science-based agricultural techniques focused on enlarging their assets and their product volume. This reflects simply the change from farming as a way of life to farming as an entrepreneurial undertaking.

This pattern of change, which became the grand theme of much classic sociology, has been described as de-peasantisation. To put it more precisely, this type of de-peasantisation, recognised as a part of modernisation, might be called farmerisation. European peasants liberated from various forms of feudal domination have become entrepreneurial farmers. However, farmerisation reflects reality as experienced by only very few societies, namely Western Europe and, to some extent, the so-called white colonies, especially North America and Australia. Some other parts of the world have tried to copy this 'western' experience with mixed and sometimes quite unexpected results. In this context one should note the peculiar developmental trajectory of Eastern and Central European societes. We agree with Daniel Chirot who suggests using a deep historical perspective to analyse this process, and stresses that:

> (...) Eastern Europe was in some sense economically backward long before it was absorbed into the broader Western world market. This backwardness had roots in the very distant past, not in any distortions imposed on Eastern Europe in the past few centuries (Chirot 1989, 3).

The peculiarities stressed by Chirot include a specific system of organisation in agricultural production, namely the system of serfdom, as well as domination of gentry over peasants and other categories in the social structure. Jacek Kochanowicz, a Polish economic historian, argues similarly in considering the problem of Poland`s fitness to become capitalist (Kochanowicz 1991). According to him, one of the most important parts of the contemporary Polish socio-economic system is the "post-traditional peasant farm" (Kochanowicz 1992). This phenomenon results mainly from late (compared to the West) modernisation, accompanied by so-called agrarian over-population, the emergence of selected markets, the dominant position of the State in the process of modernisation (see the extensive analysis by Konrad and Szelenyi 1979) and the slow process of 'peasants' transformation from serfdom or a semi-slave position to citizenship and full recognition as members of the national community. That last process was not completed in Poland until after World War II, when the Nazi occupation put peasants as well as other classes and/or strata of Polish society under the heavy threat of extinction (Przybysz 1983, Gorlach 1990, 69-75). But, as Kochanowicz says:

> The emergence of some characteristics of the modern economic system did not change the nature of peasant economy – its tendencies towards

> survival and risk avoidance. (...) But East European peasants as a social category were no longer traditional in the sense of a pre-industrial era when their farms operated on the basis of natural forces, with no connection to the world market and nation-state culture. But they did not become western-type farmers, treating their farms as capital and a means to gain some profit (Kochanowicz 1992, 157).

One might, however, point out another type of de-peasantisation observed in Central and Eastern Europe in the communist period. Under this regime de-peasantisation took the form of collectivisation. Peasants were transformed by political pressure and administrative means from the status of small owners to the status of formal co-owners of large collective farms. In fact, however, they were reduced simply to agricultural workers subordinated to and commanded by state bureaucracy. This policy was implemented in the Soviet Union from the late 1920s to the early 1930s. Then, after World War II, other countries of Eastern and Central Europe belonging to the Soviet bloc took the same route. By the early 1960s significant (if not all) parts of the agricultural sector in countries such as Romania, Bulgaria, Czechoslovakia, the German Democratic Republic and Hungary were collectivized. Economic performance varied (quite good in the case of German Democratic Republic or Hungary and much less better in the case of the Soviet Union, Bulgaria or Romania) but it is the general effect of collectivisation which is especially important. In each country, collectivisation resulted in a declining number of farms as well as a declining percentage of the labor force employed in agriculture. It also resulted in enlargement of the average farm area.

In other words the results of collectivisation were quite similar to changes in the West during the slow process of farmerisation. Collectivisation, using quite different measures, also gave an opportunity for 'economies of scale'. In this sense both types of de-peasantisation might be treated as alternative routes of agricultural modernisation.

Poland, however, is exceptional among countries of the Soviet bloc because its unsuccessful collectivisation. Attempts at Soviet-type collectivisation, as well as later state agricultural policies such as the so-called 'growth without development' of the 1960s, and 'socialist reconstruction of agriculture' in the 1970s, paradoxically strengthened the peasant character of Polish agriculture. Polish peasant farms were not collectivised but their Western-type farmerisation was blocked. As a result they became 'post-traditional peasant farms', which meant that:

> Poland remained a country of a quite traditional structure of agricultural landholdings, with few modernised farms, a strong majority of which do

not produce for the market and do not provide owners and their families with an appropriate standard of living (Kochanowicz 1992, 87).

Based on these observations, one may argue that there was a peculiar process of re-peasantisation under communism in Poland. This process, however, might not only have resulted in the survival of the traditional type of agrarian structure. We have to remember that farms operate in the context of changing rural communities. Thus another question might be raised: is this re-peasantisation process also to be observed in the changing networks, bonds and identities of rural communities?

This chapter is divided into two parts. In the first, we discuss the process of re-peasantisation, on the level of agricultural structure as well as farmers' rationalities and identities. In the second, the same process will be analysed in the context of changing rural communities and their relations with the so-called macro-social order.

From Traditional to 'Post-traditional' Peasant Farms

The History of Peasant-state Relations after World War II

Family farms operated in Poland for 45 years under a system of state-directed, planned economy, which resulted in the emergence of particular behavioral strategies among farm owners. The economy allowed some margin of free enterprise, although this varied from one period to another. At the same time it placed constant pressure on private farms, as both monopolistic supplier providing them with the means of production and simultaneously the main channel purchasing agricultural products. In other words, private farms heavily depended on state agencies which were in favour of collectivized farming i.e. state and collective farms. The situation might be described, following Marcuse (1976), as one of 'repressive tolerance' (see also Gorlach 1989 and Gorlach, Nowak and Seręga 1993). The limits of the system's repressive character were given by food shortages, while the limits of its tolerance were given by communist ideology and the pressure exercised by the political bureaucracy. Despite abandonment of the most drastic political measures against the peasantry after the late 1940s and early 1950s, the policies of 'growth without development' and 'socialist reconstruction in agriculture' of the 1960s and 1970s still shaped an unfriendly environment for peasant farms.

Our brief presentation of this system starts with the policy of agricultural collectivisation implemented in Poland following the Soviet pattern during the Stalinist era.

Beginning in 1948, this was the first experience of a communist-type agricultural policy for Polish peasants and their families. The establishment of co-operatives in the late 1940s and early 1950s was linked to certain measures which were expected to hasten the disintegration of the peasantry by setting various groups with differing socio-economic interests against each other. These groups were distinguished according to their farm size, affluence, equipment and machinery holdings, livestock, and employment of hired labor or payment of labor dues. The communist authorities emphasised the economic divisions among the peasantry, hoping to convert the potential economic tensions into open class conflict between rich and poor peasants.

A frantic effort towards collectivisation emerged as a result of a decision by the Information Bureau of Communist and Workers Parties (*Kominform*) at a meeting held in Bucharest on July 28, 1948. Between 1949 and 1952, 4234 cooperatives were established. In 1953 alone, 4363 more cooperatives were created. At the end of 1955 almost 10 thousand cooperatives existed, with about 800 thousand members, using slightly more than 2 million hectares of farmland (11 percent of total arable land in Poland) (1989, 81).

Attempts at collectivisation were accompanied by certain economic measures and by political pressures on peasants. Family farm owners were forced to pay much higher taxes than members of cooperatives or state farms: the basic land tax (*podatek gruntowy*) per hectare was five times higher for private owners than for cooperatives. After 1954, state farms were completely immune from this type of taxation. In July 1952 a law was passed on compulsory deliveries from private farms, which set delivery quotas even higher than during the Nazi occupation of 1939-1944. Farmers who were not able to meet the quotas were arrested and sentenced to jail. Between 1951 and 1955, about 200,000 farm owners were jailed (about 10 percent of all family farm owners). As Zieliński (1989, 82) writes: "Farmers became the most oppressed group under the Bierut regime".

The collectivisation campaign had a dramatic side. In some regions (Pomerania or Warsaw and Białystok voivodships) local political activists and the political police (*Urząd Bezpieczeństwa*) led violent actions against peasants who protested at collectivisation. People were beaten and their property confiscated by members of the special 'acting brigades' organised by political activists and police. In April 1951 local authorities in the Wrocław voivodship in south-western Poland, copying methods used during collectivisation in the Soviet Union in late 1920s, started to requisition potatoes from peasant farms. Some peasants were killed when protesters clashed with Soviet Army troops enforcing collectivisation (Roszkowski 1992).

The peasants' devotion to private ownership of land stiffened their resistance to collectivisation. However, resistance was also the consequence of measures aimed at the disintegration of the peasantry, which had, as a rule, just the opposite effect. The economic performance of the newly established co-operatives was highly discouraging. This was an essential factor shaping the decisions of peasants who, after all, could think in economic terms. In 1955 (the final stage of Polish collectivisation) the total income of families who had joined co-operatives was less than half that of families running their own farms. The favoured state farms (*PGR*) showed during the years of collectivisation a rapidly increasing deficit from about 2 billion zlotys in 1953 to 3,5 billion zlotys two years later (Zieliński 1989, 83).

The shaky foundations of the co-operative movement were clearly revealed by the almost immediate break-up of most co-operatives after October 1956 in the period of de-stalinisation. Of the ten thousand co-operatives in existence, eight thousand quickly ceased to exist. The remaining two thousand had broken up by 1971. Only those co-operatives which had been independently established by peasants themselves and whose economic results were good, survived. This clearly revealed that if structural changes in agriculture are not to result in lower production, they must come about as a consequence of authentic, free decisions by the farmers themselves. This new agricultural policy was introduced by the new communist party leadership under Władysław Gomułka, who preferred – as he called it – 'the specific national way toward socialism' which did not strictly follow the Soviet pattern.

However after a few years, the state farm sector began growing again to reach almost a quarter of the total area of farmland by the end of the 1960s. Policy was designed to pave the way for the future collectivisation of agriculture. It had two distinct parts. On one hand the State ensured the profitability of agricultural production. Drastic non-economic stimuli to production were avoided while supplies of fertiliser, fodder, grain and pesticides were guaranteed. Nothing was done to change the model of peasant farming, however. On the contrary, a number of factors reinforced its traditional character. For example, there were restrictions on free sale of land, which prevented the growth of large, profitable farms where modern equipment would have been easier to introduce. Sale of heavy machinery to private farmers was forbidden. Not until the end of the 1960s were state farms and agricultural circles permitted to sell off used tractors. Finally, investment in peasant farming was restricted in comparison with the socialised sector.

The vacillations and ambiguities of agricultural policy in the 1960s and 1970s are easier to understand when we realize that the policy was the result of two forces. The first was the effort by the authorities to solve problems of food supply (these, obviously, could not be solved if collective farming was fully enforced). The second was that the authorities never abandoned their vision of collective farming. The outcome was a policy of "increasing production in the peasant sector, without the development of the sector" (Kuczyński 1981, 47).

Policies which favoured productivity growth in peasant farming but without structural change eventually reaped their harvest. Peasants were well aware of the intentions, since they saw their effects daily in the form of administrative measures. From the point of view of those urging the socialist 'reconstruction' of farming in Poland, the changes described seem to have been desirable. It seemed plausible that peasant conservatism, manifested in protecting one's own plot of land at all costs, was beginning to give way to acceptance of collective forms of agriculture. This also was a basis for arguments advocating more 'furtive' methods, more subtle than the attempts at collectivisation in the Stalinist era. A situation had been achieved, at least in the mind of state officials and even some academics, where everything seemed to favor the 'socialist reconstruction' of Polish farming. As Woś (1978, 94) puts it: "The point is to create a situation whereby transition to the forms of socialist agriculture will be the only rational and fully conscious step on the part of each farmer".

Such an unstable policy created food shortages at the end of 1960s. The government tried to respond by increasing the prices of basic food commodities, which led to worker unrest, mainly in the industrial cities of the Baltic coast. After the political crisis of December 1970 in which Gomułka`s regime was replaced by pragmatic Gierek's group, the political authorities took several new measures toward family farms. These included repeal of the compulsory supply of agricultural produce, higher purchase prices by the State, lifting the ban on the sale of tractors and other agricultural machinery to private farmers, and extension of the national health service program to peasants. Problems of property ownership were also tackled, peasants being given title deeds to hold land received during the Agrarian Reform in 1944-1946. Finally, the restrictions on trading in land were reduced.

However this was not the beginning of radical changes in agrarian policy but merely, as became clear by 1974, temporary measures elicited by the food shortages and subsequent workers' revolt. Fluctuations in land policy were symptomatic of this. Between 1971 and 1973 the State Land

Fund turned over two-thirds of the land it acquired to individual farmers, then between 1974 and 1976 it turned over less than 10 percent. This struck a blow at peasants' ambitions to create large, more productive farms with high turnover.

Another restriction on free development of peasant farming was the 1973 administrative reform establishing communal councils (Village People's Councils) as basic units of local administration. The heads of the communal councils, appointed by higher State officials, were given a wide range of powers to intervene on peasant farms, to impose production demands and even to take over poorly cultivated or unproductive land. These became instruments of administrative control over private farms:

> The village universe has been divided into two foreign and hostile worlds. One, that of farming peasants, convinced of the importance of its mission of producing bread, and at the same time underestimated and looked down upon. And the other a world of bureaucratic institutions identified with the State. This line of division, despite the existence of many others, has proved to be decisive (Halamska 1981, 31).

Such policies had an obvious influence upon the consciousness and activities of peasants. The 1971-1973 'zeal' to increase production quickly vanished. When food shortages led to economic difficulties and social unrest, the State reacted with concessions to peasant farming. Administrative barriers were removed (at least officially) to the transfer of land from the State Land Fund to the peasant owners. The total transferred grew from 21,800 ha in 1976 to 113,000 ha in 1979. However, due to resistance from the local administration, the plan to sell land to peasants was not fully realized. "The reversion of policy has met with reluctance on the part of local administration; it was believed that the pro-peasant policy was only a momentary necessity, whose realization was not worthwhile since with the next swing in policy one might be held responsible for neglect of socialist reconstruction of the countryside" (Kuczyński 1981, 56).

In the last decade of the communist regime in Poland (the 1980s), contrary to pressure from the party and state bureaucracy, some pro-peasant elements in state agricultural policy became evident. Even in the period of anti-Solidarity reaction after December 1981, a constitutional amendment (which had earlier been demanded by Farmers' Solidarity) was passed guaranteeing the continued existence of private farming:

> On 20 July 1983, the Polish Parliament modified the article of the Constitution to read as follows: 'The Polish People's Republic

> considering the task of feeding the nation (...) devotes particular attention and care to the private family farms, guaranteeing their stability, supporting the growth of their production and improving their technical standard' (Lonc 1992, 220).

This seems to be proof that the Jaruzelski regime's policy towards peasant agriculture was "... of a carrot and stick variety" (Korboński 1990, 274). Martial law required peasant support against strong resistance from workers and intelligentsia. "Moreover, the Polish regime needed peasant support in order to ensure at least a minimum supply of foodstuffs as part of its attempts to stabilize the explosive political situation and to start the process of normalization" (Korboński 1990, 276). Elsewhere, Korboński (1990, 273) adds:

> The record shows that starting with the declaration of martial law, General Jaruzelski has engaged in open flirtation with the peasants. Faced with open hostility on the part of the workers and the intelligentsia, the new leader had little choice but to turn to the one segment of Polish society which did not express outright enmity to his policies. Thus, in the course of 1982 and 1983 the government's ire was a directed primarily against the workers and intellectuals and the regime clearly tried to curry favor with the peasants.

Authors considering this period of peasant-state relations stress various characteristics. Korboński (1990, 273) points to "... processes which strengthened the relative position of the private sector": slow but observable growth in the number of holdings above 15 hectares (large farms in Polish conditions), an increase in the number between 7 and 15 hectares, a sharp increase in the amount of farmland transferred from the State Land Fund, and the growing mechanisation of individual farms. Politically, the communist-controlled United Peasant Party (*Zjednoczone Stronnictwo Ludowe*) took on a new, more visible role. ZSL influenced the recognition of peasant farming in the constitution. The Agricultural Circles (Kółka Rolnicze) also played a role in price negotiations with the government. As Lonc (1992, 221) writes: "A traditional cooperative style peasant farmers' organisation, Agricultural Circles, was entrusted with the role of a trade union and of farmers' representative in price and income negotiations with government agencies".

Some opposite trends should also be considered. In 1984 and 1985 there was a significant decrease in the amount of farmland transferred from the State Land Fund to the peasant sector. Much lower use of fertilisers on individual farms, and lower deliveries of quality feedstuffs to the peasant sector by the state agencies, were also part of the picture in the mid-1980s. Despite the visible role of some official farmers' organisations, there was no official peasant involvement in politics. "...The peasants had no authoritative spokesman of their own and remained largely at the mercy of the bureaucracy which controlled just about every aspect of agricultural production and distribution" (Korboński 1990, 221).

The Legacy of Post-communist De-peasantisation

Keeping the family farm became the top priority among peasants experiencing various state policy attempts to destroy their autonomy. The forty-year struggle for private farms represented a victory for peasants over collectivisation policies. The strategy peasants established for coping with repressive state policy was a strategy for survival. In more favorable policy periods (especially the first half of 1970s, and in the 1980s), however, some of them tried to modernise their farm within the limited possibilities available.

Adamski and Turski (1990, 148) stress that:

> Intensive occupational activity among Polish peasants in the context of the existing threat toward private ownership and shortages of technical equipment might be explained by following statements: 1) the sense of threat, which became a part of Polish peasant culture, caused strong support for the defense of group interests, 2) it also resulted in strengthening attitudes towards the land as a kind of heritage and ultimate value, 3) the policy of discrimination against peasant farms periodically undertaken by the communist authorities forced peasants into intensive economic activity.

The sense of threat was quite evident in peasants' opinions on agricultural policies and their future directions and results (Halamska 1991, 49-50). It was so deeply rooted in peasant consciousness that even in the 1980s, despite formal recognition of family farms as part of the national economy, almost half of peasants asked still expressed uncertainty about the future of their farms (Adamski and Turski 1990, 142).

The limited opportunities to buy and sell land under communism resulted however in the absence of economic processes which are major ones in capitalist agriculture, such as concentration of land and production in the hands of a farming minority. The average size of a peasant farm in Poland is almost unchanged since the late 1940s (!). This is quite distinctive when compared to changes in peasant farming in some West European countries, for instance, in the period of 1960-1985 (Table 1.1).

Table 1.1 Average Size of Agricultural Holdings (in ha)

Country	1960	1985
Belgium	8.2	16.7
Denmark	15.7	31.0
France	17.0	29.2
West Germany	9.3	16.9
Ireland	17.1	22.7
Italy	6.8	8.0
the Netherlands	9.9	16.7
United Kingdom	32.0	69.4
EC-10	11.2	17.4
Poland	5.0	5.6

Sources: Alessandro Bonnano (ed) (1990), and Rocznik Statystyczny (Statistical Yearbook) (1993) GUS, Warszawa.

Farmers' consciousness and production preferences also reflect this. A significant majority of farmers used to think in terms of 5-10 hectare farms with production oriented towards mixed farming. According to previous sociological research, the preferred farm acreage in most cases reflected the regional or even local situation (Adamski 1974, 189-192, Kocik 1986, 51). In other research carried out in 1991, 56 percent of farmers interviewed pointed to the 6-8 ha farm as the 'optimal' model of family farm, which is in fact the typical size in the community in question. Only 17 percent declared that larger farms are necessary (Gorlach and Seręga 1993, 109-118).

Shortages on the agricultural machinery market, which were typical of the communist economy, encouraged certain patterns of thinking. The most important might be called a 'psychosis of possession', i.e. that farmers wanted to have all available machines and equipment even if they were not directly needed on the farm. As one farmer told us (personal communication, 1991): "When I have all the machines and equipment I might need in a particular situation I feel secure to run the farm". This expressed a general attitude typical of the 'economy of shortages' under the communist system. It was strengthened by the inefficiency and corruption of the state agricultural service system. During harvest time, for example, many farmers did not have access to agricultural services due to a shortage of equipment or spare parts for the machinery possessed by state service agencies. The poor quality of such services further encouraged farmers to prefer direct ownership of all the equipment they needed. From a purely economic point of view such 'technical self-sufficiency' was irrational when one regards the farm acreage or the level of its market orientation. However, it was rational in this peculiar situation as part of the 'game for survival'.

A particular strategy used by peasants throughout the period of communism might be called 'domestication of the communist system'. It might be visible first on the level of the complex informal relations between peasants and local officers of administrative and political bureaucracies. These contained not only typical phenomena of corruption or clientelism but also more systemic elements. Jerzy Wilkin points out that:

> Peasant farms, in fact, adopted effectively to the communist economy in Poland... However, in a few decades of communist economy slow but important changes occurred in this system as well, providing some opportunities for preserving of peasant farms. Among these changes one may point out an increasing but, of course, limited significance of market mechanisms, selective decentralization of decision-taking processes, limitations on discriminatory policies against the private sector as well as an increasing importance of economic rationality. In the case of Poland the continuing existence of peasant farms was among the reasons of such processes of change. (...) It might be treated as a kind of paradox that it was not only the communist system which controlled peasant farms but peasant farms also domesticated this system (Wilkin 1988, 31-32).

Communist de-peasantisation thus turned to be in fact a kind of re-peasantisation, at least in the case of peasant farms in Poland. This peculiar process might be observed on both structural and cultural levels. In the former case it has been visible in the shape of the agrarian structure, and the relatively low level of marketisation of peasant farms, in the latter case in the peasant interpretations of both the farm and their survival strategy. We also might argue that the typical clientelist relation between peasant and landlord was preserved, to some extent, in the relation between 'post-traditional' peasant farmers and the local state bureaucracy. In order to construct the complete picture, we now address change in the local rural community, which forms the social environment of peasant farms and farm families.

Toward a Potential Community

In the past the local community, together with family, rural household and the peasant class, was the basic element of the rural and peasant system of life (Gałęski 1965). So rural community should be included in the debate about de-peasantisation and re-peasantisation in contemporary Poland. In the sociology of the peasantry the community question is a question about social ties among peasants. Within this framework, discussion focuses on the intimacy of such bonds, and their territorial foundation. Most empirical studies of the rural milieu have concentrated more on describing the disintegration of the traditional rural community model than on analysing new forms of local system (Turowski 1971).

Three hypotheses were formulated about disintegration of social ties and the role of territory in rural life. One hypothesised a complete disintegration (decay) of rural local communities; a second hypothesised transformation of these communities; and the third their spatial liberation or de-localisation (Turowski 1977, Wellman 1979, Hunter and Staggenborg 1990). The first assumed that as a result of transformation of the industrial society, the territory loses its importance as a cause of social integration. Social norms are created mainly by the macrostructural level and mass media, and local political systems are dominated by central agencies or by national corporations. In this situation, the existence of 'community' itself, in terms of both the intimacy of ties and the integrating role of the territory, comes into question. This 'community lost' hypothesis assumes that the disintegration of communities is an unidirectional process leading ultimately to the emergence of an atomised mass society (Kornhauser 1960).

The second or 'community saved' hypothesis (Wellman 1979), assumes that the *community spirit* in rural local systems does not disappear but only changes its form. However, opinions differ on the nature of this form. Gans (1962) and Suttles (1972) stress that local solidarity can survive the challenges of industrialisation and bureaucratisation, existing in informal ties and primary neighbourhood groups, which are a source of norms and supply the individual with the basic principles needed to adapt to broader society. Others such as Young and Larson (1965), Warren (1973), Piotrowski (1975) or Turowski (1977) refer directly to the concepts of Tönnies (1988) and MacIver (1960) in suggesting that the transformations taking place in rural local communities are from traditional to associational community types, within which formal associations and organisations play the most important role and social ties in some social roles have a formalised character. They also underline that the boundaries of a community conceptualised in this way undergo some changes.

In most cases, the community becomes a supra-local structure absorbing the areas and localities adjoining it.

Finally, in the third hypothesis (Fisher 1977, Wellman 1979, 1994) primary social ties are not thought to disappear, but their territorial basis shrinks. Social contacts are not concentrated in a single place and they gradually shed their spatial restrictions. The fundamental factors sustaining this process are "a cheap and efficient transport and communication system reducing the social costs of spatial distance and facilitating the maintenance of dispersed primary ties" (Wellman 1979, 1206).

Our question is: which of these three hypotheses best characterises the changes ongoing within Polish rural local communities, in both the communist period and after? What picture emerges from the findings of empirical studies? Does it go beyond the paradigm of the Tönniesian tradition? We try in this chapter to determine the structural trends appearing in local systems. We do not pretend to be fully original; our proposals are based on theoretical suggestions from Suttles (1972), Turowski (1976), Wellman (1979, 1994), and Wilkinson (1974, 1991).

Our point of departure is a conviction that "global societies are not a simple sum of local communities" (Suttles 1972, 257). Following Suttles (1972), Turowski (1976), and Bohm (1987) we also accept that contemporary rural local systems have several integration sublevels. In the sociological literature we can distinguish four dimensions of the re-structuring of rural local aggregations, which are relatively independent of each other. We call the psycho-social attitudes within which processes of structuring and symbolic shaping of space take place 'symbolic community' (Suttles 1972, Hunter 1982). The second dimension is the primary contacts organising the 'personal community' sphere (Wellman 1979, 1994). The third is the sphere in which local and territorial aggregations organise themselves (Warren 1973, Piotrowski 1975), called the sphere of 'community organisations' (C. Milofsky 1990). Finally, the fourth dimension refers to collective activities and actions of individuals and organised groups oriented almost exclusively towards solving local problems. It can be called the 'community field' following Kaufman (1973) and Wilkinson (1991).

The first dimension treats a territory as a reflex of psycho-social experience. The local space is given as a cultural value. The main object of analysis is, thus, the experiencing of a given territory as an entity, as this finds expression in local identity and emotions. Contemporary studies tell us that rural localities lose, to a large extent, their role as reference groups. Problems in the locality remain important, but not the most important for the rural population. The recognition and respect shown by co-inhabitants towards their village are ranked below that shown toward the family or work colleagues (Boczkowski 1990).

Few respondents represent the role of inhabitant of their village or town as their most important social role (Machaj 1994). On the other hand, however, a large part of the rural population declare a strong sense of attachment to the territory that they live in (Boczkowski 1990, Wieruszewska 1988, Olubiński 1990, Starosta 1995). In the late 1980s 60-80 percent of those in rural areas and small towns identified themselves strongly with their dwelling place; in the second half of the 1990s we also found a high percentage of rural people who made such identification (Starosta 1998).

Thus, identification with one's place of residence seems to be quite strong in Poland. Of course, it is not uniform in different types of local systems. Empirical studies indicate that two groups of factors decisively affect the strength of identification with the rural local milieu: socio-economic status, and cultural factors. In culturally homogenous locations, economic status is the basic factor strengthening local identity. Identification with the role of inhabitant of a given locality is stronger, the greater the individual's chances of satisfying their needs within a given territory. Identification with rural locality seems to have a selective character. People of both high and low social status may be convinced that their chances of advancement outside a given area are insignificant, and belong to the category of those identifying themselves most strongly with their place of residence. In turn, studies of ethnic communities (Latoszek 1990) show that in situations of ethnic heterogeneity, regional cultural characteristics such as dialect, religion, or costume have a stronger effect on territorial identification than social status variables.

Satisfaction with living in a rural locality is another element of psycho-social ties. Polish studies generally show that during the present period of profound transformations the proportion of people satisfied with living in a given locality is as high as the proportion who identify with their place of residence. Indices of satisfaction with one's place of dwelling have been at a consistently high level for well over ten years. Even after the shock of the political transformation, positive indications rose in comparison with earlier years. As can be seen from survey findings published by the Centre for Public Opinion Polls, the proportion of satisfied rural respondents reached a high of 83 percent in 1993. However, whereas in the 1980s the most positive attitudes were expressed by inhabitants of small rural localities, and the most negative by inhabitants of medium-sized towns, these were somewhat reversed in 1993. By then, more people were satisfied with their place of dwelling in large urban centres with 100,000 to 500,000 inhabitants, and fewer in small localities (My Locality 1993, 66).

Applying Friedman's (1989) typology of the five ways in which territory is perceived to the findings of Polish studies in the 1980s and the 1990s, what predominated in Poland was a perception of one's locality mainly as a place of to earn a living. In conditions of cultural and ethnic differentiation, place of residence is also a source of social identity. On the other hand, one's locality is only seldom perceived as home or as one's native land. Before the transformation period the rural local community was almost never perceived as the locus of citizenship and political life or as the locus of empowerment. This confirms a hypothesis formulated by Turowski (1976, 227) that "The inhabitants identify themselves with the territorial aggregation only to a certain degree of their personality, only in certain spheres of their activity and only to a certain degree of their aspirations, feelings and goals".

The 'personal community sphere' has to do with contacts established in various social contexts. Four elements are taken into account here: the scope of contacts, measured in terms of the number of interaction partners; their social bases, as defined by the social roles of partners; their content; and fourthly their spatial and social scope (Fischer 1977). Studies carried out by Kowalski et al. (1980), and by Starosta at the end of the 1980s, indicate that the number who maintain relatively stable social contacts declines steadily. Whereas in 1977 only 3 percent of rural inhabitants studied lacked any contacts, at the end of the 1980s their proportion rose to 7 percent. In the past, social circles predominantly consisted of 5 to 6 people. At present, circles of 3 to 4 persons prevail in rural communities (Starosta 1995).

Various empirical data show that the role of the neighbourhood as a platform for social contacts diminished significantly. Whereas in the late 1970s only 13 percent of the respondents in rural communities declared that they did not maintain any neighbourly contacts (Starosta 1987), a 1988 survey showed that this figure had increased to 44 percent in Central Poland villages (Starosta 1995). In Stanek's (1995) comparative analyses of rural communities, which covered Brazil, Poland, Canada, France and Tunisia, contacts within the family diminish as well. The proportion not maintaining any family contacts in rural communities increased from 1.5 percent in 1977 (villages only) to 5 percent in 1988/1989 (villages and small towns).

During the transformation period, the workplace increased its importance considerably as a platform for establishing and maintaining social contacts. While in the second half of the 1970s only 20 percent of respondents said they had contacts with other people from the same workplace, in the late 1980s this rose to 43 percent. Social circles of employees are more often found in small and medium-sized towns than in localities which are seats of district administrations.

If we look at the content of social contacts, it appears that the number of expressive contacts grows considerably. This type of basis for contacts was mentioned by 30 to 40 percent of the respondents. Meetings connected with cultural events and entertainments, along with family gatherings, ranked second. Meanwhile, all types of services from giving advice to neighbourly assistance decreased significantly in importance. Money loans declined most. In the 1970s it was a popular basis for social contacts, which was taken up within the framework of interaction circles by 45-50 percent of rural respondents. By the end of the 1970s this had decreased to 14 percent in rural localities. In the 1990s mutual money loans as a basis for social contact between rural inhabitants hardly exist.

These data show that the basic structural category in rural localities continues to be kinship. "The system of friendship concentrates around the nuclear family" (Hryniewicz 1990, 60), which can be called a family-social circle. Simultaneously, the importance of the neighbourhood as an institution of relatively stable contacts decreases substantially, and its role is taken over by the workplace. The weakening, although still active roles of family and neighbourhood are not compensated by a growing importance of associational contacts. Thus, changes in social interactions within rural local systems do not fully accord with classical assumptions about change from family-neighbourly models of community to communities of the association type. Instead, we can observe a growing role for networks composed of family members and acquaintances, which are limited in scope.

There is no doubt that the formal-institutional basis of social ties within local systems is an important element of their structure (Young and Larson 1965, Piotrowski 1982, Miller McPherson 1990). Most formal organisations operating in the Polish rural milieu performed in the past, and still perform today, the function of local agencies of national structures. This includes churches, political parties, trade unions, and organisations providing services for rural population and agriculture.

In the 1970s most inhabitants of rural areas belonged to organisations providing services for villages and agriculture such as agricultural circles, the Voluntary Fire Brigade (*OSP*) and district cooperatives (*Samopomoc Chłopska*). Political party membership ranked second (Starosta and Majer 1982). Voluntary organisations independent of political forces represented an almost marginal sphere. After 1990, the organisational set-up underwent fundamental changes. At present, besides the Church, inhabitants of rural localities frequently belong to trade unions (Wróbel 1993) and to the Circle of Rural Housewives (*Koła Gospodyń Wiejskich*) or the Voluntary Fire Brigade (*Ochotnicza Straż Pożarna*) (Pochwala 1993).

Some trends in the structure of membership in formal organisations appeared in rural and small-town communities, which are present till today and which were accelerated by the events of the political changes, had already appeared in the 1960s, such as the steady decline in the number of persons belonging to formal organisations. Whereas in the 1980s more than 80 percent of village and small-town inhabitants belonged to at least one formal organisation (besides the Church) (Starosta and Majer 1982) this dropped to no more than 10 percent at the beginning of the 1990s (Wróbel 1993; Pochwala 1993). We should add that at the end of 1990s the number of peasants belonging to formal associations increased again to 24 percent (CBOS 1998); and that the eruption of membership in the independent and self-governing trade union Solidarity and in Farmers' Solidarity at the beginning of the 1980s represented an exception to the overall trend of declining membership in formal organisations.

Economic reforms led to the fall of cooperative organisations and other structures providing services for agriculture. After 1990, large numbers of members left Solidarity and Farmers' Solidarity. No new structures able to survive replaced these organisations. Moreover, new organisations did not obtain any more financial support from the State for their activities, which diminished their attractiveness in the eyes of local inhabitants. In the 1970s and 1980s the psycho-social ties of members to their organisations were characterised mainly by instrumentalism; no ideological ties were observed. As a result, members did not treat institutional norms with great respect. They had a selective attitude toward them, neglecting ideological content which they found incomprehensible, or interpreting it in a peculiar sense. Thus, ideological norms could not be found at the core of their mental map. Individual life problems were at the centre.

The macrosocial transformations which have occurred have not so far accelerated the emergence of voluntary local organisations as a third force operating between State and market and strengthening the subjective identity of local areas (Billis 1989). The earlier model of agency-type organisations appears not to have been replaced by a network model of local organisations. The social vacuum at the intermediate level of social structure appears to be expanding.

The final level of analysis we discuss here is that of general relations within a given area or, as described by Kaufman (1973) and Wilkinson (1974, 1991) 'the community field'. "The community field is a process of mutual interactions by means of which inhabitants express their community and territorial interest" (Wilkinson 1991, 2). Actors develop a sense of local community and transform themselves into an interest group through collective actions oriented towards a given community. They pass from the level of a *local society* as a set of individuals and structures coexisting together, to the level of 'community' as a territorial interest group. The main element of the community field is 'community action' (Sutton and Kalaja 1960), including local activities aimed at solving local problems, electoral behaviour aimed at creating local structures of power or strategic activities of local power structures seeking to develop a given locality.

It is extremely difficult to determine the extent of inhabitants' involvement in community actions as most studies are focused more on inhabitants' readiness to get involved and their declared desire to participate rather than on their actual participation. At the end of the 1970s, about 26 percent of members of rural communities from nine administrative districts in various regions of Poland admitted that they had not taken part in any community action during the preceding three years. Forty two percent took part in such actions by performing manual work, 6 percent made money contributions, and 26 percent engaged in both forms of participation (Kowalski 1982). This survey also revealed a relatively weak relationship between participation in community activity and level of involvement in organisations.

Data collected in the small town of Poddębice (Starosta) and at Łęczna and Ostrów (Machaj 1994) show that the proportion of inhabitants taking part in community activities ranged from 50 to 60 percent in the late 1980s. In Starosta's studies the proportion who participated through money donations reached 60 percent, but in the communities in the eastern region it was much lower (c. 21 percent). The proportion participating through manual work amounted to 31 percent at Poddębice as compared with 37 to 42 percent in the eastern region. Data on inhabitants' participation in community activities in the second half of the 1990s show that the number participating within rural areas has diminished to 5 percent (Starosta 1998). All these studies point to weakening involvement in community activities. Their form also undergoes change, as participation through direct manual work is more and more replaced by indirect contribution in the form of money donations.

These findings suggest that the hypotheses presented about the decay of communities, or their evolution from a primary to a secondary group or a multitude of primary groups, are limited in their scope. Communities appear to be instead evolving natural to potential communities. The territorial community spirit does not find expression in stable primary or secondary interactions. What distinguishes the structure of the contemporary territorial aggregation is the presence of four different integration levels, which are largely autonomous from one another. In terms of their temporality, they absorb inhabitants only at certain times, according to their differentiated interests or needs. In terms of their scope, comprehensive ties are being replaced by fragmentary ones. The community exists really only at particular moments and is contained potentially in behaviour models, social norms and interests resulting, as Parsons (1963) emphasises, from the territorial context of each social role. The territory is, after all, an important supplement of the social situation and as such, belongs to the social world as an ecological dimension of social phenomena. Its significance is "as a place, where something socially important happens" (Parsons 1963, 253). The local space plays, to an increasing degree, a part as a prerequisite for the mutual networking of inhabitants rather than a determinant of their social life. A durable social cohesion expressed in the principles of community and association is generally found at the symbolic level only.

Conclusion

The changes experienced by peasants have been framed by various contradictory forces and features. On one hand, some elements of private property and economy in the sector of family farming remained alive under communism and formed small islands of market logic within the overwhelming system of state-directed economy. On the other hand, many features linked peasants to the system of state socialism both on institutional and cultural pattern levels. Such linkages might be recognised as a serious disadvantage when entering a process of transition toward a market economy. Therefore peasants' contact with the harsh reality of capitalism should be seen not as an ending but, quite the contrary, as the next period of painful experience.

Four decades of political and administrative threat experienced by peasant farms in Poland resulted in particular characteristics which have become dysfunctional for them in emerging market economy. As Adamski and Turski point out:

> State policy against the economic foundations of the peasantry limited the opportunities for Western-type modernisation of this social class. However, the policy, which created the great level of uncertainty among peasants, also resulted in creating a kind of 'artificial' certainty among some of them. We have in mind some farm owners who in various periods and because of different economic, cultural and even psychological characteristics were nor inclined to engage in entrepreneurial activity on their farms but preferred to limit their operations to a subsistence scale. State agricultural policy which, for ideological and political reasons, in fact stopped the process of rational, market competition among peasants, at the same time gave a chance for traditional, subsistence farms to survive. In

> other words, it was the communist state policy which caused the reproduction of ineffective, self-subsistence farms (Adamski and Turski 1990, 148).

In a similar way, Maria Halamska (1991) stresses the 'incomplete' character of the modernisation of peasant farms. Despite the increasing level of technical equipment and the emergence of many signs of modern farming, the majority of farmers in communist Poland remained, in fact, peasants. As she says:

> Changes in the economic system of peasant farms are minimal despite the increasing level of technical equipment. The relations between the farm in-put and out-put, between amount of labour and farm produce, orientation towards markets, and vertical and horizontal integration in peasant farming have been almost unchanged throughout the whole period. Reproduction of the peasant economy overshadows its change. Despite signs of technical modernisation, peasant farms, in fact, did not change their logic of operation... (Halamska 1991, 47).

In 1996 (Cichocka 1998) only 40 percent of farms sold some part of their produce on the market. More than half own less than 5 ha of land while only 7 percent own more than 20 ha. And based on research in 1994 (Gorlach 1995), one may say that only slightly more than 20 percent of peasant farms have a strong chance of survival in the market economy. The recent process of transformation seems to be extremely difficult and painful in agricultural and, more generally, rural communities. All this reminds us of Kochanowicz's concept of the 'post-traditional peasant' farm as a major characteristic of the contemporary Polish farm economy.

The rural local community, as a highly integrated group of peasants, lost its function as a basic element of rural life. In the communist period, attempts were made to replace primary community ties by secondary relationships on the supralocal level, to coincide with administrative boundaries. The institutional system was to be the basis for the integration process and to fulfil residents' needs within the framework of several villages. This project eventually collapsed. The institutional subsystem could not fulfil the increasing expectations of rural society. Local authorities did not have sufficient support for these activities from the rural population. State policy, through planned urbanisation and industrialisation, disorganised rural communities. After the great political changes the disintegration processes caused by industrialisation and urbanisation were blocked. Rural local communities did not get from the new government sufficient institutional support to fulfil rising expectations within the framework of local systems. As a result a loose network of private contacts between family members and relatives appears as the base for the new form of community.

One can thus observe a peculiar process of re-peasantisation of the Polish countryside in the period of communism. This process is visible on two levels: on the level of farming, as an emergence of 'post-traditional peasant' farms; and on the level of rural community, as its incomplete modernisation. Under almost fifty years of state socialism, the peasantry seem to have be strengthened as a class in opposition to the state authorities. On the other hand, there are some signs of class diversification resulting both from processes of agricultural modernisation, and from increasing commuting and peasant involvement as manual workers in off-farm jobs. This means that the emergence of a post-traditional peasant class did not result in strengthening peasant rural communities. On the contrary, individualisation and privatisation of various activities became even more visible. All these characteristics, however, result nowadays in an especially difficult path to post-communist transformation in Poland. A significant majority of the peasant class seems to be marginalised because of their weak economic position based on small, traditional and non-productive farms, as well as their low skills which exclude many of them from the labour market in a restructuring economy. However a significant minority seem to be rapidly converting into typical agricultural producers fighting for their position in market economy and demanding state support. In this process they quickly adapt the methods of protest of EU farmers (Gorlach 2000). The destruction of imported Hungarian grain by protesting Polish farmers in the summer of 1998 could be seen as a symbolic farewell to a peasant culture in which grain was treated as food not as a commodity and therefore remained somewhat sacred. Destroying the grain appears as a breach of one of the fundamental taboos which formed the core of peasant culture. Thus we conclude that together these processes are leading to a last, and this time complete, de-peasantisation of Polish society in the near future.

References

Adamski, W. (1974), *Peasants and the Future of Countrysid*, PWN, Warsaw (in Polish).

Adamski, W. and Turski, R. (1990), 'Peasants' interests as a source of crisis', in Adamski, W. (ed) *Interests and Conflict. Studies on the Dynamics of Social Structure in Poland*, Ossolineum, Wrocław, pp. 127-150 (in Polish).

Boczkowski, A. (1989), 'Selected aspects of socio-spatial awerness in different local communities', in Hryniewicz, J. (ed.), *Local Communities Face the Great Changes*, University of Warsaw, Warsaw (in Polish).

Böhm, A. (1987), 'Social Structure and Local Society', in Thorniley D, (ed), *The Economics and Sociology of Rural Communities: East-West Perspectives*, Avebury, Aldershot.

Bonnano, A. (ed, 1990), *Agrarian Policies and Agricultural System*, Boulder, Westview Press, Colorado.

Bryden, J.M. (1994), 'Towards Sustainable Rural Communities: From Theory to Action', in Bryden, J.M. (ed), *Towards Sustainable Rural Communities*, University School of Rural Planning and Development, Guelph.

CBOS (1998), *Social Collective Activities of Pole*, March, Warsaw (in Polish).

Chirot, D. (ed) (1989), *The Origins of Backwardness in Eastern Europe. Economics and Politics from the Middle Ages until Early Twientieth Century*, University of California Press, Berkeley.

Cichocka, E. (1998), 'Harvest on the Rails', *Gazeta Wyborcza*, August 29-30 (in Polish).

Fischer, C.S. (1977), *Network and Places, Social Relations in the Urban Setting*, The Free Press, New York.

Friedmann, J. (1989), *Human Territoriality and the Struggle for Place*, University of Guelph, Guelph.

Gałęski, B. (1965), *Rural Sociology. The Basic Concepts*, PWN, Warsaw (in Polish).

Gans, H. (1962), *The Urban Villager*, The Free Press, New York.

Główny Urząd Statystyczny (1993), 'Annual Census', GUS, Warszawa (in Polish).

Gorlach, K. (1989), 'On repressive tolerance. State and peasant farms in Poland', *Sociologia Ruralis*, vol. XXIX, no 1, pp. 23-33.

Gorlach, K. (1990), *Peasant Question in Polish Sociology*, Universitas, Kraków (in Polish).

Gorlach, K. (1995), *Peasants, Farmers, Entrepreneurs. The Awkward Class in Post-communist Poland*, Jagiellonian University Press, Kraków (in Polish).

Gorlach, K. (2000) 'Freedom for credit. Polish peasants protests in the era of communism and postcommunism', *Polish Sociological Review*, vol. 129, no 1, pp. 57-86.

Gorlach, K. and Seręga, Z. (1993), 'From repressive tolerance to oppressive freedom: Polish family farms in transition', in Coenen-Huther, J. and Synak, B. (eds), *Post-communist Poland: From Totalitarianism to Democracy?*, Nova Science Publishers, Commack, N.J., pp. 109-120

Gorlach, K., and Seręga, Z, and Nowak, P. (1994), 'Family farms in post-communist Poland: From repressive tolerance to oppressive freedom', in Symes, D. and Jansen, J. A., (eds), *Agricultural Restructuring and Rural Change in Europe*, Agricultural University, Wageningen, pp. 153-162.

Halamska, M. (1981), 'The reasons of the emergence of Rural Solidarity', Wieś Współczesna, vol. 11, pp. 26-33 (in Polish).

Halamska, M. (1991), 'Polish peasants at the edge of XXI century', *Wieś i Rolnictwo*, vol. 73, no 4, pp. 42-55 (in Polish).

Hryniewicz, J. (1990), 'Process of Public Involvement in Urban Local Communities. Social Structuralization, Social Change and Shape of Public Institutions', in *Studia Socjologiczne*, Vol. 3-4 (in Polish), p. 60.

Hunter, A. (1982), *Symbolic Communities*, University of Chicago Press, Chicago.

Hunter, A. and Staggenborg, S. (1990), 'Local Communities and Organised Action', in Milofsky C. (ed), *Community Organisations. Studies in Resource Mobilisation and Exchange*, Oxford University Press, Oxford.

Kaufman, H.F. (1973), 'Toward an Interactional Conception of Community', in Warren, R.L. (ed), *Perspectives on the American Community*, Rand McNally, Chicago.

Kochanowicz, J. (1991), *Is Poland Unfit for Capitalism?*, Program on Central and Eastern Europe Working Papers Series, #10, MA, Harvard University, Center for European Studies, Cambridge, Mass.

Kochanowicz, J. (1992), *Discussion on Theory of Peasant Economy. Peasant Farm in Economic Theory and Economic History*, Warsaw University Press, Warszawa (in Polish).

Kocik, L. (1986), *Peasant Family in Poland in the Process of Modernisation*, Jagiellonian University Press, Kraków (in Polish).

Korboński, A. (1990), 'Soldiers and peasants: Polish agriculture after martial law', in K. E. Wadekin (ed), *Farming in the Soviet Union and Eastern Europe*, Routledge, London and New York.

Kornhauser, W. (1960), *The Politics of Mass Society*, Routledge and Kegan Paul, London.

Kowalski, M. (1982), *Differentiation of Village Inhabitants' Social Activity in From Studies of Social Structure*, University of Łódź Publications, Łódź (in Polish).

Kowalski, M. and Majer, A. and Piotrowski, W. and Starosta, P. (1980), *Problems of Formation of Residential-Service Centres in the Light of Social Attitudes of Rural Population. Synthesis of Sociological Studies*, PR-5 III/120/78, IKŚ, Warsaw (manuscript) (in Polish).

Kuczyński, W. (1981), *After a Great Leap*, PWE, Warsaw (in Polish).

Latoszek, M. (1990), 'Ethnicity as a Factor of Local Community Structure', in B. Jałowiecki (ed), *Region-Town-District*, University of Warsaw, Warsaw (in Polish).

Lonc, T. (1992), 'Social and economic adjustment and policies in a period of crisis and emergence from it. The experience of Poland in the 1980s', in A. Hakkinen (ed), *Just a Sack of Potatoes? Crisis Experience in European Societies. Past and Present*, SHS, Helsinki, pp. 215-226.

MacIver, R. and Page, C.H. (1959), *Society. An Introductory Analysis*, Mac Millan, London.

Marcuse, H. (1976), 'Repressive Tolerance', in Connerton (ed), *Critical Sociology. Selected Readings*, Penguin Books, Harmnondsworth, pp. 301-329.

Mariański, J. (1983), 'Changes in People's Religiousness in Rural Community', in Piwowarski W. (ed), *People's Religiousness*, Wydawn. Wrocl. Ksiegarni Archidiecezjalnej, Wroclaw (in Polish).

Miller McPherson, J. (1990), 'A Theory of Voluntary Organisation', in Milofsky, C. (ed), *Community Organisation Studies in Resources Mobilization and Exchange*, Oxford University Press, Oxford.

Milofsky, C. (ed, 1990), *Community Organisation Studies in Resources Mobilization and Exchange*, Oxford University Press, Oxford.

'My Locality. Perception of Changes at Local Level' (1993), *Serwis informacyjny* 6/1993, Warsaw, CBOS (in Polish), p. 66.

Olubiński, A. (1991), 'Identifaction with One's Own Locality and Self-fulfilment Possibilities of Small-Town Inhabitants', *Studia Socjologiczne*, vol. 1-2, (in Polish).

Parsons, T. (1963), 'Principa l Structures of Community', in Parsons, T. *Structure and Process in Modern Societies*, The Free Press, Illinois.

Piotrowski, W. (1975), *Formation Processes of Surpralocal Rural Community in Rural District. Its Functions and Development Conditions*, OBN, Olsztyn (in Polish).

Piotrowski, W. (1982), 'Socio-Spatial Structures in Rural Districts', in Pióro Z. (ed), *Space and Society*, Ossolineum, Wroclaw (in Polish).

Pochwala, A. (1933), 'Civic Society in a Rural District', *Wies i Rolnictwo*, vol. 1.

Przybysz, K. (1983), *Polish Peasants under Nazi Occupation 1939-1945*, LSW, Warszawa (in Polish).

Roszkowski, W. (1992), *History of Poland 1914-1991*, Warsaw, Wydawnictwo Naukowe PWN (in Polish).

Smelser, N. (1964), 'Toward a theory of modernisation', in A. Etzioni, E. Etzioni (eds), *Social Change*', Basic Books, New York and London.

Stanek, O. (1995), 'Territorial Identity and Relationship to Local Community in Rural Aggregations', in Starosta, P. (ed), *Territorial Aggregations and Social Ties*, University of Łódź, Łódź (in Polish).

Starosta, P. (1987), 'Neighbourly Interactions and Relationships in Rural Milieu', *Wies i Rolnictwo*, Vol. 1(54) (in Polish).

Starosta, P. (1993), 'Social Interaction with Mass Organisations in Rural Communities', in Gorlach, K. and Seręga, Z. (eds), *The Polish Countryside in Socio-Political Transformation Processes*, Jagiellonian University, Cracow (in Polish).

Starosta, P. (1995), *Outside the Metropoly. Rural and Small-Town Local Communities and Models of Macrosocial Order*, University of Łódź, Łódź (in Polish).

Starosta, P. (1998), 'Inhabitants' Feeling of Social Bonds with Rural and Small Town Communities', in Granberg, L. and Kovách, I. (eds), *Actors on the Changing European Countryside*, Institute for Socialogical Sciences of the Hungarian Academy of Sciences, Budapest.

Starosta, P. and Majer, A. (1982), 'Participation in Formal Organisations and Social Differentiation of Rural Population', in Bokszański Z. (ed), *From Studies of Social Structure*, University of Łódź, Łódź (in Polish).

Suttles, G.D. (1972), *The Social Construction of Communities*, The University of Chicago Press, Chicago, p. 257.

Sutton, W.A. and Kalaja, J. (1960), 'Elements of Community Action' in *Social Force*, Vol. 38.

Szelenyi, I. and Konrad, G. (1979), *The Intellectuals on the Road to Class Power. A Sociological Study of the Role of the Intelligentsia in Socialism*, Harcourt Brace Jovanovitch, New York and London.

Tönnies, F. (1988), *Community and Association*, PWN, Warsaw (in Polish).

Turowski, J. (1965), 'Transformations of Traditional Rural Local Community', Roczniki Socjologii Wsi, Vol. 4.

Turowski, J. (1971), 'Development of Studies in the Field of Rural and Urban Sociology during 25 Years of the Polish People's Republic', in Kryczka, P. and Szwengrub, L.M. and Turowski J. *Rural and Urban Sociology in Poland*, IFiS PAN, Warsaw (in Polish).

Turowski, J. (1977), 'Local Community', in *Studia Socjologiczne*, vol. 3 (55) (in Polish).

Warren, R.L. (1973) (ed), *The Community in America*, Rand McNally, Chicago.

Wellman, B. (1979), 'The Community Question, the Intimate Networks of East Yorkers', *American Journal of Sociology*, vol. 84, p. 1206.

Wellman, B. (1994), *Personal Communities. Some Basic Characteristics in Changing Values and Attitudes in Family Households with Rural Peer Groups*, Social Networks and Action Spaces, FAA, Bonn.

Wilkin, J. (ed) *Peasant Farm in State Socialism Economy. Basis and Effectiveness of Adaptative Mechanisms*, Warsaw University Press, Warsaw (in Polish).

Wilkinson, K.P. (1974), 'A Behavioral Approach to Measurement and Analysis of Community Field Structure', *Rural Sociology*, vol. 39:2.

Wilkinson, K.P. (1991), *The Community in Rural America*, Greenwood Press, New York p. 2.

Woś, A. (1978), 'Class structure and stratification among the rural inhabitants and their evolution in the process of construction of the developed socialist society', *Studia Socjologiczne*, no 2 (in Polish).

Wróbel, S. (1993), 'Political Involvement of Local Communities during the Period of Socio-Political Changes' in *Local Communities and Local Democracy Development*, Domino, Katowice (in Polish).

Young, R.C. and Larson, O. (1965), 'The Contribution of Voluntary Organisations to Community Structuree', *American Journal of Sociology*, vol. 71.

Zieliński, M. (1989), 'Peasant war in Poland', *Res Publica*, vol. 5-6 (in Polish).

2 De-peasantisation of the Hungarian Rurality

IMRE KOVÁCH

Introduction

In the middle of the 20th century some 50 percent of Hungarians worked in agriculture as landed or landless peasants or wage-workers (or members of their families) while in 1998 7.2 percent of the total labour force were registered as full-time farmers or agricultural employees. This trend and other structural changes clearly show that the historical peasantry, which had formed a dominant social strata demographically within the Hungarian society, no longer exists.

In the introduction to this volume the term 'Green Ring' is used to comprise those countries around Europe which followed different pathways to modernity from the early industrialising countries, and which have been characterised by a strong agricultural presence in the recent past. In Hungary until the 1990s, as in other 'Green Ring' countries, agriculture and the peasantry had a remarkable position in the economy, culture and, more indirectly than in Nordic countries, also in politics.

The Hungarian economy and society have never been amongst the most backward in Europe but in no period could it achieve the development level of core regions. The duality of a close connection with European Centres and the backwardness of the entire economy (and especially agriculture) in its relation to the level of development of the core, early industrialised regions, was exacerbated in recent centuries by the crucial problem of modernisation, which multiplied and strengthened the points of connection. This backwardness and constant pressure to modernise led to a series of state interventions into agrarian structures and to trends towards de-peasantisation – the decreasing economic and social importance, and finally disappearance, of the peasantry. Up to the post-modern changes of the 1990s, 'peasant' and 'rurality' were integrated within national symbols and images, and the peasantry was regarded as a source or custodian of national culture and the demographic basis of the nation which implies that de-peasantisation is a culturally meaningful process in the Carpathian Basin. This paper focuses on both the long drawn out trends and much more recent phase of rural development in discussing de-peasantisation and its broad impacts in Hungary.

Understanding 'Peasant' and 'De-peasantisation'

If the peasantry had conflicting positions, demographic domination and a source of national culture on one side and almost absent political representation on the other, how can we understand the terms 'peasant' and 'de-peasantisation' in the Hungarian context? Does 'peasant' have a different meaning from that in other Green Ring countries?

The Hungarian term '*paraszt*' (peasant) in the vernacular speech has a complex sense with negative connotations and a sometimes pejorative meaning, so it was slowly and occasionally replaced by other names. Different landowning and agricultural worker groups identified themselves as '*kisgazda*' (smallholder) or '*gazda*' (master) or '*földműves*, *földmunkás*' (land labourer) in the first half of 20th century but '*paraszt*' remained the general term for millions when, as Ferenc Erdei, peasant-born sociologist and theorist, argued, Hungarian society had a dual (feudal and capitalist) structure. Later, during communism, '*paraszt*' was a collective term meaning the opposite of urban dwellers, and a kind of synonym for somebody who occupied a low position in the social hierarchy (as in Poland; see Introduction to this volume) or whose way of life represented an inferior culture. Peasant, as the name of a social category characterised by involvement in small scale agriculture, lost its specific meaning after collectivisation and is recently thought of as a cultural label for something which is not modern or post-modern.

The etymological history of terms connected with non-urban dwellers or food production workers became more complex after the 1960s. 'Villager' or 'countrysider' denoted the rural population or those part-time farmers who lived in cities. The term 'co-operative peasantry' (and less frequently, state farm workers) was a category name in bureaucratic and statistical terminology, but the word 'peasant' was not in usage to identify food producers. 'Peasant' was succeeded by 'member of co-operative' or by other labour force terms such as 'skilled', 'semi-skilled' or 'unskilled worker'. The structural changes in agriculture, the productive coexistence of big co-operatives, state farms and part-time farming, and the emergence of the 'second economy' complicated the terms used in everyday language. Part-time farmers, employed in the socialist/state sector of the economy and engaged in commodity, spare-time farming, were identified also under labour force categories and never under 'peasant' or 'farmer'. If somebody referred to their agricultural production they used the phrase 'small scale producer'.

External terms (appearing in socialist ideology and state/official documents and romantic images of the peasant considered by the intelligentsia as a source of national culture on one hand, and treated as some kind of second class citizen or recently as non-European on the other) have been discordant with the internal terms of food producers themselves. The new actors in family agricultural production, after the fall of communism and the restoration of private land property rights, are 'agricultural entrepreneurs' sometimes and, much less often, 'farmers'. 'Peasant' nowadays is more synonymous with a lower cultural level than an unambiguous social category. The shift of meaning from a social to a cultural sense entailed a re-interpretation of rural, rurality and also peasant. Images connected to rural and 'peasant' have shifted from their former positive socio-cultural meaning to a situation where they have more problematic and negative connotations. These rather positive images have been replaced by more post-modern constructions, a mixture of new and positive meanings and negative connotations, with the 'rural' and 'peasant' viewed as a problematic outcome of processes of modernisation and Europeanisation.

The social sciences provided varying meanings for the concept of peasant. Ethnography, statistics, history, agricultural economics and sociology applied it differently, insisting that its use roughly corresponds to analyses of late feudal and capitalist modes of production. 'Peasant' and its synonyms (agriculturist, primary producer) in this tradition represent food producers who live in rural settlements. Ferenc Erdei's non-positional interpretation of 'peasant' classified rural occupations (with the exception of landlords, intellectuals and officials and tradesmen but including artisans and landless workers) within the peasantry which he understood as a social stratum or estate excluded from the capitalist class structure.

Analysers of the socialist era concentrated on theories developed to describe complex rural changes and did not formulate new concepts of 'peasant'. As they saw it, collectivisation ended the existence of the historical peasantry and theoretical reflections (proletarianisation, peasant-worker theory, interrupted embourgeoisement, see Introduction) were addressed to post-peasant rural transformations. A common understanding of these theories is that the strong rural or agricultural presence in

Hungarian society gradually ceases in the process of socialist modernisation and in other words de-peasantisation, de-agriculturalisation and de-ruralisation are definitively connected processes. This perspective on the disappearance of the peasantry presumed that socialist modernisation and urbanisation repress rurality; part-time farming was viewed as successful resistance by rural people against industrialisation and collectivisation. Szelényi's thesis on the four destinations of the transforming rural class structure integrates proletarianisation, peasant-worker and interrupted embourgeoisement theories. For Szelényi, rural changes are mainly transformations of rural class structure and this dominates over any complexity in the disappearance of the peasant state and any linkage of subsystems. Peasant transformation is seen as a one-dimensional social change. Embourgeoisement primarily means a trend-like continuity in the social position of middle peasants through agricultural small scale production (part-time family farming). Its essence is that middle and rich peasants' descendants were able to preserve or regain their social position which have been relatively independent from the redistributive-bureaucratic dimensions of social reproduction. He considered that the permanent co-existence of dual economy and dual reproduction order is most important feature of Hungarian social and economical structures in capitalism as well in socialism.

In the 1990s, after the fall of communism, privatisation, land restitution and economic liberalisation were the key influences on rural restructuring. The dual economy and dual reproduction order came to an end but rural embourgeoisement (farmerisation) has been limited to narrow social strata. As far as rural embourgeoisement is concerned, entrepreneurial activity and the slow transformation of family farming to become part of the market economy have not brought a break-through. Post-socialist privatisation and economic liberalisation have brought about not only rapid de-agriculturalisation of the rural economy and partial renewal of family farming, but also renewal of forms of subsistence farming and peasant survival strategies. The main actors of part-time family farming of the 1980s lost the struggle in the period of transition for social and power position, or else they have invested their successfully accumulated economic and social capital in non-agricultural sectors.

But how should we understand Hungarian rural development in the context of de-peasantisation? Does de-peasantisation replace the concept of embourgeoisement? Has there ever been a Hungarian model of transformation of the peasantry from the time of the 'agricultural miracle' to the post-socialist transformational crises and attempted Europeanisation of the countryside? To answer these questions, in the rest of the introduction I look at how to identify differences and similarities of transformation of peasants in Hungary in comparison to other Green Ring countries.

First, I look at historical-structural peculiarities of Hungarian rural structures. Some individual characteristics of the spatial structure of food production were a consequence of the long Ottoman occupation of 40 percent of the territory of the medieval Hungarian state (the best arable land area on the Great Hungarian Plain!) Hundreds of thousands of landowners and agricultural producers moved to market towns at that time and together with other social groups such as liberated peasant-soldiers or peasants with communal nobility they constituted privileged free peasant strata in the feudal system. The rural population (agricultural producers) remained the majority within Hungarian society until the 1950s. In spite of accelerated communist urbanisation and industrialisation, the settlement structure changed only slowly in the 20th century. Hungary is considered under-urbanised (Szelényi 1988) since the urban population was growing faster than urban industrial jobs. The proportion of the rural population in the total population decreased from about 60 percent to 36 percent by the late 1980s, but the question is to what extent was that decrease a consequence of the bureaucratic re-designation of settlements from villages to towns, rather than of real urbanisation. Until 1994, the population of settlements officially registered as villages declined; afterwards it started to increase from 34.3 percent in 1994 to 37.4 percent in 1998. There is no clear definition of rural, and elsewhere I have argued that about 50 percent of Hungarians live in rural settlements, that is, farmsteads, villages, or towns with a strong rural character (Kovách 2001). The formation of a mixed farming structure (big units or landlord estates and small-holding peasant farms) that had come into being from the second half of the 19th century in some respects survived even the distribution of land in 1945, collectivisation, and the post-communist reconstruction of agriculture, all of which involved radical transformations of landed property patterns.

Second, I stress the importance of politics in the making and unmaking of peasants. In tracing the process through which de-peasantisation became more complete, political scholars stress the decisive role of reforms and interventions of state policy into rural economy. In the 19th and 20th centuries rural development was a consequence of successive attempts to modernise and restructure rural economy and society. These included emancipation of the serfs and creation of peasants in the 19th century, then the protection of big estates, introduction of a war economy and increased state control during World War II, and four radical reforms – land redistribution in 1945, two collectivisations (an unsuccessful one from 1948 to 1958 and a completed one from 1958 to 1963), the socialist modernisation of the 1960s and 1970s, and reprivatisation, land restitution and attempted Europeanisation of rural development policy after 1989. These interventions have been of a rather aggressive nature particularly in the second half of the 20th century; they left practically no space for autonomous civil institutions, the participation of local residents or organic processes in rural transformation.

Third, the socialist collectivisation should be re-interpreted form the viewpoint of de-peasantisation. Alanen, Varis, Meurs and Begg, and Majerova (all in this volume) call attention to the appearance of some features of peasant economy and survival strategy. Family farms, in the post-socialist era, also seem to adopt elements of peasant food production. Szelényi also found a mysterious, strong correlation between family entrepreneurship before socialism and the market orientation of family agricultural producers in late socialism. The fast revival of some components of a peasant way of life suggest that collectivisation, the destruction of private ownership in land, has not eliminated totally what we understood as peasant, and it is useful to emphasise again that shifts can occur independently in the different dimensions of de-peasantisation.

The 'Hungarian model' or 'agricultural miracle' marked a particular project within socialist rural modernisation. One building block of this was a unique symbiosis between co-operatives, state farms and small-scale (part-time) family farms in which family farms, both in rural and urban areas, produced some 30-50 percent of total agricultural products. An estimated 40 percent of family farms included subsistence production but almost two thirds, or about 60 percent of the total households, had been involved in commodity production. The market orientation of family farming rapidly increased from the late 1960s to the early 1980s, when 61 percent of food produced by family farming was sold on the market. This proportion is higher than it has ever been in peasant history. How can we interpret this Hungarian model of socialist modernisation which destroyed the historical peasantry by collectivisation and simultaneously all-powered market oriented entrepreneurial mini-farms to reappear?

Fourth, an unusual symptom of post-socialist rural development is, surprisingly, that while agricultural production has been continuously shrinking and rurality de-peasantised (disappearance of the historical peasantry), the number and ratio of rural inhabitants within the total population started to increase by 0.7 percent per year. De-agriculturalisation, de-peasantisation and de-ruralisation are disconnected and de-peasantisation has not resulted in the depopulation of rurality. Something has been occurring in rural Hungary which cannot set in process the Pan-European rural exodus.

The thesis of embourgeoisement might be considered the most investigated model of Central European socialism. It should be understood as about the organic evolution of (rural) society versus power. As Szelényi recalled, this was the common meaning of Hungarian 'embourgeoisement' and Polish 'civil society' terms. Addressing embourgeoisement as an essentially one-dimensional social change, it may be viewed positively on breaking out of state socialism. This assumes that the process of embourgeoisement may be one way of achieving a lasting peasant history in socialist circumstances, and that many traditions could survive the 'great transformation' of the 20th century. However after the collapse of socialism the Hungarian road of embourgeoisement did not prove to be viable. Rural transformation in the 1990s clearly shows that de-peasantisation is negative social change or at least an unbalanced, transitory period, versus the concept of positive embourgeoisement. Many peasant traditions became extinguished from rurality. Indeed there is a contradiction in post-socialist Hungarian rurality: embourgeoisement failed in the end but this did not at all start rural migration to urban centres as happened in other Green Ring countries. This is not a success story: many rural people lost accumulated social and economic capital and their ability to resist any set of economic, political or social pressures has been undermined. In late industrialised regions within the European Union, the loss of the peasant state set off massive outmigration from rural areas, and qualitative positional restructuring such as farmerisation or the raising of skill levels. The peasant entity is replaced by some other kind of viability. In Hungary (Central Europe) the size of the rural population remained relatively stable but peasant capacity for survival or for economic accumulation went to pieces. The new countryside is a mixture of new actors and economic activities but with no strong connection to the historical peasantry. Dead-end socialist modernisation, and the necessities and deficient political decisions of post-socialism, are responsible for this loss.

Fifth, in the 1990s re-peasantisation is a rather negative term, both in vernacular speech and in scientific terminology. It means rebirth of poor peasant practices, habits of mentality and cultural norms.

There are two important implications of the above assertions. First I must take a stand on what I mean by peasant, second on how I interpret de-peasantation. As a working definition, the historical 'peasant' can be described by private property, dominance of agricultural production, pluriactivity in food production (weak specialisation of production), duality of subsistence and commodity production, capacity to survive and resist any pressure with their focus on social reproduction and economic accumulation, constrained social mobility, and community with a peasant-type public sphere. 'Historical' means that this working definition of peasant is set in a particular time and related to a determined historical period, in Hungary from 1848 to 1958, from the emancipation of the serfs to the start of the second collectivisation. With regard to the definition of de-peasantisation, I share the Introduction's approach: de-peasantisation is a multidimensional process of societal change in which different constituents of the definition of the historical peasantry are transformed or disappear. In the rest of the paper I look at peasantisation and de-peasantisation as a historic process, on the lines suggested in the Introduction including the three different levels of multidimensional change – cultural, structural and societal de-peasantisation.

Structural De-peasantisation

The Making of Peasants

Hungarian agriculture, having been integrated into the economy of the Habsburg Empire as a primary food producer internal colony, rested on feudal structures until the middle of the 19th century. Alongside noblemen's large estates with their system of enforced labour, serfs lived on subsistence farming engaging partially in commodity production. The feudal system was based on the work of the serfs who, after the abolition of feudalism and serfdom in the 1848-1849 revolution, became wage-workers and free peasants within two decades when some half of agricultural land were allotted to them. This beginning to the making of peasants was associated with the extension of capitalism. The peasant families practised subsistence farming with a weak commodity orientation. Peasant labour capacity was for mostly absorbed by their small-scale farming. There was also a shortage of capital, despite the fact that the transformation of the banking system and the development of credit-facilities did not leave agriculture untouched either. In the last third of the 19th century, when the demand for agricultural products rose both abroad and on the domestic market, Hungarian agriculture could meet this challenge only by developing its extensive forms. The crucial question for expanding production was access to land. Under the technical conditions of the age both the cultivation of plants and livestock-breeding required large tracts of land and there was simply not enough. This is why – with the passage of time – the struggle for land became an acute political issue. The population grew continuously during the whole 19th century, while the extent of land which could be legally possessed did not keep pace with this after the middle of the century. Thus owing to the conflicts between the small-scale and the large-scale enterprises demographic problems manifested themselves towards the end of the century and the consequences included emigration, settling down in towns and birth-control. Up to the end of the capitalist period of Hungarian agriculture, i.e. until World War II, the size and requirements of the market demanded only to a small degree the development of a competitive composition of products, including the production of mutually exclusive products.

The large estates remained unchallenged, defended also by the system of payment in kind. The inalienability of the large estates was decreed by law, preventing thereby the decline of the economic and political power of the landholder aristocracy (Tóth 1981) At the beginning of the last century the large feudal estates began to be transformed into real large-scale enterprises. The first phase in this process was the concentration of territory, in the course of which those lands were gradually expropriated and taken back by the landlord which had been previously allotted to the serfs for use.

Only the large enterprises could specialise, although in some regions some small-scale enterprises also managed to change the composition-mix of their products. The traditional composition of products, in which specialisation was an exception to the rule, retained its dominance for a long time. Neither did the process of work nor the organisation of the enterprise change significantly. The proportion of products which were self-consumed was extremely high; even in the 1930s and 1940s this was estimated to be approximately 70 percent of all agricultural production. (Gunst 1987) The traditional and less specialised composition of products and the predominance of subsistence production mainly characterised small-scale enterprises, which did not specialise in production but produced various kinds of agricultural products. Their main purpose consisted not so much in selling products on the market as in securing self-provision and absorbing labour-power in a useful way. The formation of a mixed farming stratum, together with the development of the large-scale enterprises, greatly increased the tensions in the political struggle for the possession of land between the small and large-scale enterprises. Production in the self-provisioning type of peasant holding was guided primarily not by criteria of profitability but by the need to absorb as high as proportion of the labour-power of the family as possible. Self-consumption required the cultivation of arable lands. The redundant labour-power, resulting from technological progress and over-population, could have been absorbed by those branches of agriculture, which require relatively little land. In the production of the small-scale enterprises there were some signs of specialisation, but the subsistence character of farming and the resulting composition of products and labour-utilisation remained unaltered practically throughout the whole period of capitalist agriculture. Small-scale peasant enterprises were characterised by the predominance of corn-growing and, for reasons of security, by an extremely mixed composition of products on lands which were not sown with corn. Commercial relations between small-scale and large-scale enterprises were weak and occasional, requiring mediators. The large-scale enterprise succeeded in organising within its own framework the production of those articles, which, for various reasons, could not be bought from industry. The small-scale enterprises organised instead the regular exchange of services. The large-scale enterprise was able to process its own produce and could also transport it directly either to a processing factory or a commercial firm. The small-scale enterprise sold its products at fairs, often with the help of merchants. The large-scale enterprises conducted commercial transactions primarily with urban society, whereas the small-scale ones maintained economic relations with the population of the villages. Their transactions with the town-dwellers were mediated by agents. This contributed to the preservation of peasant traditions to a significant degree. Owing to the market conditions, the structure of land property and the division of large-scale and small-scale production, the commercialisation of agriculture was not realised fully until World War II. The majority of small-scale agricultural enterprises concentrated to a significant degree on production for self-consumption. In the majority of peasant units more than 50 percent of the products were used by the producers themselves or went into barter. Only a very narrow stratum was in a position to regularly guide its production according to the market and to take part in financial transactions. The number of farms which specialised in producing vegetables or fruits and whose main purpose was selling their products on the market, was estimated to be approximately some thousands. Thus the absence of specialisation, besides the unfavourable proportion of those products which were put on the market.

The structural problems of agricultural production are reflected in the fact that while the proportion of those who lived on agriculture practically stagnated (48 percent), the contribution of agriculture to the national income had decreased by World War II to 37 percent, almost reaching the proportion of industrial production (36 percent) which provided jobs only for 23 percent of the population (Berend-Ránki 1972).

In 1940 4,538,900 people lived on agriculture, close to half the total population. The social structure of the agricultural population displayed vast inequality. The number of landowners with holdings larger than 100 hectares was 32-33,000. The majority owned plots smaller than 10 hectares. In the 1930s some 700,000 peasant small-scale enterprises were operating, but the number of small-scale enterprises exceeded that of the strictly peasant-owned holding, because the large masses of agricultural and fewer industrial wage labourers, totalling almost 2 million people, also ran agricultural small-scale farms (Kovách 1999a). Peasant society existed in a state of accumulating crises.

Extension of the Peasantisation Project: Land Reform in 1945

The new parliament just after the war passed a law on land distribution. In addition all large estates, exceeding in size 100 hectares, were to be nationalised without compensation. Later decrees enabled the large estates to be expropriated up to the limit of a 57 hectares, with compensation. In the case of peasant holdings the limit was set at 114 hectares; the rest of their lands were taken away from them and the promised compensation was never fully paid. Three point two million hectares, 35 percent of the total land, were nationalised for the purpose of land distribution (Donáth 1977). The lands to be expropriated were to make up 35 percent of the territory of the country. Sixty percent of the nationalised lands were allotted to individual peasants, the rest remained in the hands of the state. Land and/or building sites were given to 810,000 people. Of the 730,000 claimants some 660,000 were able to get land. Fifty-two percent of agricultural labourers, 47 percent of domestic servants, 44 percent of peasants, owning only dwarf holdings, and 75 percent of smallholders received nothing. The average amount of distributed land was 3.2 hectares (Donáth 1969) Some ninety percent of those who were given land by the state were agricultural labourers, domestic servants or peasants with mini holdings; they received 93 percent of the allotted land. Simultaneously with the land reform and partly in order to remedy the problems deriving from it, the government initiated a programme of new settlements. The average size of the allotted plots was 2.9 hectares, so commodity production could not increase its proportion. In 1941 only 21 percent of the population ran peasant farms, this rose to 43 percent by 1946 (Szakács 1996).

The redistribution of land was an attempt to extend the peasantisation project, therefore it was implemented primarily according to social and political considerations. The new peasant farms were incapable of commodity production and the number, even the proportion, of those medium-sized peasant holdings which would have been capable of intensive commodity-production decreased. As early as 1952, 68,000 peasants gave back their lands to the state. In the post-war period, and especially from 1947/48 when the communists came into power, the redistributive nature of state agricultural policy became stronger. The land reform, however it realised the peasants' demand for land, was one of the most redistributive interventions in the rural economy. The communist state introduced a markedly controlled system of agricultural policy, depriving peasant farms of profit – there was hardly any profit produced in agriculture – but also of agricultural natural products (by expropriation, taxes, prices, indirect measures, legal procedures and physical force) (Pető-Szakács 1985, Berend-Ránki 1972, Szakács 1996). The land reform associated with emerging state intervention and control could not create an extension of the peasantisation project.

De-peasantisation by Collectivisation

A new wave of political (communist) intervention started in 1947/48 with the first aggressive collectivisation and although it was aborted, about 400,000 peasants migrated to cities. The second attempt from 1958-1963 led to almost full collectivisation of agriculture, some 93-94 percent of previously independent peasants joined – more often than not under external pressures – the new co-operatives (Harcsa-Kovách-Szelényi 1998a, Szakács-Petö 1985, Kovách 1999a, b) The dominance of the co-operative form of cultivation was further enhanced by the fact that almost half a million active earners left agriculture between 1959 and 1963. The number of individual farmers dropped to one-tenth of the figure of 1958 by 1963, to about 150,000 people. The unexpected rapid success of the second wave of collectivisation may be explained by several causes (Donáth 1972, Petö-Szakács 1985, Berend T. I., Harcsa-Kovách-Szelényi 1998 and Kovách 1999a, b).

After 1945 there was constant migration from the sphere of agriculture. To the enforced collectivisation in the early 1960s the response was also migration: some 500,000 people left the villages for urban centres; however, the pace of industrialisation and urbanisation in this period was much slower than it had been ten years earlier. In the 1960s and 1970s another type of migration was movement from small villages into larger ones, agricultural market towns. The decade of the 1970s and partly also the first years of the 1980s were years of intensive development of regional centres, in the course of which the settlements that had qualified as large towns received significant financial grants for development (Vági 1982). The population of some county capitals and regional centres rose in a decade from 20-30-40,000 to 100-150,000 mainly by in-migration from smaller settlements. It was primarily the young labour force which left villages (Andorka-Harcsa 1982). The importance of regular monthly, weekly or daily commuting also increased (Andorka-Harcsa 1982, Harcsa 1996) In the 1970s some 1 million people, half of the Hungarian worker class, could be registered as regular commuters (Böhm-Pál 1984). As a result of these processes of migration the ratio of urban and rural populations as well as food producers changed significantly by the 1980s: the rural population fell to around or under 40 percent of the whole population of the country.

Socialist Modernisation

In the peak periods (1968-1982) from the middle sixties to 1989, Hungarian agriculture belonged to the most productive in the world and the rural society and economy underwent intensive modernisation (for details see Swain 1985, Szelényi 1988, Harcsa, Kovách and Szelényi 1998a, Meurs 1999, Kovách 1999b). From 30 to 50 percent of agricultural production was small-scale and family based (Juhász 1982, Bell 1984, Szelényi 1988, Kovách 1999a, 1999b). Collectivisation and rapid agricultural development launched restratification processes in rural society. From the middle of the 1960s human investment was made in agriculture, big groups were trained at the universities and high schools as agronomists, lawyers, engineers, veterinarians, economists to work in agriculture and more and more young people left secondary school as skilled workers. Full employment of the rural population, the new forms of economic activity in the first, socialised and the second economy, the controlled worktime at workplaces, the close relations between educational level, job and incomes, and the attractiveness of the new way of life opened up new channels for social mobility and at the same time pushed large numbers to leave their or their parents' peasant condition. According to figures on the peasantry, the rural population had the most intensive social mobility (Andorka-Harcsa 1982, Harcsa 1996). In the fifties and sixties 900,000 to 1 million people migrated into the cities from villages. The fact of migration did not lead to the urbanisation of the majority of those who had come to live in towns. However, these processes of migration and especially the practice of commuting had a significant social effect: a certain exchange of values between cities and villages. Those who commuted between town and village became acquainted with the urban way of the life, and took home some of its elements, owing to which a new structure of consumption came into being in the villages, leading to the formation – together with other factors – of part-time farming and the industrial second economy. Increasing agricultural and industrial incomes, commodity-oriented, intensive part-time farming and alternative forms of the second economy led to rural social transformation (Andorka-Harcsa 1982, Juhász 1982, Szelényi 1988, Kovách 1988, 1991).

The investments in family farming remained limited by the end of the communist era and the rural households invested their increasing incomes, their profits from part-time farming and the second economy, into the human sector, mainly into the education of their children and into consumption, inducing new waves of social mobility and rapid transformation of rural living conditions and way of life. At the time of the political transformation in 1993, the Hungarian rural population lived in the best housing conditions in post-communist Central Europe (Kovách, 1997). The former monolithic structure of a peasant family, dominated by the husbands who, having a central position in the household economy, made the decisions on all the most important questions like investment, economic and social strategy, cracked and eroded. In families oriented to consumption and social mobility, the wives determined or took part in forming the strategies for better education, prestige consumption, social representation and manifestation of their new social status.

The economic activity of the family became subordinated to consumption purposes as the capacity of wives to determine family life strengthened. By the 1970s, the historical peasantry had disappeared from the society. Local communities, rural civil society and rural individuals lost their political autonomy, which had existed, although dominated by bigger landholders, before the collectivisation; when their parties, associations and cultural societies were liquidated, they lost any real representation in political life. Local civil society lost its economic basis, relations to private property, and also adapted to the intensive rural social restratification. Necessity maintained close economic connections between rural households (such as labour exchanges), but the formerly strong co-operation and solidarity dissolved.

The multigenerational, large families of the peasantry which had been the basis of local civil society atomised, and the connections of nuclear families with their kin became occasional and casual, as the various forms of the peasant social life became empty. The de-peasantisation of Hungarian rurality started decades before the economic success of collectivised Hungarian agriculture, but the politically created collectivisation, the new double structure of the economy and rural society, social mobility, and changes in the way of life or more generally the new system of social reproduction speeded it up.

The New Rural Regime: Farmerisation, Remaking the Peasants or Colonising the Countryside?

After the change of regime in the 1990s the new political elite began to build up a new rural regime. The transformation of property rights and farming structures, and economic liberalisation, radically reformed socialist structures. Arable land was re-privatised (for details see Harcsa-Kovách-Szelényi 1998b, Kovách 1999b, Kovách 2001). Some 2 million people were entitled to restitution and 1.5 million households had become landowners by 1996. Ninety percent of arable land was privatised but the national average plot size for lands acquired by restitution was only 4.4 hectares per household. The peasant land ownership and production structure could not be restored by restitution (Kovács K. 1994)

The organisational structure of agriculture keeps on changing. Private production is becoming dominant in agriculture but the farming structure is heterogeneous (re-established co-operatives, family farms, new enterprises, part-time farms) (Harcsa 1999, Kovács and Bihari 1999). While the registered number of full time food producers in the co-operative and family farming sectors together amount to some 300,000, nearly 1 million households run part-time farms (see interpretation of term in societal de-peasantisation section of this paper). In terms of number of participants, the primary form of agricultural production is still part-time farming. Between 1992 and 1999 some differentiation emerged among the top family farms (see Harcsa 1994, 1995, Harcsa-Kovách 1996, Kovách 2001 for details). A weak stratum of family farms were able to develop into real enterprises (Western farms), even under the adverse economic conditions of the period, when gross agricultural production dropped by 1998 to 60 percent of what it had been in 1988, and in the same period the numbers in agricultural employment shrank to one third, not more than 10 percent of rural dwellers! Farmerisation is an uneven project just as remaking the peasantry is.

Economic reforms are linked with a dramatic re-stratification of rural society. With the appearance of massive rural poverty and its new forms, identified by several researchers as a rural underclass (Csite and Kovách 1997, Ladányi-Szelényi 1996, Laki 1997), new actors and networks came into existence in the process of democratisation and the establishment of a new rural regime including new administrative and economic elite groups, experts on the new development policy. The monolithic socialist power structure has been replaced by multi-actor networks. New leisure and producing classes claim the use of rural resources. Some rural regions, mainly in Western Hungary, are being industrialised or re-industrialised by multinational and Hungarian firms through a putting-out system, the reorganisation of commuting and the foundation of new factories. Yearly some 10,000 migrants leave big cities for rural residences. Pensioners and other losers in the post-socialist transformation hope to have a better chance for survival, while the urban upper middle classes and non-Hungarian citizens move to leafy suburban settlements and remote villages. Pioneers of new producer roles, developers of rurality (Csite 1998), and members of religious sects are invading villages. Rural tourism, from 'vine routes' to hunting, attracts millions of visitors. Bruckmeier (in this volume) suggests using concepts of 'colonisation of the rural' and 'post-traditional rurality' to describe the post-modern nature of such processes. As I consider it, de-traditionalisation is rather a cultural than a structural term, although it is also true that theorists such as Sztompka interpret de-traditionalisation as both cultural and structural phenomena. Elsewhere we have argued for using the concept 'pluralised rural' (Csite and Kovách 2000). By this we mean no more than that (i) the Hungarian countryside dominated by food production exists no longer, the rural economy is pluralised by other sectors of the economy; (ii) rural society also has a more pluralised structure than ever before; (iii) new rural images as well as new actors and policy networks are appearing in the 1990s; (iv) rural problems and the definition of rurality are more focused in policy and in the mass media. Hungarian rurality is more pluralistic than Bruckmeier interprets post-socialist rurality to be, and it is indeed an uncomfortable task to try to predict the directions of processes of de- or re-peasantisation.

Societal De-peasantisation – the Meanings of Small Scale Farming in Socialism and Beyond

In the late period of socialism, more than 80 percent of Hungarian households had additional sources of income beyond their regular wage from different sectors of the second economy. Small-scale family farming was a most popular form in which around 60 percent of the total adult population participated. Ninety percent of rural dwellers and 30 percent of urban inhabitants ran some kind of farm. Small-scale family farming had a massively increasing commodity character but the importance of self-provisioning also remained stable.

Did this small-scale farming contribute to de-peasantisation if it preserved many attitudes and practices of peasant farming? What was the meaning given to these farming practices by food producers? One word of caution: Eastern analysts, including myself, apply the term of part-time farming to Eastern family agricultural production very carefully. Peasants and farmers in the West may obtain incomes from other sources but their off-farm work is a contribution to the survival of farming activity. In Hungary, by the early 1960s, almost every member of the labour force was employed in the socialised sector (rural dwellers in co-operatives, state-farms, industry and service sectors) and those involved in small-scale agricultural production ran family farms outside their working hours. While the most significant characteristic of Western part-time farms is their focus on economic accumulation (or the survival of commodity farming), Hungarian family part-time agricultural production rather served the purpose of extended social reproduction. This should not be identified with a Chayanovian state of peasant farming, the condition of peasantry in Hungary surpassed this by the beginning of the 20th century. But how can we understand the social function of small-scale food production in socialism, which has been viewed by most interpreters as a way of resistance against the pressures of collectivisation and proletarianisation? I argue that the societal meaning of socialist family farming is de-peasantisation. From a Chayanovian perspective, de-peasantisation is the destruction of family-oriented farming in which peasant practises are part of a strategy for social reproduction rather than outcomes of a search of profit. The Hungarian peasant economy in the first half of the century was in a transitory situation from family to market and economic accumulation oriented farming. The key goal of peasant farming was investment of accumulated profit into the economy, above all into land property. They preserved their status within the peasant stratum by very productive farming.

Did socialist modernisation reproduce this transitory phase and preserve a peasant state on a different level? The paradox of Hungarian family farming is that while market orientation continuously strengthened from the late 1960s to the early 1980s and households supplemented their incomes to an increasing extent they invested profit into consumption rather than into extended economic reproduction. This can be considered as a reverse of the Chayanovian model: intensifying small-scale farming practises in socialist circumstances are explicable as part of a strategy for extended social reproduction which reoriented cultural norms and disarranged the historical frames of peasant state.

In the arguments of socialist politicians and proletarianisation theorists this process of de-peasantisation and its accompanying social-positional mobility as well as the modernisation of rural life circumstances are primary evidence for the legitimisation and viability of the socialist model and of collectivisation. I am obliged to call readers' attention to the fact that while the socialist attempt undoubtedly modernised agricultural production and the standard of living rose higher than ever before, the capacity of rural dwellers to survive dramatically weakened. This was the price for modernisation. Post-socialist liberalisation and the contradictory making of a market economy at the same time as the Europeanisation of rural policy (Kovách 2001) met low levels of participation and interest, and even mass demonstrations of resistance by top farmers and representatives of co-operatives.

Attempts to conceptualise societal de-peasantisation on these lines might be more explicable if we look at some characteristics of family agricultural production. Small-scale farms functioned under various forms of limitation and control. Private ownership and purchasing of land were regulated and strictly limited by law. Trading in the means of production and other materials for industrial use were hindered by the gap between the prices of agricultural and of industrial products. Limits on agricultural loans and terms of repayment resulted in a permanent shortage of capital. The infrastructure available for small-scale farmers was obsolete to the extent that the majority of family farmers had almost no information on the world market: nonetheless their products were not destined only for regional or local market. Agrarian small-scale production was regulated so that the government and the trading companies could isolate small producers from the market. The state guaranteed prices and in fact COMECOM provided unlimited demand. Because of many factors – limited availability of capital, land and means of production, the inability of the market to function adequately, the hindrances imposed by the infrastructure and political uncertainty – only a small portion of the income earned by small scale farming was reinvested in the family agricultural farming. The limited income and profit of small-scale farming was steadily being used up to eliminate the gap between urban and rural living standards.

In cases where income was earned by not selling products directly on the market, economic decisions concerning the volume and intensity of small-scale production had to keep in mind marketing efficiency as well as demand, and many aspects of income-earning activity within various sectors of the economy were taken into consideration in order to enforce the interest of the family agricultural producer. Besides the ability to market a product, small-scale producers had to consider fluctuations in the non-agricultural labour force, and the skills and means of production available on their farms, as well as making money off the farms and finally the constraints of saving a sizeable portion of their profit for consumption.

Therefore, the part-time character of small farming also regulated and limited the marketing of their agricultural products. These peculiarities listed above led to a paradox of small scale production that, in spite of its increasing commodity and market orientation, farmerisation (specialisation of production) as one positive alternative to de-peasantisation remained an isolated phenomenon in the socialist era as it has done also in the post-socialist period despite economic liberalisation and the restoration of a market economy. 'Pluriactivity' is a generally typical symptom of family farming as well as of rural and urban economy. This is also a term which should be handled carefully in the Central European context. It means that the majority of economic actors (not only farmers!) gain their income from several different sources. The economic consequence is the capacity of the masses to survive, and from another viewpoint, difficulties in specialisation of production and in various spheres of social life (most problematically in the representation of political interests).

Family farmers modernised their residences, obtained consumer goods and could promote the educational mobility of their children, but they reproduced their social status by small scale farming. They accumulated a kind of capital in the economic sphere which did not prove to be re-convertible. They could not culturally convert accumulated economic capital which was needed for the reconstruction of their social position. Societal de-peasantisation, extended social reproduction, is a partially attained process with the consequence of farmerisation reaching deadlock, as well as mental and cultural heterogeneity.

Cultural De-peasantisation – De- and Re-traditionalisation of Rural Hungary

Rural, or actually peasant, culture is understood as a basis and source of national culture in 19th and 20th century Hungary, as in most Green Ring countries. The rural was integral to national images and symbols. From the variety of rural images historically available, two components survived romanticism and emerged as dominant in the 20th century: the articulation of relationships between national and rural/peasant imagery, and criticisms of the rural idyll (Kovách 2001) De-peasantisation is not an isolated structural change; it has a broad cultural impact on differing segments of Hungarian society and on the remaking of post-socialist and post-modern national symbols and images. The decisive cultural component of de-peasantisation is de-traditionalisation, which in the context of Hungary in the 1990s might be interpreted in varying ways.

First I look at how the values of rural people have been de-traditionalised with the disappearance of historical peasants. As was briefly explained in the section on societal de-peasantisation, in the mode of social reproduction in the age of late socialism economic accumulation was subordinated to consumption goals and this together with the erosion of institutions of peasant society directly led to decomposition of the values of the historical peasantry before post-socialism. What was learned by rural people during the 1990s is mostly the set of values of post-modern consumer society. But, as I expound elsewhere:

> The 1990s is not only the period of the spread of consumer culture. The disintegration of traditional peasant communities, which started in the previous century, had not been completed when the introduction of the socialist system began. The values and patterns of behaviour of local societies have had rather strong community control until recently. This community control was always the means by which peasant traditions could be defended in the face of the alien and dangerous outside world. The force of these external influences can only be questioned by those who have not experienced the pressures to conform, where in a community of any self-respect one could not dress and consume, greet, celebrate and mourn just as one pleased until a few years ago. This was not an archaic, or ridiculously obsolete world and set of values, and its use of objects met all the expectations of its time at the end of the century. Its survival without the old economic background (with economic relationships based on private property) is likely to be due to the fact that rural people could still count on their ability to be protected by values regulated by community control. This was the only force keeping together local society in rural settlements during the period of rapid economic modernisation, when the functions of big families, the church, the school and other institutions of control and co-ordination disappeared (Kovách 2001).

In my view, the disappearance of the values of the historical peasantry regulated by community control, or de-traditionalisation, means the end of the long de-peasantisation process. We may learn from Polish examples that local community life in post-socialist Europe is far from dead (Starosta 1998, Gorlach and Starosta in this volume) but at least in Hungary this is a new structure with no strong connection to historical peasant communities. There are quite few traditions of the historical peasantry which continue in existence and retain a strong intimate link with the present, but on the other hand we may be witnessing a rural renaissance. The dance-house movement is more popular than before (new) old local dishes have been cooked, wines have been bottled, the State Folk Ensemble is re-designing historical dances and songs, a new local elite has found legitimisation in the process of remaking local traditions and the tourist industry creates and re-creates rural values (Borsos et al. 2000). Rural images gained new importance after the introduction of the new system of regional/rural development. Competition for development resources prompted local elites to search for ways of putting into action new discursive strategies for the creation and elaboration of local/rural images. New, powerful actors of rural development, designers and intellectuals are strengthening their own class position in refashioning the rural (Kovách 2001). The emerging importance of discourses on rural image making, the 'intellectualisation' of rural development, ultimately may lead to a state of reflexive modernity. In vernacular speech, 'rural' and 'peasant' became devalued in meaning but new consumer and leisure classes are appearing in the countryside discovering the treasures of rurality. The pluriactive form of economic activity, which was a strategy of historical peasants before, has been adopted by urban dwellers, pushing them closer to the peasant way of life and making them more open to a rural/peasant past. 'Contrived' or 'invented' traditions are in the process of making.

It is a difficult task to guess what is traditional and what has been de-traditionalised. The key question of cultural de-peasantisation is, if the 'intimate link' which connects the present, or even more the future, to the past is interrupted, what are the social and economical consequences of this in the period of emergent reflexive modernity? As might be considered in other Green Ring countries, especially in the Nordic regions, the making of modern nationhood was based on a peasant heritage with enviable efficiency. As I understand it, cultural de-peasantisation in Scandinavia and maybe in Ireland and in some Mediterranean regions has proved to be an organic 'bottom up' project. Making the modern gained the interest and participation of peasants and post-peasant groups. In Hungary, cultural de-peasantisation is in a rather transitory period but for time being outsiders, rural elite groups, invaders and colonisers are much more active in culturally refashioning farming life than descendants of the historical peasantry.

References

Andorka, R. and Harcsa, I. (1982), 'A községi népesség társadalomstatisztikai leirása' (Social statistics of Hungary's rural population), in Vágvölgyi (ed), *A falu a mai magyar társadalomban*, Akadémiai Kiadó, Budapest.

Bell, P. (1984), *Peasants in Socialist Transition*, University of California Press, Berkeley.

Berend, T. I. (1990), *Hungarian Economic Reforms, 1953-1988*, Cambridge University Press, Cambridge.

Berend, T. I. and Ránki, G. (1985), *The Hungarian Economy in the Twentieth Century*, St. Martins Press, New York.

Böhm, A. and Pál, L. (1982), *Bejáró munkások* (Commuters), Társadalomtudományi Intézet.

Borsos, E., Csite, A. and Letenyei, L. (2000), *After change of regime*, BKE-Számalk, Budapest.

Csite, A. (1998), 'Constructing the miserable countryside', in L. Grandberg and I. Kovách (eds), *Actors on the changing European countryside*, Institute for Political Sciences, Budapest, pp. 231-257.

Csite, A. and Kovách, I. (1997), 'A falusi szegénység' (Rural poverty), *A falu*, vol. 1.

Csite, A. and Kovách, I. (2000), *A falusi társadalom átalakulása* (The transformation of rural society), Unpublished manuscript.

Donáth, F. (1969), *Demokratikus földreform Magyarországon 1945-1947* (Democratic land-reform in Hungary, 1945-47), Közgazdasági és Jogi Könyvkiadó, Budapest.

Donáth, F. (1977), *Reform és forradalom. A magyar mezögazdaság strukturális átalakulása* (Reform and revolution. Structural change in Hungarian agriculture), Akadémiai Kiadó, Budapest.

Gunst, P. (1987), *A paraszti társadalom Magyarországon a két világháború között* (The peasant society in Hungary between the two world wars), MTA Történettudományi Intézet, Budapest.

Harcsa, I. (1994), *Parasztgazdaságok, mezőgazdasági vállalkozók* (Peasant farms, agricultural entrepreneurs), KSH, Budapest.

Harcsa, I. (1995), *Farmerek, mezőgazdasági vállalkozók* (Farmers, agricultural entrepreneurs), KSH, Budapest.

Harcsa, I. (1996), 'A paraszti népesség társadalmi mobilitása' (The social mobility of the peasantry), in Für, L., Orosz, I. and Romány, P. (eds), *Magyarország agrártörténete* (History of Hungarian Agriculture), Mezőgazda, Budapest.

Harcsa, I. (1999), 'Farmers and agricultural enterprises in Hungary', in P. Starosta, I. Kovách and K. Gorlach (eds), *Rural Societies Under Communism and Beyond. Hungarian and Polish Perspectives*, Lodz University Press, Lodz, pp. 80-94.

Harcsa, I. and Kovách, I. (1996), 'Farmerek és mezőgazdasági vállalkozók' (Farmers and agricultural entrepreneurs), *Társadalmi Riport*, Tárki, Budapest.

Harcsa, I., Kovách, I. and Szelényi, I. (1998a), 'The Hungarian agricultural miracle and the limits of socialist reforms', in I. Szelényi (ed), *Privatising the Land: Rural Political Economy in Post-Socialist Societies*, Routledge, London, pp. 21-43.

Harcsa, I., Kovách, I. and Szelényi, I. (1998b), 'The Crisis of Post-Communist Transformation in the Hungarian Countryside and Agriculture', in I. Szelényi (ed), *Privatising the Land. Rural Political Economy in Post-Socialist Societies*, Routledge, London, pp. 214-245.

Held, J. (1980), *The Modernization of Agriculture - Rural Transformation in Hungary (1884-1975)*, Columbia University Press, New York.

Juhász, P. (1982), 'Agrárpiac, kisüzem, nagyüzem' (Agrarian market, small producers and latifundia), *Medvetánc*, No. 1, pp. 117-139.

Kovách, I. (1988), *Termelök és vállalkozók* (Producers and entrenrepeurs), Társadalomtudományi Intézet, Budapest.

Kovách, I. (1991), 'Rediscovering Small-Scale Enterprise in Rural Hungary', in S. Whatmore, P. Lowe and T. Marsden (eds), *Rural Entreprise. Critical Perspectives on Rural Change*, David Fulton Publishers, London.

Kovách, I. (1994), *Embourgeoisement Theories and Processes*, Politikatudományi Intézet.

Kovách, I. (1997), *Quality of Life in Rural Settlements in Six Post-socialist Countries*, Institute for political Sciences, Budapest.

Kovách, I. (1999a), 'Hungary: co-operative farms and household plots', in M. Meurs, (ed.), *Many Shades in Red. State Policy and Collective Agriculture*, Rowman and Littlefield, London and Boulder, pp. 126-151.

Kovách, I. (1999b), 'The lessons of history for future rural development: agriculture, rural society and state policy in 20th century Hungary', in Starosta, P., Kovách, I. and Gorlach, K. (eds), *Rural Societies Under Communism and Beyond: Hungarian and Polish Perspectives*, Lodz University Press, Lodz, pp. 17-44.

Kovách, I. (1999c), 'The post-socialist Hungarian countryside: transformations in 1990s and the alternatives', in Starosta, P., Kovách, I. and Gorlach, K. (eds), *Rural Societies Under Communism and Beyond. Hungarian and Polish Perspectives*, Lodz University Press, Lodz, pp. 361-372.

Kovách, I. (2001), 'Leadership, local power and rural restructuring in Hungary', in Halfacree, K., Kovách, I. and Woodward, R. (eds), *Leadership, local power and rural restructuring in contemporary Europe*, Ashgate Publishers, forthcoming.

Kovács, K. and Bihari, Z. (1999), 'State and co-operative farms in transition. The Hungarian case', in Starosta, P., Kovách, I. and Gorlach, K. (eds), *Rural Societies Under Communism and Beyond. Hungarian and Polish Perspectives*, Lodz University Press, Lodz, pp. 115- 151.

Ladányi, J. and Szelényi, I. (1996), *Making of a Rural Underclass; the Gypsy Village Ghetto and the Poor of Csenyéte*, Unpublished Manuscript.

Laki, L. (1997), 'A magyar fejlődés sajátszerűségeinek néhány vonása avagy a polgárosodásból kimaradó társadalmi csoportok' (A few specific features of Hungarian development/or the social groups left out of embourgeoisement), *Szociológiai Szemle*, No. 3, pp. 67-91.

Meurs, M. (ed, 1999), *Many Shades in Red: State Policy and Collective Agriculture*, Rowman and Littlefield, London and Boulder.

Petö, I. and Szakács, S. (1985), *A hazai gazdaság négy évetizedének története 1945*-1985 (The history of Hungarian economy, 1945-1985), Közgazdasági és Jogi Könyvkiadó, Budapest.

Starosta, P. (1998), 'Inhabitants' feeling of social bonds with rural and small town communities in Bulgaria, Poland and Russia', in Granberg, L. and Kovách, I. (eds), *Actors on the changing European countryside*, Institute for Political Sciences, Budapest, pp. 104-131.

Swain, N. (1985), *Collective Farms which Work*, Cambridge University Press, Cambridge.

Szakács, S. (1996), 'A földreform és a kisüzemi mezőgazdaság' (The land reform and small-scale farming), in Für, L., Orosz, I. and Romány, P. (eds), *Magyarország agrártörténete* (History of Hungarian Agriculture), Mezőgazda, Budapest, pp. 437-485.

Szelényi, I. (1988), *Socialist Entrepreneurs*, University of Wisconsin Press, Madison.

Vági, G. (1982), *Versengés a fejlesztési forrásokért* (Competition for development sources), Akadémiai Kiadó, Budapest.

3 The Changing Role of Agriculture in the Czech Countryside

VĚRA MAJEROVÁ

Introduction

The countryside and farming have long been tied by numerous links. The importance of farming for rural areas varied throughout history. Modern equipment and technologies combined with the development of trade may make rural areas independent of agriculture and its importance may continue to decline in the future. However, it is hard to imagine a society completely dependent on outside food supply, while it is impossible to see a society ignoring its natural environment in the cultivation and preservation of which farming has always played both the positive and negative roles. Agriculture is understood here to imply the people who have invested their skills, experience and feelings in farmwork, and in the improvement of the rural areas where they live and engage in agricultural as well as non-agricultural activities. These people live in a network of social ties which have, more or less, allowed them to adopt a certain type of social and economic behaviour. Further development of the countryside is not sustainable without a working social structure. This chapter focuses on how the farming population's working and living conditions have been evolving and, through the prism of sociology, tries to identify correlations between the long-term evolution of certain elements which have influenced the social structure of the countryside and farming.

Historical Background

During the time of the Austro-Hungarian Empire (1867-1918), which included the Czech lands, farming was an important sector not only for the countryside but also for the existence and stability of the Empire itself. Unfortunately, this importance was almost always related to the Empire's expansionist and defensive policies, when agriculture kept the war machine running. Both the Empress Maria Theresa and her son, the Emperor Joseph II, understood the importance of farming economics very well. The Theresian and Joseph's Cadasters gradually brought order into land ownership, although primarily for tax reasons. Their objective was to stabilise rural settlement, to found new villages in place of the by then disappearing manors and divide large farmsteads to increase the number of settled farming families who would later become the founders of the new villages.

The abolition of serfdom in 1781, together with the Tax and Levies Patent promulgated in 1789, which regulated the manors' obligations to land rulers and the church, were highly beneficial for the countryside and farming. The rural population was allowed to pursue different types of living (e.g. crafts or trade) whilst farming dominated in many areas. Only a limited number of artisans (wheelwrights, blacksmiths, saddlers, carpenters, joiners, bricklayers, basket weavers, etc.) could earn their living in a village, very often in symbiosis with farming itself. It was often difficult to reach markets for both the farm produce and non-farming products. Agriculture therefore also helped to stabilise social groups of other professions, frequently constituting the economic backbone of the whole region.

The end of the 19th century saw a rapid development of co-operative farming, initially supported by credit facilities in reaction to an agrarian crisis caused by overproduction. The agrarian crisis affected the Czech lands heavily. The social structure of the Czech village became sharply differentiated, land and property was held by the nobility (mainly those who flourished after the battle of Bílá hora) and large estate owners (often foreigners), on one hand, and a mass of small producers who worked small holdings on the other. These small farmers were unable to compete with large, better equipped farm businesses and the crisis caused them to be mercilessly liquidated. Social dissatisfaction had a nationalistic undertone, too.

It is difficult today to try and gauge the real social status of the farming profession at the turn of the 19th century. This can be deduced however from literary renditions of rustic stories[1] historical treatises and works of political science.[2] There is no doubt that independent landlords showing good results, farmers endowed with organisational abilities and leadership qualities, as well as those perceived and termed as 'wise Czech rural (amateur) scholars and chroniclers', enjoyed a high social status in their villages. A much worse position was held by members of the poor strata – the landless, farm workers and servants, and part-time labourers. Regular or seasonal employment in nearby or even more distant settlements and emigration overseas were a frequent solution to the untenable living conditions in villages at the turn of the century. Soil tilling was often left to women and older children, while the father of the family looked for a livelihood outside agriculture.

As industry developed at the beginning of the 20th century, farming began to lose its role as a major source of income in rural areas, but the cruelty of World War I served as a reminder that hunger could be one of the most powerful political tools, both during the war and in its aftermath. The Czech lands gained independence in 1918 when the Czechoslovak Republic was founded. Its first days are evocatively described by Ferdinand Peroutka (1933).[3] They brought with them a score of various radical intentions to improve society, including the 1919 Land Reform.

Despite its outcome being quite different from the original intentions, this Reform distinctly changed the face of the Czechoslovak village because it helped to strengthen the middle class of peasantry. Changing social stratification was one of the main aims and ideas of this land reform. A relative consolidation of Czechoslovak villages continued until World War II. 'Relative' because, once again, both towns and villages experienced a setback caused by the Great Depression of the 1930s. Nevertheless, co-operative farming went on developing very strongly;[4] great attention was devoted to education tailored to the needs of the countryside and agriculture; a young agrarian intelligentsia loudly demanded its place in political life and the Agrarian Party offered an attractive ideology.

During World War II agriculture resurfaced as a strategic industry on whose operation both the individual and the State depended for their survival. The German occupying power was well aware of how important a stable source of raw materials, food and money was and, therefore, subjected agriculture to rational administration by the occupants rather than suppressing it. So-called 'farming surpluses' were redistributed to Germany and credit co-operatives were directly subordinated to the Reichsbank. The damage inflicted on agriculture and the village was more of a moral than material nature.[5]

Having outlined the key developments that marked the formation of the pre-war Czech countryside and farming, the phenomena and processes which from the sociological perspective appear to be the most significant can be summarised as follows. The agrarian crisis was an important milestone, causing acute extremes in the social structure. The situation prevailing at the end of the 19th century was unfavourable to the countryside and migration increased. The growing interest in co-operative farming was a social phenomenon which, to a degree, helped to stabilise the countryside and farming. From the sociological perspective this process may be perceived as the emergence and stabilisation of rural co-operation patterns based on democratic principles. World War I dealt a hard blow to the Czech countryside. Thousands of young men in active age either returned from battlefields in bad health or did not return at all. The most significant experience of WWI was an extremely difficult situation in food supply. Hunger during and after the war underlined the importance of farming.

Interventions in economic and social life affected more institutional aspects of the new Czechoslovak Republic in 1918. The need to change the rural social structure prompted an adoption of a radical Land Appropriation Law in 1919 and Land Allocation Law in 1920, which notably changed the social stratification of the village. Great attention was given to raising the social significance of the Czech countryside and farming to a higher level. That is why the Czechoslovak Academy of Agriculture was established as early as 1924.

The Agrarian Party played a special role in political life at that time. Its position was very strong, despite the agrarian ideology being split into right and left-wing factions. The development of the agrarian movement continued from the end of the 19th century. Its political expression had roots in the merging of the economic and cultural values of both town and countryside. The abandoning of aristocratic society was real and evident, mainly because of the foreign origin of the aristocracy. Rural society reflected different lifestyles, and perceived the influence of urbanisation as a destructive factor for rural society. The social base of the Czech agrarian movement consisted of a large number of peasants. At first, they defended economic interests, later also political ones (Kubálek 1998, Čmejrek 1998). The farmer had his place in society; his existence was associated with the vision of national prosperity.

Collectivisation – a Fundamental Change in the Perception of Agricultural Workers as a Social Group

Collectivisation after 1949 caused one of the most radical upheavals in the Czech countryside and farming within their entire development history. Collectivisation was in a way an organised concentration of production forces. A kind of concentration appears in all societies during modernising processes. Land and capital concentration is a necessary attribute of a market-orientated economy, for only strong and professionally managed enterprises are able to withstand competitive pressures, compounded by natural factors (poor crops, droughts, natural disasters, etc.) to which even large-scale farming is susceptible.

West European agriculture has undergone similarly complicated processes of concentration and specialisation. However, respective governments tried to reduce their impact on small disappearing producers as far as possible, granting various special-purpose subsidies in respect of production changes, retraining both the farmer and his family, maintaining the minimum living standard, etc. These processes were accompanied by human tragedies, bitterness and hopelessness. Any loss of independence can cause depression and farming is no exception since it is almost never a mere profession but a way of life for the whole family. But the difference between forced, centrally controlled collectivisation on one hand and a complete disappearance of production due to competitive pressures on the other is very noticeable indeed.

East European collectivisation models were inspired by the violence of the Soviet pattern. A short time after the war, as early as the beginning of 1946 at the 8th Congress of the Communist Party of Czechoslovakia, the communists set forth guidelines concerning the co-operative system in general and co-operative farming in particular.[6] They had tried to push through a unified co-operative system before the war, but with no success at all. Eventually, they managed to do so in July 1945 when the Central Co-operative Council became the highest-level organisation to which nine new independent co-operative associations were subordinated.[7]

Agricultural policies were launched by extensive personnel purges. The unification of the co-operative system marked the end of all the previous co-operative institutions (namely Kooperativa, Moragra, Zemka, Peru, and other marketing societies); the entire purchase of farm produce and its processing were subjected to a monopoly.[8] Credit co-operatives were made directly responsible to the State Bank of Czechoslovakia, thus becoming part of the State's financial sector and being deprived of all the scope of independent operations. The co-operative system as a democratic form of rural co-operation was rapidly heading towards extinction.

Today, historians are beginning to describe the elements of the collectivisation process as including: the most efficient means of pressure comprising class-based scheduling of produce supplies and purchases, forced sales of machines, a ban on employing labour, forced threshing and sale of grains, forced leasehold, forced land swaps, the seizure of textile and footwear rationing vouchers, a ban on hog killing for one's own needs, and many other restrictions.[9] The economic and social pressure exerted on peasants grew.

In February 1949, the National Assembly passed Act No. 69 on Standard Agricultural Co-operatives whereby all the co-operatives hitherto existing in a village were amalgamated or transformed into a single co-operative, while the formation of additional Standard Agricultural Co-operatives (SACs) was launched. These newly formed SACs did not operate according to the pre-war principles of co-operative farming; rather, 'Standard Statutes' of 1949 were imposed on them. The Statutes had been approved by the Ministry of Agriculture and regulated in detail the terms of reference of the co-operatives and their management practices.

The forcibly collectivised co-operatives failed to show good results for a long time. The co-operative peasants' standard of living was low and it was therefore decided not to seize their crofts (private backyard plots) in order to allow some compensation for their inadequate financial income. It was only the 1960s which brought about a partial consolidation for the countryside. The persecuted generation tried to forget, the next generation was starting a new life in the village under different circumstances. The wrongdoings inflicted on their fathers had lost some of their edge, life went on, down a different path.

The sociological aspects of the era can be summarised as follows: the post-war Czechoslovak village encountered terror, lawlessness and violence to an extent not previously known (Kaplan 1993, Jech 1993). The rural élite of able and industrious agricultural workers was systematically liquidated in economic and social terms. Whole families were affected, including the children whose access to education was restricted. Massive propaganda ridiculed farmers, frugality was interpreted as niggardliness, employing labour as exploitation.

Co-operative work on common democratic grounds disappeared, the new type of co-operative imposed different rules. Personal responsibility gradually vanished, to be replaced by collective irresponsibility compounded by feelings of injury and injustice in the wake of land and property confiscation. An outflow of labour into industry led to depopulation of villages, the farming profession lost prestige, a solely agricultural education was imposed as punishment on the children of unreliable class enemies. The Communist Party of Czechoslovakia together with the Ministry of Agriculture took over the political representation of agricultural workers and steered the process of collectivisation.

All democratic civic groups and political parties of the pre-war type vanished from the villages, to be replaced by the National Front and a system of special-interest organisations having no real political power. Remuneration based on a 'work unit' in poorer areas did not even provide for an average standard of living. Additional work on their crofts put burdened villagers and peasants with a round-the-clock workload, stifling in them all interests other than preserving one's sheer subsistence. Cultural traditions were replaced by collective entertainment models, cultural heritage disappeared.

Public opinion was manipulated. Only rarely did real farmers appear in the media, which promoted role-model milkmaids or 'Collectives of Socialist Work'. The farmer gradually turned into a surreal character within the society, depicted as a tractor driver in overalls with a broad smile, while in real life resembling a shabby labourer in mud-covered rubber boots. The seeds of ridicule, misunderstanding and contempt were sown.

Village for Life or Recreation?

Collectivised mass-producing agriculture had its advantages as well as disadvantages. The positive features stemmed from land and capital concentration and opportunities to deploy equipment and technologies leading to economies of scale. The disadvantages primarily reflected the society's social conditions. Throughout the existence of socialist agriculture, differences between so-called State Farms and SACs persisted, whether they were similar in size or not. Alienation was always deeper at State Farms than at SACs, despite land ownership in the latter being hypothetical only. Co-operatives nevertheless showed a more stable social structure as well as better on-the-job relations and results, even under comparable natural conditions.

The Soviet occupation of Czechoslovakia in 1968 and subsequent 'normalisation' (the communist terminology for bringing the country back under Soviet control) did not affect the countryside too much, as the centres of intellectual resistance which sprang up were mostly in larger cities. The hardening of the political and economic course – strictly orientated towards the Soviet Union – was reflected in structural changes in the early 1970s. These changes concerned both SACs and State Farms – their acreage was markedly expanded through amalgamation.[10] At the same time, a massive replacement of management personnel took place.

To be fair, amalgamated enterprises showed relatively good economic results for a certain period of time. In fact, a similar trend towards concentration and specialisation in agricultural production was then under way in advanced capitalist countries, only the ways and means of fusing agricultural units to form large farms differed. However, the differences between state and co-operative farms had existed all along.

Historically, state farms provided higher benefits to their employees. In the 1970s, the government's agricultural subsidy policies changed, the conditions of social consumption in the state-owned and the co-operative sectors levelled out and the standard of living improved. Actual differences between state farms and co-operatives gradually adjusted, specifically in salaries, the duration of maternity leave and conditions of social and health insurance. Other benefits were provided for agricultural workers and inhabitants of the villages – for example farm-crèches, nurseries, kindergartens, child care leave for mothers and summer vacation camps for children. Also provided were canteens, transport to and from work-places by farm buses, recreation facilities, organised leisure time activities such as sports and cultural events, etc. Most leisure time in villages was devoted to building, modernising and renovating houses and caring for gardens and private plots. Rural household amenities gradually improved.

At the same time the urban population discovered the value of spending weekends in the countryside, sometimes in a second home. Besides the purely material advantages of having a second home in the countryside, this trend also reflected social conditions. Owning a cottage offered an opportunity to escape into one's own secluded world, sometimes to return to the ancestors' way of life or to discover nature and rural communities. However, Czechoslovakia's socialist, mass-producing agriculture reached its peak in the second half of the 1980s, whereupon it began to decline, in economic terms. In 1989, political changes took place in Czechoslovakia.

The main social changes occurred in workplaces, in leisure time, the mobility of the population and their values. De-peasantisation continued in the various spheres of economic and social life. The level of qualifications of agricultural workers increased, as did mechanisation and technological advances in agricultural production. These processes led also to less favourable results – alienation from nature and a worsening of the farming population's state of health (an increase in the occurrence of injuries, allergies, occupational diseases). At the same time a system of extensive care for employees was provided by agricultural enterprises.

The importance of television in the village increased in relation to the decrease in importance of earlier rural patterns. The togetherness of neighbours was diminishing, external services replaced mutual help between rural inhabitants. Consumerism (i.e. the purchase of status-enhancing goods) was on the way up, urban patterns influenced village lifestyles (dress styles, household furnishing, sports etc.). At that time, the three most characteristic items of rural household assets were a motor car, colour television and a freezer. Life in villages adapted to that in the towns and its typical rural features disappeared. The inflow of holiday-makers (including cottage owners) into villages increased in recreational localities.

Villages, especially smaller ones, grew older. Young people could get education without any restrictions but employment in agriculture was losing attractiveness for them in comparison with life in town. Gradual depopulation affected most those small villages with a low level of infrastructure, as only older people remained there. Agriculture gradually became an attractive profession only for families living near agricultural enterprises, for women with children, for skilled men with a technical education who could achieve a good income in mechanised professions. Social programmes run by agricultural enterprises, the system of recruitment and stabilisation allowances, opportunities for cultivating private plots and gardens, accumulating family income, assistance provided by relatives and neighbours, limited spending opportunities and traditional frugality – all of this made agricultural workers a relatively affluent social group.[11]

The social standing of farming professions increased to various degrees according to local circumstances although it depended considerably on the proximity of infrastructure and commuting possibilities. The limited opportunities for upward career mobility in an agricultural enterprise were reflected in the state of general dissatisfaction, which intensified in the Czechoslovak society before 1989. Then, with the political changes of 1989, the prospects of weakening the barriers that prevented people from applying their skills and abilities raised hopes for improving their well-being; also a restitution of land and property began to be expected with heightened interest.

Winners and Losers – or Cheated by the Velvet Revolution of 1989

The first news of property restitution was anticipated with hope, although many people were not clear what the restitution would imply. Attitudes to privatisation were very positive in general although considerable reserve persisted in specific cases. The first research conducted after 1989 (Hudečková 1991, Majerová 1990, 1991, 1992) indicated that agricultural workers were satisfied with the changes occurring at that time.

Free elections, the end of the Communist Party's hegemony, departure from the Marx-Leninist ideology, abandonment of the practice of filling managerial posts based on Party membership and ranking, the re-emergence of religious life, political rehabilitation and opportunities for free decision-making were ranked highest of all in the study. However, doubts were evident as to whether civil and political rehabilitation would actually happen and whether formerly persecuted persons would be able to find their place in political and social life.

Such doubts were most frequently voiced in connection with the aggressive way former Communist Party 'nomenclature cadres' made their presence felt – in many agricultural enterprises they remained on in their positions. The idea that these people might make up for the wrongdoings they had inflicted in the past on others and become guarantors of democratic development appeared rather unbelievable to many villagers.[12] At this stage of the country's transformation, economic objectives included a complete diversification of production units, the maintenance of economies of scale derived from mass production while making its organisational structures operate more efficiently, creating the preconditions for a natural restructuring of production which would take into account soil and climatic conditions, and shortening the path from agricultural producers to consumers.

At the same time, legislative preconditions were being prepared for launching work on privatisation proposals for state-owned organisations and transformation proposals for agricultural co-operatives, i.e. the governments of the respective Republics decreed how compensation for livestock, equipment and inventories was to be calculated.[13] It was not easy to clarify ownership titles either. A considerable part of the Czech border territory, where agricultural production constituted the primary or major source of living, was a strategic area which had been affected by war, desertion/expulsion, resettlement and then collectivisation. Identifying logical sequences in transfers of property rights was extremely complicated in both legal and logistical terms and highlighted how sensitive the human aspect of the transformation was. The entire restitution process lasted longer and proved to be more complicated than originally expected; some of its aspects still remain open now.

In 1991 agricultural enterprises still continued their operations in the original scale, structure and orientation, but preparations for systematic changes were already under way. These preparations were both official (drafting the procedures for disaggregating enterprises into smaller units, collecting and preparing documentation and accounting statements for privatisation, settling restitution claims etc.) and unofficial (drafting legal and economic documents for transactions involving transfers of state-owned and co-operative property into private hands) in nature. The absence of laws to protect the legality and transparency of the privatisation process, inefficient overseeing of the process, networking, and the well-oiled organisation and unscrupulousness of former powerholders all made it easier for some individuals to enrich themselves beyond their due share while casting a shadow of doubt on some aspects of the privatisation process.[14]

Some forty years have passed since the beginning of collectivisation. Among the original farmers and land property owners, only very few could be counted on as farm workers. However, in the context of restitution claims, heirs without any direct relationship to land cultivation proliferated; their numbers were estimated at 3.5 million. This structure of eligible claimants also determined the way in which restituted property was disposed of: sale, donation, lease, realisation of investment for an existing business or to start one's own business could all be considered. Visions of returned property would be used usually reflected the personal and family situation. Men were more inclined towards engaging in business, including private farming, while women preferred property sale or lease. On the whole, however, much vacillation prevailed in relation to property because the fact itself of property restitution to the original owners or their heirs felt unusual and was outside any previous experience. This initial uncertainty was stronger in workers with a lower level of education, senior citizens, women and, in geographic terms, the Slovak part of the Czechoslovak federation.[15]

In respect of employment, a sizeable shift in opinion could be remarked. Fear of losing one's job, infrequent before this, grew increasingly acute. The first major wave of dismissals swept over retirement age employees; some administrative and white-collar workers were dismissed, or later left of their own accord. Gradually, agricultural enterprises divested themselves of their non-farming operations. Losing a job turned out to be a very serious problem. Some agricultural workers felt totally helpless, others tried to find jobs using various known and new ways and means – through job centres, advertising, enquiries among relatives and friends, commuting or starting their own business, including private farming. Retirement and, for women, staying at home as housewives, also had to be considered. Manual workers with low qualifications faced the greatest handicap; they found themselves in a situation where their basic material and social certainties were jeopardised. They were aware of their inability to meet the requirements for a substantial improvement in the quality of their work, for upgrading their qualifications, or retraining. They were poorly informed about legislative regulations and opportunities for applying their skills elsewhere. They characteristically expressed strong dissatisfaction with the situation then prevailing in agricultural enterprises, criticism of managerial personnel, lack of confidence in the future, refusal to accept any changes (with the exception of replacements in managerial posts) and demands to keep the same job, in the same company, under the same terms and conditions.

Efforts to deal with problems in employment and living conditions increased with the level of education. The best aptitude for adaptation was shown by workers in mechanised professions (tractor and combine harvester operators, drivers, repair and maintenance workers) whose professional orientation increased their employment opportunities both inside and outside agriculture, and also by some white-collar workers who, thanks to their managerial experience, were able to find their way through the maze of legal regulations and intensively sought pragmatic ways out of their new situations.[16] The new structure of agricultural production units in which workers could seek employment is described in Table 3.1.

Table 3.1 The Size of Agricultural Enterprises (to 31.12.1993)

Enterprise	Acreage in thous. ha	Percent	Average acreage in ha
Enterprises arising from transformation of co-operatives	2279	53.2	1357.0
Enterprises arising from privatisation of state farms	566	13.2	599.0
Private farmers	780	18.2	15.0
- under 1 ha	15	0.4	0.5
- 1-10 ha	83	1.9	5.6
- 10-50 ha	173	4.0	24.6
- 50-100 ha	89	2.1	72.7
- over 100 ha	420	9.8	397.3
Private sector together	3625	84.6	67.0
Private sector together without farmers to 1 ha	3610	84.3	135.0
Other enterprises (school farms. residual farms etc.)	657	15.4	-
Total	**4282**	**100.0**	-

Source: *Green Report of Ministry of Agriculture. Czech Republic, 1994.*[17]

The agricultural workers who chose private farming differed substantially from the majority of farming population. Return on investment was, and still is, uncertain. Some private farm businesses set out on the path of intensive large-scale production on rented land supported by modern technologies, other private farmers opted for extensive production on small holdings.[18] The diversification of private agricultural production has developed into what today are relatively clear contours. Although the average 15 hectare private farm is not large, if we include corporations the average holding per farm is 67 hectares, and after deducting some 28.000 one-man units working up to one hectare of land it transpires that the resulting average cultivated area of 135 hectares markedly exceeds the size of private undertakings in the EU countries where the average area is 14.7 hectares.[19]

The changes brought about by transformation and privatisation in agriculture can be summarised as follows. Agriculture is losing its position of an industry enjoying permanent attention. Large-scale agricultural production is experiencing economic problems which limit the development of social programmes for the countryside and farming. The restitution process has been a completely new experience. Rural society looks to property restitution with anticipation. However, in the transparent environment of villages under strong social control, any untoward scheming in restitution and privatisation can be seen more clearly than elsewhere. Not all injustices have been corrected. The slogan 'Love and truth will win over lies and hatred' (buzzword of the 1989 Velvet Revolution in Czechoslovakia) can elicit bitterness at times. Some wrongdoings have been redressed, others have yet to be resolved; in some cases, old wounds may have been re-opened.

The loss of economic and social certainties is accompanied by disillusion with political development. The growing electoral preference for the Communist party may raise concern. Agricultural workers face unemployment and have considerably worse conditions for dealing with it than urban population. The transformation and privatisation of agricultural enterprises is producing both vertical and horizontal social mobility. Considerable changes take place in the hierarchy of local structures, particularly in places where the agricultural enterprise constitutes the community's economic and social pillar. The standard of living of farmers and of the rural population is becoming significantly more differentiated.

Creating social structures on the principles of popular democracy is difficult for many reasons: strong traditions, wealth of experience, plentiful finances and material means are all lacking. The rural population is scattered and malfunctioning telecommunications together with deteriorating and increasingly expensive transport services cause almost total isolation of some rural areas. The political representation of the countryside is mediocre. Rural problems are almost permanently absent from the media.

It is apparent that agriculture as the primary source of income is economically sustainable only in areas with favourable natural conditions.[20] The transformation of agricultural enterprises and their basic re-categorisation is coming to an end, marked by all of the above problems and implications. The countryside of today, its economic and social structure, faces additional important changes.

The European Prospects

Such information as is available about the conditions of the Czech Republic's future integration into the European Union does not allow us at present to identify which parts of the Common Agricultural Policy will be beneficial or otherwise to Czech farmers. Some of the economic and social processes already under way in Europe are identical with those in the Czech Republic, and are not associated solely with either the capitalist or the socialist system of production. Examples of these are:

- the decline in the importance of agriculture as the primary source of livelihood in areas with unfavourable conditions;
- the increasing level of education of the population, including the rural population;
- the weakening interest among young people in farming professions;
- the de-population of smaller villages;
- the deterioration of the natural environment;
- the increased occurrence of 'civilisation diseases';
- the growth of social tensions in areas suffering from a high rate of unemployment.

Nowadays Czech agriculture is experiencing inhibited growth and its poor competitiveness expresses itself in the differential development of different regions. Depression in living standards and uncertainty about the future development of rural areas lead to growing dissatisfaction and to some social tension in the villages. It is clearly not possible to return either to older conceptions of peasant farming or to socialist large-scale production. The attempt at renewal of private farming after 1989 was only partly successful from an economic point of view, especially in these cases where farmers left off peasant methods of farming and tried to use the managerial practices developed in large-scale farming.

However, socialist large-scale farming was constructed as a model of highly productive agriculture linked to an adequate market space. There is no such space for present-day agriculture. Moreover, the earlier relative self-sufficiency of the villages has disappeared. They are now much more connected with towns, not only in terms of their communication systems, but also by the mass media. Actions shown on TV influence people more that those that take place in their own neighbourhood. Various aspects of environmental as well as country protection become very important. It is well known that only some resources can be reproduced. Their permanent exhaustion and deterioration is a hazard for the future of humanity. The social environment also changes. Every generation has its own experience and interpretations. Some values may continue to be shared, but a return to past is impossible. In any case, contemporary interpretations of the past carry features of idealisation and ideologisation.

Looking objectively at the common European future of rural areas, the main task for the Czech countryside will be the gradual achievement of a balance between the various regions, especially from the point of view of employment. This means that agriculture need not be everywhere the main source of employment and incomes. If agriculture is not the main one than we need to find other sources of employment. Natural resources, cultural heritage, closeness to important communication systems, recreation possibilities, local advantages of various kinds can all be utilised. Good migration and immigration policies will be essential. We consider that migration processes will continue. They are a significant part of social development and their strength and structure will play a crucial role. At the same time, settlement policy must also be developed, as the Czech Republic is more and more also a destination and not just a transit country for migrants.

The Czech village needs also to strengthen its social prestige in the mass media. In contemporary Czech society, which is so strongly influenced by the media, news of the countryside and its problems is rare. Social invisibility helps to support the centralist approach of the state and puts a brake on the construction of a democratic civil society in the countryside. Though the rural population is a majority in the whole Czech population, it finds itself in the position of a minority. Loss of the peasantry and its ideology is a historical reality and a return to past history is not possible. However, emancipation of the rural population should not be connected only with the peasantry and agricultural production. Economic and social self-confidence could have other roots. The power of the so-called 'new rurality' can spring from a recognition of the immense advantages found in local society, which has close to personal knowledge of all the important actors in local life, and from a renewed trust in its own powers and capacities for collaboration.

Acknowledgements

The first part of this chapter has been published in Czech under the heading '*Měnící se role zemědělství v trvale udržitelném rozvoji venkova*' in the Proceedings of a scientific conference on The Czech Society at the End of the Millennium organised by the Faculty of Social Sciences of Charles University in Prague on March 12th to 14th 1998 on the occasion of the 650th Anniversary of Charles University.

Notes

1 E.g., B. Němcová, K. Světlá, J.Š.Baar, J. Holeček, K. Rais, A. Jirásek and other authors.

2 E.g., Krofta, Kamil (1919), *Přehled dějin selského stavu v Čechách a na Moravě*, Praha; Pohl, J. (Doberský 1948), *Vylidňování čs. venkova v Čechách 1850-1930*, Praha; Říha, O. (1948), *Hospodářský a sociálně politický vývoj Československa*, Praha; Volavka, A.(1948), *Sto let zemědělské politiky*, Praha; Vrba, R. (1889), *O zachování stavu rolnického*, Praha; Žalud, A. (1919), *Česká vesnice*, Praha.

3 Peroutka, Ferdinand (1933), *Budování státu*, Part I. Fr. Borový, Praha, p. 241.

4 Helešic, F. (1997), *Právní úprava družstev ve vyspělé Evropě a u nás*, Law School of Charles University, Praha, pp. 11-14.

5 "There is no need to idealise the Czech farmer and look in retrospect for his patriotic deserts during the War. Besides industrious, brave and self-sacrificing people there were those who without any scruples engaged in profiteering and collaborated with the occupants. It is difficult and pointless to express mathematically which of those deeds predominated."

6 The key policy of co-operative farming was formulated as follows: "To provide for effective support to co-operative farming and its full democratisation and simplification so that co-operative farming can become an effective tool in the restructuring of the entire distribution system, as well as for production cost cutting in agriculture" *(Protocol of the 8th Communist Party Congress*, p. 145).

7 Association of Co-operatives for the Elevation of Agricultural Production; Association of Warehousing and Related Agricultural Co-operatives; Association of Agricultural Engineering Co-operatives; Association of Credit Co-operatives; Association of Production and Labour Co-operatives; Association of Small Trades Co-operatives; Association of Housing Co-operatives; Association of Community and Cultural Houses.

8 Co-operative societies of blue-collar workers, artisans, small trades and agriculture.

9 Above all Kaplan, K. (1993), *Nekrvavá revoluce*, Mladá fronta, Archiv, Praha, pp. 224-225.

10 At the beginning of collectivisation in 1950, an average SAC had 359 ha of farmland and still in 1960 by when average acreage had increased to 420 ha of farmland an

average SAC equalled in size one community's area (there were 10,816 SACs for 19.343 settlement units in Czechoslovakia). Before 1970, amalgamations went on relatively slowly (farmland area averaged at 625 ha per SAC). The figures are 1.325 ha in 1975; 2.476 ha in 1980; and 2.507 ha in 1982 remaining so until 1990 (2.504 ha). In: Karlík, J. (1991), *Vnitřní a vnější aspekty přechodu agrárně potravinářského komplexu ČSFR na tržní mechanismus*, Academy of Agricultural Sciences, Vol. 144, Praha, p. 7.

11 Generalisation based on final reports produced by VÚEZVž (Research Institute of Agricultural Economics and Nutrition) in Prague between 1977 and 1988: Social position of agricultural youth; Social position of women working in agriculture; Comprehensive care for the handicapped; Comprehensive care for retired agricultural workers in the system of social planning; Position of selected social groups in the context of the concentration and specialization processes in agricultural production; Structure of income and expenditures of co-operative peasants' households; Trends in household income and amenities; Formation and differentiation of the socialist way of life of agricultural workers; Social development in the agro-food complex; Trends of selected components of socialist agricultural workers' standard of living; Proposal for improvements in the working and living conditions of women active in agriculture in the 8th and 9th Five-year Plan periods.

12 Majerová, V. (1990), *Důsledky současných společenských změn v sociální oblasti v zemědělství a na venkově*, VÚEZVž Final Report, Praha, pp. 44-45.

13 Lipták, M. et al. (1991), *Strategie transformace APK a její konkretizace v jednotlivých organizačních formách*, VÚEZVž Final Report, Praha, p. 2.

14 Majerová, V. (1990), *Důsledky současných společenských změn v sociální oblasti v zemědělství a na venkově*, VÚEZVž Final Report, Praha, p. 46.

15 Majerová, V., Němcová, E. (1992), *Ekonomické a sociální důsledky transformačního procesu pro pracovníky zemědělských podniků*, VÚEZVž Final Report, Praha, p. 29.

16 Majerová, V. (1992), *Sociologický pohled na změny v československém zemědělství po r. 1989*, Sociology of Agriculture 1/1992, p. 37.

17 Green Report (1994), Ministry of Agriculture, Czech Republic, p. 9.

18 In figures: 53.7 percent of private farmers worked less than 1 ha (0.5 ha on average), 28.6 percent up to 10 ha (5.6 ha on average) and 9.8 percent up to 30 ha (18.6 ha on average). Larger farms took a much smaller share, 3.7 percent of private framers had 30 to 50 ha (40.2 ha on average), 2.4 percent 50 to 100 ha (72.7 ha on average), 1.2 percent 100 to 250 ha (186.4 ha on average), 0.3 percent 250 to 500 ha (336.4 ha on average), 0.2 percent 500 to 1.000 ha (689.1 ha on average) and 0.1 percent over 1.000 ha (1.704.7 ha on average). Green Report (1994), Ministry of Agriculture, Czech Republic, p. 178 (A7) and the author's research.

19 When making comparisons with the EU countries, a difference must be made between family and non-family farms. This is not easy in the Czech Republic, either, because private farms 'up to 1 ha' and 'over 1.000 ha' may be both family and non-family farms if the relevant criterion is the share of family members working it and the share of hired labour. It is, however, obvious that with increasing size of private farming the share of rented land increases. Large private farms (over 50 ha) are very likely to be – in many economic terms – more like incorporated companies working on soil.

20 Generalisation based on final reports produced by VÚEZVž (Research Institute of Agricultural Economics and Nutrition) in Prague between 1989 and 1995: Consequences of current changes in the society for the social sphere in the countryside and farming; Qualitative analysis of social structures in APK; Transformation of social structures in agricultural enterprises; Economic and social consequences of the transformation process for agricultural workers; Private farmers – a newly rising social group of agricultural workers in the Czech Republic.

References

Čmejrek, J. (1998), *Agrární aspekty formování demokratického politického systému*, Inner material, PEF ČZU Praha, p. 31.

Green Report (1994), Ministry of Agriculture, Czech Republic.

Hudečková, H. and Lošťák, M. (1991), 'Ochota k soukromému hospodaření na půdě', *Sociologie zemědělství*, 1/1991.

Jech, K. (1993), 'Vyloučení selských dětí ze zemědělských škol (1951-1953)', in *Stránkami soudobých dějin*, ÚSD, pp. 123-141.

Kaplan, K. (1993), *Nekrvavá revoluce*, Mladá fronta, Archiv, Praha.

Karlík, J. (1991), *Vnitřní a vnější aspekty přechodu agrárně potravinářského komplexu ČSFR na tržní mechanismus*, Academy of Agricultural Sciences, Praha, Vol. 144.

Krofta, K. (1919), *Přehled dějin selského stavu v Čechách a na Moravě*, Praha.

Kubálek, M. (1998), 'Postavení agrární strany v politickém systému první republiky', in *Sborník Agrární perspektivy* VII., PEF ČZU, Praha, pp. 926-932.

Lipták, M. et al. (1991), *Strategie transformace APK a její konkretizace v jednotlivých organizačních formách*, VÚEZVž Final Report, Praha.

Majerová, V. (1990), *Důsledky současných společenských změn v sociální oblasti v zemědělství a na venkově*, VÚEZVž Final Report, Praha.

Majerová, V. (1992), 'Sociologický pohled na změny v československém zemědělství po r. 1989', *Sociologie zemědělství*, 1/1992.

Majerová, V. (1995), *Private Farmers – Newly Rising Social Group of Agricultural Workers in the Czech Republic*, RSS CEU, Praha.

Majerová, V. and Němcová, E. (1992), *Ekonomické a sociální důsledky transformačního procesu pro pracovníky zemědělských podniků*, VÚEZVž Final Report, Praha.

Peroutka, F. (1933), *Budování státu*, Part I. Fr. Borový, Praha.

Pohl, J. (Doberský) (1948), *Vylidňování čs. venkova v Čechách 1850 – 1930*, Praha.

Volavka, A. (1948), *Sto let zemědělské politiky*, Praha.

4 Rural Change in Bulgarian Transitions

BOB BEGG AND MIEKE MEURS

> Winter had begun. The first snow of the year had fallen. It covered the muddy streets of the village, softening the nakedness of the fruit trees which stood lonely in the bare yards. The wind from Mt. Vitosha came whistling down the empty lanes, sending both man and beast to seek protection and warmth. The fall sowing of wheat was in. The fields, wind-swept and snow-laden, were deserted, seemingly forgotten by the men and women who had spent so much time with them during the past months. One ordinary day followed another (Sanders 1949, 1).

Introduction

Irwin Sanders' study of change in one Bulgarian village between 1934 and 1945 emphasises the continuity of rural life, even during this period of change. Much more radical changes occurred in the Bulgarian countryside in the second half of the 20th century than those occurring from 1934-1945. In the 1950s, the majority of agricultural land was transferred from private to state control, and in the 1970s and 1980s small-scale industry was located in the countryside, creating a class of rural semi-proletarians. The 1990s have again brought abrupt re-organisations, as private property in agricultural land has been returned to rural households and parts of rural industry and the central state have collapsed. Referring to the most recent transition, Francis Fukuyama (1989) declared the "end of history", presenting a substantially differing perspective than Sanders', a version of history emphasising extreme separatism or discontinuity with the past.

In this chapter, we examine the most recent changes in Bulgarian rural life from the two perspectives presented above: history as 'separate', as a disjunction from communism, and history as organic, "a legacy of complex symbiosis" (Todorova 1996, 48) of aspects of the communist and post-communist periods. We do not argue for the correctness of either perspective, rather we illustrate ways that the social life of Bulgarian villagers[1] has been altered substantially by recent changes and, at the same time, exhibits strong continuities with both the communist and pre-communist past. By drawing on both the 'separatist' and 'organic' perspectives, we come close to what might be seen as an emerging 'third way' in the debate on historical perspective.

This third position, advocated by Katherine Verdery and Michael Burawoy (1996), suggests that emerging institutions and behaviours are neither entirely new artifacts of the new conditions nor pure continuities from the past. Instead, they are new behaviours which are possible and desirable due to the new conditions but which are not constructed from whole cloth. They are adaptations of old behaviours and institutions. In the case of Bulgarian villagers, relative isolation has meant that change emanating from the centre does not always result in significant new behaviour at the local level. In some ways, Sanders' romantic vision holds. All three historical perspectives yield important insights, but in the case of rural Bulgaria particular weight on continuity may be justified.

We make our case by focusing on the Kurdjali and Smolyan *okrugs*, in the Rhodope region of Bulgaria. By focusing on these two rural districts in which we have done extensive fieldwork and data gathering over the past five years, we are able to provide a high level of detail. This will assist in illustrating both continuities with Bulgaria's recent past and radical change from that past. Although our example is one of rural transition from a collective to a private, market-based economy, the argument we wish to make is more general. The emphasis one places on historical continuities or discontinuities plays an important role in the understanding of change in any context (Todorova 1996, 51-52). Self-conscious and balanced use of the historical methods is therefore central to developing a deep understanding of emerging institutions and to making effective policy.

Three Approaches to Explaining Change

Separatist or Revolutionary Interpretations

Todorova (1996), in discussing the legacy of the Ottoman Empire in the Balkans, identifies the two differing historical perspectives on transition. The recently dominant 'separatist' perspective for studying transition minimises the influence of a displaced regime. In this case, the historian interprets change as a disjunction from the immediate past and a connection to an idealised historical (or even ahistorical) past. The earlier regime (Ottoman, communist) is seen as an aberration imposed on a legitimate, historical society. Once liberated, that society is able to return to a natural state, rapidly throwing off the imprints of the recent, parasitic culture. Here 'natural' has two meanings. In the case of the Ottoman Empire, the separatist perspective draws connections to an historical or Utopian Slavic golden age. But in studies of the transition from communist institutions, economists and historians view change as movement toward the diverted, natural order of private property and markets.

The ideological potential of such a perspective is clear. In the case of the Ottoman Empire, the nationalist ideology driving the politics and culture of the emerging Balkan states minimised the impact of the Ottoman Empire on indigenous culture and, quite frequently, demonised 'the Turk'. Historiography here involves selecting strands of history and culture for which the 'external' influence can be isolated and minimised. It stresses disjunction rather than continuity.

In the case of the demise of Balkan communism, most economic studies have taken the 'separatist' approach, assuming a disjunction from the communist era. In much of this work, history becomes utopian: the formerly communist states are returning to a natural order and look forward to reclaiming their place in the greater capitalist and European community of nations. Rural producers, as a result of the new policies, are becoming property holders and moving inexorably toward modern, market-oriented private production. History is 'recent' history, with baselines for measuring change set at 1989. Burawoy and Verdery (1999) call this approach to understanding transition the 'revolutionary' path, in which the totalitarian regime is extinguished and the world of capitalist, liberal democracy emerges from the shock therapies of structural readjustment leaving only 'rubble' of communism.

Other versions of the separatist approach may also emphasise links to far earlier times, especially pre-communist Bulgaria, while downplaying the impact of recent history. Thus, Begg and Meurs (1998), point to the nature of pre-communist land holdings in shaping the current patterns of agricultural land restitution and the current willingness of Bulgarian farmers to remain in cooperative farms. Meurs and Morrissey (1995) look to pre-communist village cultural practices in an effort to understand ecological conditions and change within Bulgarian farming villages after 1989.

Organic or Evolutionary Interpretations

The second perspective Todorova (1996) labels 'organic'. From this perspective, historians view transition as a broad and complex symbiosis of cultures with emergent properties. Here, the Ottoman rule and an earlier Slavic culture do not simply coexist for 500 years, they influence each other and the future.

An organic approach to studying the recent transition also exists. In discussing emerging literature on transition, Burawoy and Verdery (1999, 4, 6) call this the 'evolutionary' approach: "Those who saw, beyond and within the party state, an autonomous realm of society regard the latter as laying the foundations for a post-communist order. They see continuity where (revolutionary) theory sees collapse". Such an approach not only recognises the persistence of institutions and the institutional elite as part of path dependence, but is also more apt to understand the communist regimes as complex civil societies "percolating outside (and even within) the formal apparatuses of power". In this approach, formal restructuring of the state has been accompanied by strong continuities in the actual institutions of the state and civil society. In other examples, Meurs and Begg (1998), in reviewing the post-1989 transition in agriculture in Bulgaria, stress the continuity of private production from the command to market economies (see below). Begg and Pickles (1998) examine the influence of communist era industrial structure and regional policy in shaping the regional economy of Smolyan in the 1990s. In both cases, state inaction and persistence in local power structures and traditions combine with new conditions to create a complex synthesis. Old institutions and behaviours are influenced by and interwoven with new ones.

Todorova is clear in saying that either the 'separatist' or the 'organic' perspective may produce good history. The point is to be conscious of one's perspective, the benefits of each, and the appropriateness of each for a given problem. A potential ideological bias sometimes drives the use of the separatist perspective in studying transition economies. Fukuyama (1989), in heralding the transition as a final victory for capitalism and liberal ideology, the end of history, set the tone for a great deal of research. In the case of developments in the rural community, the actions of the new democratic, capitalist state are seen to radically transform rural life toward the fore-ordained end-state of capitalist agriculture. This tendency ignores important continuities in private and co-operative agriculture. The separatist perspective urges a focus on the changes in overt policies of the new, reformed state, when local institutional legacies or state inaction may also be powerful forces in shaping rural life.

A Third Way?

Burawoy and Verdery (1999, 2) find both the organic and separatist approaches inadequate. Completely rejecting the notion of an extinguished past, they see the notion of persistence as also flawed:

> …we challenge those analyses that account for the confusions and shortcomings of the transition process as 'socialist legacies' or 'culture'. Repeatedly, we find that what may appear as restorations' of patterns familiar from socialism are quite different: *responses* to the new market initiatives, *produced by them, rather than the remnants of an older mentality*. In other words, we find what looks familiar has causes that are fairly novel. Both innovation and reversion are responses to unstable environments, at least as much as they are evidence of socialism's legacies or its culture (Italics added).

Their distinction is important, but we argue that it may overstate the case. Of course the actions of villagers taken today are in part responses to new circumstances, but the nature of those responses is heavily conditioned by the material landscape, the slow demise of socialist institutions, and the nature of Bulgarian agriculture that had emerged after forty years of negotiated compromise with socialism. Creed (1998) in his ethnography of the village of Zamfirovo, makes the point that the advent and progress of co-operative agriculture has been one of negotiated compromise between the ideology of the state, the necessities of production, and the ability of the peasant to 'domesticate' revolution. He argues that:

> Socialism created a distinctive historical context. The social relations of socialism, the systematic interdependencies of the economy, and the cultural identities developed over forty years cannot be erased as easily as Marxist-Leninist political philosophy or party institutions. This context is the one in which transition proceeds; there is no other place to start, and there are no other materials to use (1998, 278).

We will emphasise that in rural areas, households retained a certain autonomy through the almost 40 years of communist industrialisation. The centre made less effort to integrate rural than urban households into central planning and made compromises which often left rural households with plenty of latitude for the continuation of (slightly adapted) pre-socialist subsistence agricultural practices. The recent radical changes at the centre may have had only limited impact on the subsistence of rural households. Rural industry, which has always remained only a second 'leg' of the household economy, has taken on new characteristics but also remains surprisingly unchanged, at least in the case examined below. Thus while we would agree with the Third Way's rejection of either a purely organic or purely separatist approach, we do not believe that it is appropriate to insist on framing current rural practices as 'new' adaptations to current conditions. Instead, we would argue that in the case of rural households significant weight should be given to continuities.

In considering the impact of the recent changes on rural households, then, we are offered three perspectives. A separatist perspective would emphasise the radical impact of collapsing state structures and re-ordering of property rights in bringing forth new behaviours and institutions. In particular, it might predict the quick emergence of a new class of private, commercial farmers. An organic perspective might emphasise the tradition of dependence of rural households on collective forms of organisation, and predict that these structures and household dependence would play an important role in institutions and behaviours displayed in the 1990s. The third way might predict that households would draw on production traditions in order to form new responses to new conditions. What we suggest is that all three versions can illuminate important aspects of the recent period. However, because of the limited integration of rural households into structures of the state and centre, in some areas households may experience only impact from the radical restructuring of the state. In this case, we recommend using a kind of emended third way – one which more heavily weights the continuities in ongoing adaptations.

Recent Change in Rural Bulgaria

In this section we examine three examples of change in rural Bulgaria since 1989, using both the 'organic' and 'separatist' lenses. What we try to illustrate is the way that the same change can be described so as to emphasise either continuity or discontinuity with the recent past. How we understand the impact of recent changes in the state of rural life thus depends greatly on the selection of data and on its specific lens used in its analysis.

The Emergence of Private Agriculture

For rural households, perhaps the most important consequence of the recent transition has been the restructuring of agriculture away from state-run co-operatives and toward market-based private farming. A separatist historical perspective on this development would stress the extent to which private production has replaced state-run, co-operative production. From this perspective, it is important that by 1996 for Bulgaria as a whole 62 percent of the value of crops came from private production. For some crops, the success of privatisation appears even more impressive: 97 percent of both potatoes and tobacco were privately produced (NSI 1997, 116). These data tell a clear story of radical agricultural restructuring, and the changes can be linked to the dramatic collapse of the socialist state. Once the newly liberal state ceased to restrict private households to production on tiny 'household plots' and relinquished its monopoly on grain and vegetable distribution, private producers responded to market forces, rushing to expand production.

From another perspective, however, the changes may not appear so dramatic. If we examine the transition in agriculture from an organic perspective, we note that the cropping patterns and activities of the rural producers have not changed as much as the previous numbers indicate. Although co-operatives are no longer state-run, the majority of agricultural land remains in co-operatives and many farms remain substantially unchanged since.[2] While 62 percent of production was produced by private producers in 1996, only 39 percent of arable land was privately held (NSI 1997, 114-115). This compares to 13 percent of agricultural land that was already under private control (in self-sufficiency plots) in 1986 (Meurs 1998, 53). The data on land use suggests much less change than does the production data.

If we look more closely at what is meant by 'private' and 'collective' in the two periods, the distinction between past and present is further muddied. In the late 1980s, socialist state officials responded to agricultural stagnation by implementing various forms of quasi-privatisation. Provisions were made to encourage increased production for sale on private plots, including the extension of agricultural credit and the sale of small machines to producing households. Beginning in 1982, an accord (or contract) system was also introduced. Under this system, part of collective farm work was subcontracted to groups of workers, families, and later to individuals, who rented machinery and land from the collective farm for a fixed period of time (usually one or two years) and contracted with the farm for the supply of other inputs. Sometimes these contract groups would contract to fulfil a specific task, such as weeding a given area, but increasingly the contracts were for final production, making the groups owners of their products. By 1988, 78 percent of those working in agriculture worked under such accords. Twenty percent of agricultural workers were organised into family-based accords, 30 percent in-group accords, and the remaining 28 percent in different types of individual accords (NAPC 1988). The extent of subcontracting of collective farm work varied significantly by crop. One hundred percent of tobacco was organised under accords, as was 80 percent of vegetables, 70 percent of meat and milk, and 62 percent of grain (NAPC 1988). With this in mind, the complete privatisation of tobacco and potato production by 1996 can tell a story of continuity as well as one of radical change.

Table 4.1 Percent Private Production for Various Bulgarian Crops, 1985-1996

Year	Vegetables	Potato	Wheat	Cattle
1985	38	52	13	34
1986	36	52	15	31
1987	41	64	14	32
1988	43	62	13	32
1989	42	54	13	36
1990	48	63	14	35
1991	60	81	29	50
1992	73	93	30	64
1993	82	95	28	79
1994	87	98	43	86
1995	89	97	48	85
1996	91	97	41	86

Sources: NSI 1991:236, NSI 1994, 232, NSI 1997: 117,124

The effect of using the organic lens, then, is to see that some of the changes attributed to market reforms were already emerging prior to the 'transition'. Private plots were responsible for an increasing part of production, and a form of tenant farming, which gave substantial control to family groups, was an important part of co-operative farming. These early roots of private production are reflected in Table 4.1 (above). Although the private share of production of basic crop groups shows a marked jump in 1991, private producers accounted for a significant and increasing share of output of many crops by 1985.

Further, as Figure 4.1 shows, even Table 4.1 overstates the extent of change, since part of the rise in private share is due to the collapse of output on co-operatives as they were freed from state control and re-organised. The actual increase in private output is minimal in all product groups but wheat. For many rural households, the changes since 1989 may have had little impact on the amount and type of private agricultural production.

The choice of unit of analysis can also influence whether the history of agricultural restructuring is written as one of continuity or change. If we examine the Rhodope region, we can illustrate even stronger continuities in agricultural organization and production than appeared at the national level. For this illustration, we draw on data from the Rhodope *okrug* of Smolyan, because this *okrug* has published the most complete set of district-level data.[3]

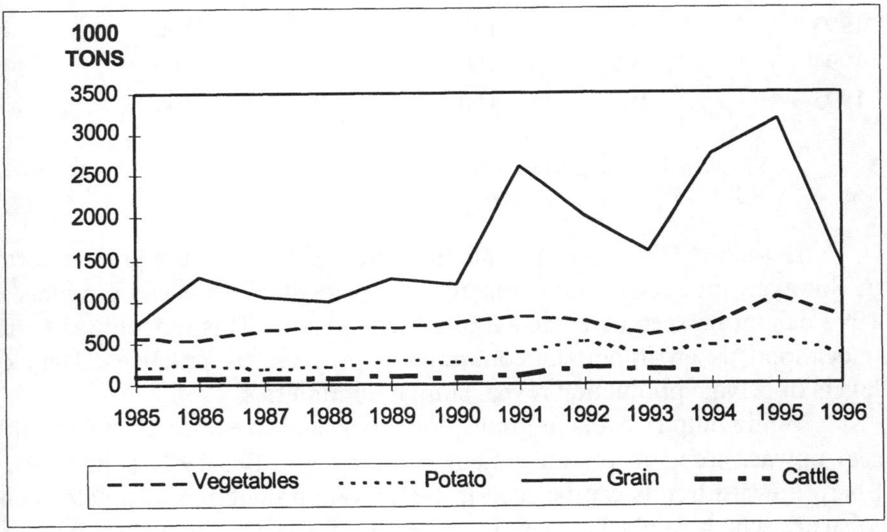

Figure 4.1 Private Production of Selected Bulgarian Crops, 1985-1996

Smolyan is a mountainous region where the collectivisation of agriculture posed particular problems. Because mechanisation, and even access to fields, was difficult, socialist collective farms allowed local households to employ unusually large amounts of 'marginal' land as 'private plots'. On the remaining land, the state-run collectives concentrated on a few products, mainly tobacco, livestock and certain vegetables. As a result, in 1985 private individual households produced nearly 100 percent of grains, beans and most types of fruit in the Smolyan region. While the state did produce the majority of tomatoes and potatoes in 1985, reforms of the late 1980s reduced the state role in these areas too. By the end of the socialist period, private households produced 87 percent of tomatoes and 53 percent of potatoes (Table 4.2).

Table 4.2 Percent Private Production for Bulgarian Crops, Smolyan*, 1985-1996

Year	Rye	Maize	Tobacco	Potato	Cattle
1985	98	100	0	40	69
1990	98	100	0	53	76
1991	99	100	93	71	85
1992	100	100	96	98	97
1993	100	100	96	100	99
1994	100	100	98	100	NA
1995	100	100	98	100	NA

* Smolyan does not produce wheat
Source: NSI 1996

If we keep these antecedents in mind, the fact that the private sector in Smolyan produced approximately 100 percent of all these products in 1997 cannot be seen as evidence of radical change. This outcome is better understood as an important continuity with previous practices. Data on levels of private production reveal similar continuities.

While output levels in grains, tomatoes, and livestock fluctuate, little real upward trend in private output can be seen after 1989 (Figure 4.2). Sharp upward trends can be seen in the private production of potatoes and tobacco, which might be seen as evidence of a sharp distinction between the production practices of the two periods.

These trends too may illustrate continuity with the socialist period, since potatoes were by far the most important crop for private plots under central planning. With the expansion of production opportunities after 1990, households drew on it in this tradition. The production of tobacco was a legal monopoly of the state prior to 1991, and no tobacco was produced privately. But even so, individual households under the direction of the collective farm often grew it on dispersed plots.

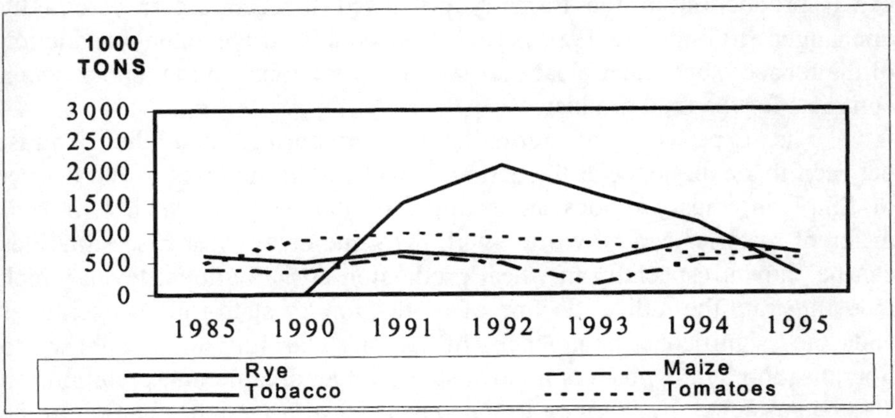

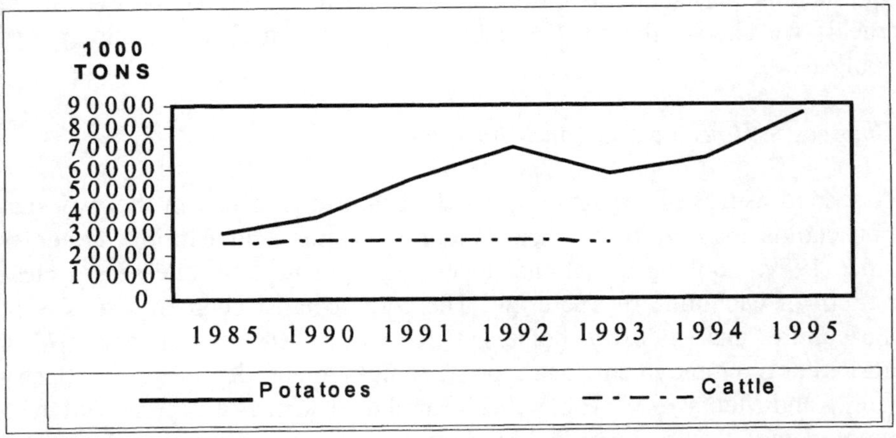

Figure 4.2 Private Output of Selected Products in Smolyan, 1985-1995

By 1988, 100 percent of tobacco was produced under the accord system. When the production of these households was finally recorded as their own after 1991, they often remained in an essentially unchanged situation, producing on contract to the (still) state-run tobacco monopoly (Field Notes 1993).

These continuities can be linked to continuities within the apparently restructured state. Despite efforts at mass privatisation of state assets and at exposing remaining state firms to competition, certain old interest groups have proven particularly difficult to dislodge. Those affiliated with the state tobacco monopoly are among these. Old institutions and networks serve to prevent restructuring of pre-1989 practices in the tobacco sector, leaving actual production in the formally privatised tobacco sector essentially unchanged from the pre-1989 period. Households do the actual production of the tobacco, but under close supervision by the tobacco monopoly which contracts for the final product.

The experience of agricultural restructuring, and the contrast between this experience at the national level and at the level of the district of Smolyan, thus provides an example of the way in which two very different histories can be written with the same data. Clear discontinuities can be shown, especially in wheat production at the national level, which may illustrate the radical decline of production by state-run co-operatives and the significant emergence of private farming in response to liberalisation by the post-communist state. But an organic interpretation can also be sustained, by pointing to the expansion of private production in the late socialist period and its limited expansion since 1991. This is even more true if we choose the rural Rhodope region of Smolyan as our unit of analysis.

Changes in Household Labour Allocation

A second aspect of the ongoing rural transformation lies in the way that households have adjusted their labour allocation and earning strategies since 1989. Looking at national level employment data, one sees a clear pattern of economic restructuring. The core socialist economic sectors of agriculture and industry collapsed following 1989, with declines in agricultural employment outstripping employment declines in industry (Begg and Meurs 1998, Table 10.2). Rural areas actually experienced even more change than is suggested by this data, since industrial conglomerates chose to close rural branch plants before laying off workers in the urban enterprises.

As the old industrial and agricultural core declined, employment patterns restructured, with a larger share of employed workers shifting into sectors more characteristic of modern, post-industrial economies – commerce and transportation. By 1996, nationally representative survey data suggests that the share of workers in manufacturing had fallen from 35 percent to 25 percent and the share in agriculture had fallen from 17 percent to 12 percent. At the same time, the share of workers employed in the commerce sector rose from 7 percent to 11 percent (Meurs and Giddings 1999).

These data suggest a radical restructuring away from agriculture and a significant impact on labour allocation in rural households. As with the data on agricultural restructuring, these data on employment can be linked to the radical restructuring, and scaling back, of the state. Once socialist planners were no longer at work transferring resources into the inefficient core of agricultural and industrial firms, these collapsed. Entrepreneurs rushed into the void left by the state and, responding to market signals, shifted resources into the underdeveloped areas of commerce and transportation.

If we again focus on rural households in the Rhodope region, we can paint a picture of much stronger continuities, however. In a 1996 survey of 85 rural households, the share of households reporting at least one member employed in agriculture actually increased slightly from 1986 to 1996, from 43 percent to 45 percent, despite the sharp decline in this sector (Survey Data, 1996). This is partly due to the decrease in the total number of households with working members. The share of households with a member employed in industry declined from 15 percent to 10 percent. Overall, in contrast to changes at the national level, the share of households with a member employed in these old core sectors fell only very slightly, from 57 percent to 55 percent. As at the national level, the share of households with a member employed in commerce rose, from 5 percent to 8 percent.

Of course, the nature of employment changed as privatisation proceeded. In 1986, 76 percent of households reported that all members worked in state enterprises and only 1 percent reported that all members worked in the private sector. By 1996, the distribution had become more equal, with 25 percent of households reporting all members working in state firms and 22 percent reporting all members in the private sector (Survey Data, 1996). But strong continuities persist in the degree of household reliance on employment in the old core sectors of agriculture and industry. At the same time, households also continued to allocate significant amounts of labour to subsistence agricultural production. Seventy-one percent of households reported growing at least half of their food needs in 1996 (Survey Data, 1996).

Yet another story of continuity and change in employment can be told if the Rhodope data on rural households is disaggregated by ethnic group, separating the Turkish minority (which actually forms a majority of the population in some parts of the Rhodope region) from ethnic Bulgarians. For the Bulgarian population in this region, strong discontinuities can be shown in household labour allocation, whereas Turkish households seem to illustrate much greater continuities. Many Bulgarian households in the Rhodope region suffered long term unemployment after 1989, something unheard of under central planning. The number of employed members of the surveyed Bulgarian households fell from 73 to 33 from 1986-1996, as 30 household members reported that they were unemployed over the whole year of 1996. For Turkish households, the number of employed members actually increased over this period, from 77 to 84 (Survey Data, 1996).

Of the Bulgarians who remained employed, many had shifted to jobs in different sectors. Slightly over 40 percent of both ethnic groups worked in agriculture in 1986 (Table 4.5). By 1996, the share of employed Bulgarian household members working in agriculture fell to 24 percent, while the share of the employed Turks working in this traditional sector rose to 55 percent. Overall, employment opportunities in commerce did not expand in the rural Rhodopes, but Bulgarians shifted into services, displacing Turks. The share of employed Bulgarians in this sector rose from 25 percent to 30 percent, while the share of Turks fell from 26 percent to 23 percent (Survey Data 1996).

Even the significant changes in Bulgarian employment can be understood in terms of late socialist precedents, however. In 1986, Bulgarians were highly concentrated in the socialist economic core-state sector jobs. Ninety-five percent of Bulgarian rural households received income from state firms, compared to only 79 percent of Turkish households (Survey Data, 1986). Within the dominant agricultural sector, Turkish households were much more likely to be employed in tobacco production than Bulgarian households (Survey Data 1996). These historical precedents meant that after 1989, the collapse of the state sector in the rural Rhodopes resulted in much higher unemployment among Bulgarian households than among Turkish households. And Turkish households, which had already been producing tobacco under the accord system, were better placed to expand the production of tobacco and to undertake other independent agricultural production once agricultural markets were decontrolled.

Again, the continuities and change in employment restructuring can be linked to subtle continuities in the role and character of the post-socialist Bulgarian state. As ethnic Bulgarians were squeezed out of the collapsing state industries, the service sector offered the main new source of Bulgarian jobs, despite the fact that this sector did not account for a growing share of all jobs. The majority of these jobs remain in the state sector (NSI 1996, 43), which has maintained a fairly constant level of employment. In displacing ethnic Turks from these jobs, Bulgarians drew on their traditionally closer association with the Bulgarian state and the unequal benefits, especially in terms of education, which they have drawn from it. Despite the state's declining resource base and economic role, it continues to be an important source of power and benefits for many who had good connections prior to 1989.

Restructuring Rural Industry: Apparel

The history of the rural Rhodope *okrugs* of Smolyan and Kudjali is not only agricultural. Although each was largely marginal to the building of state socialist industries, in the 1980s both saw substantial investment in assembly operations. As part of a series of regional policies for border, mountain, and ethnic regions branch plants and rural workshops were established in both regions. Textiles, leather, and apparel all grew more rapidly in the Rhodopes than in the nation as a whole. In our last example of the interpretation of change as organic and separatist, we will look briefly at changes in the apparel industry in these counties.

From 1980 to 1990, West Germany responded to market pressures in its textile and apparel industries by 'off-shoring' labour to Eastern Europe (Dicken 1998, 307), with "the most rapid growth in clothing production in Eastern Europe (occurring) in Bulgaria and Romania" (Dicken 1998, 291). The Bulgarian state firms which received the contracts were organised as a distributed system of production, with central factories in the towns and small village workshops (*tsehs*). Subsequently, Bulgaria slowly became a part of the international production network supplying department stores, mass merchandisers, and discount chains in Europe and the United States. As a result, the apparel sector grew rapidly between 1980-1990 (Table 4.3).

Table 4.3 Index of Apparel Production, 1980-1996 (1990=100)

1980	1989	1990	1991	1992	1993	1994	1995	1996
66	90	100	87	74	62	75	63	71

Sources: NSI Statisticheski Godizhnik 1991, 168, 1995, 23; Statisticheski Spravochnik 1997, 90

As the central state and its industrial sector collapsed after 1989, the sector declined precipitously. Between 1990 and 1993 production declined by 48 percent (Table 4.3) and between 1990 and 1996 employment dropped by 50 percent (Table 4.4). Workshop closure occurred more quickly than production declined, as a result of rationalisation and concentration of production (Begg, Meurs, and Pickles, under review). Some apparel companies were able to diversify and even extend certain product lines, however, even as they closed out-lying village workshops.

Table 4.4 Employment in Apparel, 1989-1996 (1990=100)

1980	1989	1990	1991	1992	1993	1994	1995	1996
64448	80467	83259	67483	57354	54024	48535	47014	41381

Sources: NSI Statisticheski Godizhnik 1991, 147, 1995, 89; Statisticheski Spravochnik 1997, 42

Despite the setback, assembly operations remain much more important in the rural Rhodopes than in the nation as a whole. Collectively textiles, leather, and apparel made up 7 percent of Bulgarian manufacturing employment in 1995 (NSI 1996, 89), but about 30 percent of manufacturing employment in the southern Rhodope region (NSI 1996a). In understanding the impact of changes on rural actors, it is important that these assembly operations in these counties were and remain largely rural and female (Table 4.5).

Table 4.5 Employment in Textiles, Apparel, and Leather Goods, 1992

Region	Employment	Manufacturing %	Female %
Kurdjali Region	5818	39	80
Kurdjali Rural	3528	57	83
Smolyan Region	6439	27	80
Smolyan Rural	3443	55	84

Sources: Socialno-Demografski Harakteristiki na Nacelenieto I Hilnishten Fond: Kurdjali, 139; Smolyan, 168. Statistcheski Sbormuk: Kurdjali 1996, 37; Smolyan 1994, 40

Since the mid-1990s, however, there has been a remarkable resurgence in the rural Rhodopes of what has been called *maquila* production (Pavlinek 1998). Not only have state firms garnered new foreign contracts, but new workshops have opened under diverse forms of ownership. Former state managers using old state equipment have secured foreign contracts (Classic Mode, RS Club). Greenfield operations, again staffed by former apparel workers, using high-tech imported equipment, produce for the domestic market (*Slavi*). Foreign joint ventures operate in former village workshops doing *maquila* assembly for foreign firms. Village co-operatives have opened workshops in unused village buildings and contract either with domestic or foreign firms (Beli Vir, Mladost). Finally, old state firms from the industrial core are contracting with village workshops in a variety of ways (Artex).

In calling textile assembly in the Rhodopes *maquila* production, a name that is in many ways apt, the impression is conveyed of the *tsunami* of global capitalism overwhelming the bucolic world of the Bulgarian peasant. That image ignores the past. Table 4.6 indicates a more organic transformation. Old forms, old equipment, old workers are engaged in a variety of forms of foreign and domestic production that draw from and mimic the late socialist period while responding and adjusting to new market incentives. Indeed, the vast majority of the apparel workshops rely on the same rural women who learned the trade during the late command era. In many cases the workshop, equipment, and management personnel remain unchanged.

This is not to minimise the influence of the changes. Many of the firms are now private enterprises producing for export. At the same time, the relative wage in apparel for the region moved from 120 percent of the national apparel wage in 1990 to 65 percent in 1993, only 37 percent of the average national manufacturing wage for that year (Begg and Pickles 1999). But the set and setting for the rural worker remain the same. Nor would the opportunity have arisen but for the movement to assembly of EU garments in the region during the late communist period.

Change and Rural Life

In our emphasis on continuities in rural practices, we do not want to deny the significant impact which the collapse of the central state and introduction of markets and private property have had on rural households. There have clearly been very significant impacts on rural households. These include the steep rise in unemployment, collapse in industrial wages, decline in state pensions, and the de-collectivisation of agriculture, all of which have had a radical, negative impact on rural households. Our point is that excessive emphasis on these radical changes can lead to the uncritical adoption of a separatist perspective, which we believe is inappropriate.

Drawing on detailed micro-level data, we hope to have shown that a significant part of rural household economies remains relatively unchanged despite the transition, in part due to the limited collapse of certain state structures at the local level. The persisting parts of the household economy provide a kind of 'leg' on which households can stand as other institutions collapse. They provide a known way to secure a very basic standard of living. But they often do not appear to be in the process of expanding or adapting significantly in response to the processes of privatisation and market development – private commercial production and industrial restructuring do not appear to be taking place on a large scale. Further, we want to emphasise that where and who you are plays an important role in whether changes appear more appropriately framed in the separatist or organic framework. Wheat producers, for example, provide a case for understanding institutions radically reformed. Turkish tobacco workers in the Rhodopes, however, appear difficult to characterise in that light.

Conclusion

We can see the withdrawal of the state and the intrusion of market relations as producing radical economic changes in rural households since 1989, or we can find in households a continuance with a set of economic activities which is relatively unchanged from those of the socialist period. These outcomes may be explained by a radical shift from socialist, totalitarian to weak, liberal state, or by the ongoing actions of entrenched interest groups and long-lived local agents. The way we choose to tell the story is influenced by the choice of historical 'lens' – separatist or organic. Both stories obviously contain grains of truth, valuable for developing a more complete understanding of the developments in rural life. The third approach, proposed by Burawoy and Verdery (1999), which emphasises that even the apparently unchanged activities are in some sense new in their changed context, provides additional insights.

In arguing for a greater relative emphasis on continuity when studying rural Rhodope households, we hope to provide an approach more appropriate to the study of rural Bulgaria. But we are not suggesting that rural life is unchanged. The Sanders quote with which we began this piece could easily be construed as romanticising a notion of the 'eternal peasant'. There is nothing romantic, however, about the grinding poverty which many Bulgarian villagers currently endure, nor have forty years of central planning left village ways unchanged. Still, there are constraints and echoes of a former age embedded in the furrows of the small fields, which are now ploughed again with a horse and which now generate products for trade through old barter mechanisms (Cellarius, 2000). How the practices of the emerging private *maquila* firms or newly commercial agricultural producers affect these practices remains to be seen. But, for now, we can say that for many village households, 'one ordinary day follows another'.

Notes

1 There is no commonly used equivalent for peasant in Bulgarian.

2 See Begg and Meurs, 1998 for a more complete description of the process of restructuring of collective farms.

3 The Rhodope okrug of Kurdjali would perhaps have made a more interesting example, since agriculture makes up a much more important part of the economy of Kurdjali, but published agricultural data for Kurdjali is not disaggregated into private and cooperative production.

References

Begg, R., Meurs, M. (1998), 'Writing a new song: path dependency and state policy in reforming Bulgarian agriculture', in Ivan Szelényi (ed), *Privatizing the Land: rural political economy in post-communist societies*, Routledge, London, pp. 245-270.

Begg, R., Pickles, J. (1998), 'Institutions, Social Networks and Ethnicity in the Cultures Of Transition: industrial change, mass unemployment and regional transformation in Bulgaria', in John Pickles and Adrian Smith (eds), *Theorising Transition: The political economy of post-communist transformations*, Routledge, London, pp. 115-146.

Begg, R., Pickles, J. (1999), *Ethnicity and Globalizing Production in a Border Region: FDI and Ishmolay in the Bulgarian Apparel Industry*, manuscript under review.

Begg, R., Meurs, J., Pickles, J. (1999), *Regional Economic Change and Ethnicity in Post-Communist Transitions: The Case of Southern Bulgaria*, manuscript under review.

Buckwalter, D.W. (1995), 'Spatial Inequality, Foreign Investment, and Economic Transition in Bulgaria', *The Professional Geographer*, vol. 47(3), pp. 288-298.

Burawoy, M., Verdery, K. (1999), 'Introduction', in Michael Burawoy and Katherine Verdery (eds), *Uncertain Transitions*, Rowman and Littlefield, New York, pp. 1-18.

Cellarius, B. (2000), 'You Can Buy Almost Anything with Potatoes: an examination of Barter During Economic Crisis in Bulgaria', @ *Ethnology*, vol. 39 (1,2), pp. 73-97.

Creed, G. (1998), *Domesticating Revolution*, Penn State Press, University Park.

Dicken, P. (1998), *Global Shift*, 3rd Edition. Guilford, New York.

Field Notes (1992-1997).

Fukuyama, F. (1989), 'The End of History', *The National Interest*, vol. 16, pp. 3-18.

Meurs, M. (1998), 'The Alchemy of Reform: Bulgarian Agriculture in the 1980s', in Ivan Szelényi (ed), *Privatizing the Land: rural political economy in post-communist societies*, Routledge, London, pp. 43-61.

Meurs, M., Begg, R. (1998), 'Path dependence in Bulgarian agriculture', in John Pickles and Adrian Smith (eds), *Theorising Transition: The political economy of post-communist transformations*, Routledge, London, pp. 242-261.

Meurs, M., Morrisey, M. (1995), 'State Policy and Agricultural Ecology in the Great Transformation: lessons from the Bulgarian experience', *Review of Radical Political Economics*, vol. 29 (2), pp. 70-95.

Ministry of Regional Development and Construction (1996), *Regional Policy in the Republic of Bulgaria*, Sofia.

NAPC (National Agro-Industrial Union) (1988), *Chelni posttizhenia po prilagane na akorda y selskoto stopanstvo*, NAPC, Sofia.

National Institute of Statistics (1991-1995a), *Statisticheski Godizhnik* (Statistical Yearbook), National Institute of Statistics, Sofia.

National Institute of Statistics (1995b), *Staticheski Sbornik Smolyan 1994*, Territorial Statistical Office, Smolyan.

National Institute of Statistics (1996a), *Socialno-Demografsski Harakteristicki na Nacelenito Hilishten Fond Statistichesko Izdatelctvo Petchatnusta, 1996 Reyon Smolyan*, National Institute of Statistics, Sofia.

National Statistical Institute (1996b), *Socialno-Demografsski Harakteristicki na Nacelenito Hilishten Fond Statistichesko Izdatelctvo Petchatnusta, 1996 Reyon Kurdjali*, National Institute of Statistics, Sofia.

National Statistical Institute (1996c), *Statistical Reference Book of Bulgaria*, National Institute of Statistics, Sofia.

National Statistical Institute (1997), *Sraristicheski Spravochnik* (Statistical Yearbook), National Institute of Statistics, Sofia.

Pavlinek, P. (1998), 'Foreign Investment in the Czech Republic', *The Professional Geographer*, vol. 50, pp. 71-85.

Sachs, J. (1994), *Poland's Jump to the Market Economy*, MIT Press, Cambridge.

Sanders, I. (1949), *Balkan Village*, University of Kentucky Press, Lexington, Kentucky.

Todorova, M. (1996), 'The Ottoman Legacy in the Balkans', in L. C. Brown (ed), *Imperial Legacy*, Columbia University Press, New York.

5 Soviet Patrimonialism and Peasant Resistance during the Transition – The Case of Estonia

ILKKA ALANEN

Introduction

Estonia's first independence between World War One and World War Two (1918-1940) can be regarded as a true republic of small farmers. A radical land reform had created a fairly homogeneous ownership structure and a relatively large class of small farmers, which also wielded great political influence. The farms were cultivated by literate peasants oriented to commodity production. At that time, Estonian agriculture was in many ways more modern than agriculture in most other countries reckoned today within Eastern Central Europe. A comparable level of modernisation was only achieved in the Czech Republic and Latvia (see Lieven 1993, Raun 1989).

Soviet occupation was followed by forced collectivisation and almost fifty years of Soviet rule. Estonia's independence was restored in 1991 after the collapse of the Soviet Union, and the restoration of independence was followed by an agricultural reform carried out against the wishes of the majority of the rural population – just like Stalin's collectivisation. The goal of the reform was the dissolution of *kolkhozes* and *sovkhozes*, and the transfer of their resources to smaller farms. The goal was set to establish new smallholdings by restoring the land ownership of the first period of independence and by returning all nationalised property (cattle, buildings, machinery, etc.) to their entitled owners. This reform resulted in vast material and mental losses, and furthermore, it did not prevent a large-scale farm system based on former collective farms from becoming the core of the present agricultural structure in Estonia. The establishment of a large-farm system as a consequence of the privatisation of *kolkhozes* and *sovkhozes* was the outcome of a reaction against the pro-small-scale government policy favoured by the agrarian middle class (the well-educated specialists and skilled workers).

I interpret this reaction primarily as a rational attempt a modernisation (Parsons 1961 and 1971), and not as a counter-revolution tinged with nostalgia for Soviet times (see Alanen 1998a and b; 1999). Although the majority of people wanted to have their old family farms back, the polls conducted prior to privatisation confirm that only a small minority were willing to start farming the restituted land (Alanen 1999). However, according to interviews conducted in Estonia, antipathy towards the Soviet system imposed from above was expressed by all social classes toward large-scale farms, and Soviet power with its *kolkhozes* and *sovkhozes* was generally regarded as a manifestation of the Russian occupation. Also the vast majority of *kolkhoz* and *sovkhoz* leaders advocated Estonian independence, although some factions of the nationalist movement labelled them 'red barons', referring to old ethnic German landlords who had lost their centuries-old position as a result of the first land reform conducted during Estonia's first independence.

The right wing of the nationalist movement (Estonian Congress) took power right after the declaration of new independence in 1991. Its pro-small-scale agricultural policy reflected the nationwide enthusiasm for restoring ownership relations that had prevailed during Estonia's first independence, since they were widely considered the only legitimate basis for the reconstruction of social relations in general. Parallel to this there was the desire to overthrow *kolkhoz* leaders, since the Estonian Congress considered *kolkhozes* and *sovkhozes* bastions of communism in the countryside. During Soviet times the Estonian agricultural sector had served as a kind of a testing laboratory for All-Union reforms. A republic-wide agricultural reform was carried out there starting in the late 1970s. The idea of the Estonian reform was to offer relatively wide economic autonomy to the collective farms. The reform resulted in increased effectiveness and production; the difference was great compared to other Soviet republics, where effectiveness was impeded by the political control of economic decision-making. Estonia did not undergo the agricultural stagnation that characterised other parts of the Soviet Union in the 1980s; instead, farms grew stronger, and positive development continued on some of them right up until the turbulent times of the new independence. Also the agrarian population grew increasingly wealthy: out of all Soviet republics only in the Republic of Estonia was the living standard of the agrarian population higher than that of the urban population (Palm 1992). Soviet farms not only paid good wages but they also provided farming plots for their employees, and the cultivators of private plots were able to gain additional income by selling their produce, which was usually marketed through the collective farms. *Kolkhozes* and *sovkhozes* also provided a variety of social services (e.g. kindergartens, firewood supplies for the elderly, and allowances to single parenting mothers). They also sponsored numerous cultural activities (e.g. musical life, theatrical performances, and other local cultural events). During the Gorbachev era, the successful Estonian model was extended into other parts of the Soviet Union, but these farms were not delegated enough autonomy, which is why the reform attempts failed. Still, the Estonian agricultural experiment contributed to the technocratisation of everyday work at the expense of the ruling Communist Party.

After privatisation, preservation of large-scale farms was not synonymous with the preservation of *kolkhozes* and *sovkhozes*, instead, the proponents of large-scale farming aimed at developing the former collective farms further and making the changes required for adaptation to the capitalist market economy. However, Soviet times and their associated mind-set and moral norms undoubtedly had an influence on the struggles that were fought for large-scale production, just as the long Soviet era unavoidably influenced people's expectations of how the new enterprises would operate. The prevalent mind-set and the associated moral code might be called the 'Soviet community spirit'. This chapter focuses on this community spirit, particularly its manifestations during agricultural decollectivisation, which will be analysed with the aid of James C. Scott's (1986) conception of the 'moral economy of peasantry'. Some of the distinctive features of the Soviet community spirit are its special form of patrimonialism, including, for example, the idea of the farm leadership as the 'workers' representatives' (this symbolised the social cohesion of the Soviet times) and on the whole as 'representatives of the people' (cf. the Introduction to this volume).

Patrimonial Bureaucracy on Soviet Collective Farms

Estonian *kolkhozes* and *sovkhozes* were a part of the All-Union agro-industrial complex and hence a part of the Soviet economy. The Soviet system was basically a variant of a barter economy: exchange, investments in production, etc. were the result of a complicated negotiation system rather than of a straightforward command structure. Production goals were nevertheless assigned from above (the Politburo and central planning ministries) down to the bottom, where production plants like *kolkhozes* and *sovkhozes* were located (Clarke et al. 1993).

The leaders of the *kolkhozes* and *sovkhozes* were at the core of the agro-industrial complex, but in some cases their role was much more important than that. Their office and power was not limited to agriculture only, but they were also vested with political power at the local level as well as regionally. The leaders of all the largest farms were public figures widely known throughout the Soviet Republic of Estonia. The Soviet large farm system was not a variety of feudalism; nevertheless, it personalised many social relationships and functions in an analogous way. The external relations of a collective farm with the local administrative body (Village Soviet) and the district level (Rayon Soviet), and also with various parts of the agro-industrial complex, were deeply personalised:

> If you needed something, let us say that you needed machinery, the *kolkhoz* did not contact EPT [Eesti Põllumajandustehnika, the sole supplier of agricultural machinery in Estonia during the Soviet era], instead, Mr X phoned Mr Z or some other person. And they wanted to pull a fast one . . . Such and such a price was agreed upon, if it was in somebody's personal interest (Chairman of a *kolkhoz*).

Negotiation skills (and the associated ability to create an extensive network of relationships within the agricultural complex; i.e. 'social capital') were thus one of the most important characteristics required of an able collective farm leader (cf. Clarke, 1993). Within any Soviet farm – *kolkhoz* or *sovkhoz* – the responsibility for decision making fell upon the collective farm leader, who was appointed by the state. Yet, in *kolkhozes* the primary decision-making body was in principle the General Assembly of all *kolkhoz* members, which elected the Kolkhoz Council, which in turn appointed the Kolkhoz Board of Management. In practise, both *kolkhoz* and *sovkhoz* leaders were elected in the same way; i.e. the *kolkhoz* administrative bodies were compelled to execute a decision that had been made at a higher level. (Immediately before the end of the Soviet era the official legal procedure prescribed by law was, however, utilised to accomplish a number of local 'little revolutions' inside *kolkhozes*.) Such being the case, the use of a general label, such as 'Soviet farms', without making a distinction between the two different legal categories is justified given the subject matter of this chapter.

The Board of Management was technocratic in the sense that typically the most important managers of *kolkhoz* production departments (brigade leaders, managers of cowshed complexes, manager of the construction unit, chief agronomist, etc.) were members of the Board, even though there were also some other members. In practise, each department manager (and all other members of the Board) had to enjoy the confidence of the *kolkhoz* chairman; and if this trust was lost, the department manager usually lost his position as a department manager in addition to his post in the Board of Management. All the Boards of Management were made up of a closed core and an outer rim. The closed core consisted of trustees of the chairman (in *kolkhozes*) or the director (in *sovkhozes*), who in general were close friends of the chairman also outside the Board. This group was capable of monopolising all available power at will, and they were largely able to filter the information presented to the other Board members. As a rule, this situation remained unchanged until the collective farms were eventually broken up.

The leading role of the Communist Party was recognised at all levels of the Soviet system. The decisions of the Communist Party were usually executed to the smallest detail, for example in the drawing up of All-Union production plans. At the level of an individual farm the situation was not as simple as that, at least not in the Brezhnev era and after. The practical value of Party-centredness diminished in farm-level everyday decision-making after the Soviet system had been firmly established, i.e. once its infrastructure, educational system, distribution and processing systems for foodstuffs and means of production had been organised. This development inevitably led to an increase in the expert authority of the *sovkhoz* director and the *kolkhoz* chairman.

At the local level, even the Communist Party was largely dependent on the leader of the collective farm. Although the vice-chairman of a *kolkhoz* was the official Party representative (watchdog of the Party line) in the enterprise, and he was also the secretary of the farm's Party organisation, the chairman was 'de facto' the person acting in the name of the Party. This held particularly true if the *kolkhoz* chairman had already managed to establish his position in the agro-industrial complex (and possibly had received a variety of formal recognitions from the Party), and if he was determined enough. In such cases a farm leader was often capable of subordinating, to some extent, the Party to defend the interest of his farm. Or, as one former chairman phrased the issue: "He could control the activities of the Village Soviet either through the Party Commission [at the local level] or through the Party Committee [at the rayon level]". The Party was virtually invisible in the management of everyday practical affairs. Obligatory Party rhetoric was reserved for meetings with a larger attendance, where the exposition of the Party line was mandatory, but few people paid any attention to the speeches. To give an example, if a young Party official who had attended Leningrad Communist Party Academy started early in his career to use Party rhetoric in daily farm activities or in some local meeting, he would have been considered stupid.

Before the Party lost all of its power, it became abstracted to the All-Union level and its sphere of influence only reached down to the Republic level, finally stopping at the rayon level, and never reaching the farms. This long and unnoticed process of agricultural technocratisation had preceded the rise of the nationalist movement, and in the end of 1980s the Communist Party finally yielded power under pressure from the nationalist movement. When the leading position of the Party was officially denounced in Estonia just before the collapse of the Soviet rule, the occurrence – according to our interviewees – went unnoticed at the local level. From the perspective of ordinary people, the Communist Party simply withered away without any dramatics, and without arousing any strong emotions.

One of the hallmarks of the Soviet system, the absence of competitive markets (although the world market, as well as the informal and 'black' markets had a minor influence), also encompassed the labour market. All this is readily understandable within the framework of a 'shortage economy' (Kornai 1980). An economy of shortage was a structural feature of the so-called state-socialist countries, and it forced corporate leaders in these countries to allow for bottlenecks in the supply of vital resources by amassing large stockpiles in advance, since the supply could be cut off at any moment. Labour was one of these scarce resources. Although the employment system in socialist countries was formally based on wage-labour, the scarcity of labour prevented competition and the creation of a true labour market in the western sense of the term. This fact had visible consequences also at the local level. Although the Soviet system could be considered relatively authoritarian in terms of workplace democracy, at the floor-level the management lacked effective means of carrying out production-related decisions due to the scarcity of labour. This held true also in the sector of agriculture. Moreover, *kolkhozes* and *sovkhozes* were obliged to hire unqualified and uneducated employees to provide for seasonal peaks in labour demand. Hence, the permanent staff of *kolkhozes* and *sovkhozes* included many employees who in any western country would have been unemployed or institutionalised – alcoholics and people with a wide array of other problems. A related phenomenon observable in all Baltic republics were the 'migrating birds'. These people, whose professional skills and work motivation were low, travelled around the countryside looking for work. In many cases their background involved social problems from the larger cities, and they hoped to find a solution for their problems in the country. This labour reserve was not much respected, yet the shortage of labour was such that apartment buildings were constructed to attract them. They did not, however, always receive "a permanent job", said a manager in charge of recruitment and housing at an Estonian *kolkhoz*.

Bad work discipline at Soviet farms by no means denoted that all workers were idlers, hard drinkers, etc., but the glaring presence of such employees is a telltale sign of this special structural characteristic of shortage economies. Consequently, employees were seldom given notice for excessive drinking, etc. Likewise, if one of the *kolkhoz* or *sovkhoz* department managers fell into the Board's disfavour for a weighty reason (e.g. came under suspicion of misappropriations or turned out to be an alcoholic), he was usually demoted to the ranks of ordinary employees instead of fired. Also petty thieving was a rule rather than the exception in *kolkhozes* and *sovkhozes*, but such small-scale thieving did not shake the economic foundations of any production units. There were also semi-legitimate means of making use of *kolkhoz* resources. The best employees were the most bold in utilising these resources for their own good, e.g. tractor drivers often worked with a *kolkhoz* tractor on a private plot, and the plot owner paid a fee directly to the tractor driver, not to the *kolkhoz*. In some cases, the *kolkhoz* leadership openly authorised this kind of behaviour, in other instances it was merely tolerated, since these practises were against Soviet laws.

In fact, the entire *kolkhoz* system at the level of basic production was a symbiotic unit, where large-scale farming and small-scale plot farming complemented each other (Alanen 1995, 1998a). This symbiotic relationship, which was complementary and economically beneficial to both parties, had originally developed spontaneously during the post-Stalin era. It arose from the inability of the Soviet government and Soviet farms to control the co-operation between private plot farmers and tractor operators, but later *kolkhoz* leaders, and Soviet leaders more widely, came to realise its obvious advantages, and developed the system further with their own goals in mind (Alanen 2001a).

In some cases the *kolkhoz* leadership was on the verge of losing control of their employees. Some of the interviewees told us about a few instances where workers had simply stopped working: not to be out on strike, but most often to do something else they deemed more pleasant. For example, an entire construction brigade sometimes stopped work at a construction site to start drinking. In this particular case the *kolkhoz* leadership debated the issue in several meetings of the Board, but they came to the conclusion that legal action or the dismissal of these employees would not have brought any improvement in the work morale of the construction brigade, since all other construction units were facing more or less the same problems. In the end, the *kolkhoz* leadership suspended the payment of wages in the hope that 'hunger' would drive the employees back to the construction site. The wealthiest collective farms were able to hire better-educated labour, and were thus more capable of avoiding the extreme phenomena described above.

The leader of a Soviet farm was ever-present, not only did he attend Board meetings, but often he personally oversaw various everyday work-related situations too. However, the arena where his personality was most prominently displayed was the Kolkhoz Board of Management. The faculty members at the Estonian Agricultural University, who knew the farm chairmen well, had come to the only half-joking conclusion that

Kolkhoz Board members "strongly resemble their leaders in outward appearance".

In line with Weber's (1978) ideal type of patrimonialism, Soviet patrimonial leaders were at first glance characterised by omnipotence. The status of the *kolkhoz* chairman was once described in the following terms: "Everything he said had to be done. He forbids, commands, hangs and shoots people" (a former brigade leader, quoting the traditional Estonian phrase "*keelab, käseb, poob ja laseb*"). However, any *kolkhoz* chairman was liable to lose his power if he transgressed the norms governing his behaviour (this was due to the structural factors described above and is consistent with Weber's ideal type). We must also consider the advanced technocratisation of Estonian agriculture. Unlike typical Weberian patrimonies, the authority of Soviet farm leaders was not based on tradition, but on expertise. Hence, *kolkhoz* chairmen did not inherit their position; rather they attained it by education and by merit. Chairmen generally avoided situations where their lack of knowledge of a specific subject would have become apparent; this view is supported by the following amusing encounter related by one of our interviewees:

> The chairman promised to bring the hay to the cowshed. As it was raining outdoors, he wanted to dry the hay with a fan. I was against it and we had a fight. I threw him out of the cowshed and locked the doors. The next day the chairman came here and behaved as if nothing had happened. He had realised that he had been wrong.

And the interviewee was not referring to any sweet-tempered collective farm leader since this particular chairman was regarded as more of an authoritarian figure. The interview material also includes a case where farm specialists had been able to have a chairman or director discharged on the grounds of lack of professional expertise by appealing to higher officials at the rayon or republican level. This is more in line with the patrimonial concept of authority than with theories of all-pervasive totalitarianism (for details on the inadequacy of theories of totalitarianism, see Smith 1999, 16-25), since a farm leader could lose his authority anytime or could even be fired if his actions were not well-founded professionally.

It is a matter of common knowledge that the basic character of Soviet society was manifest in numerous arbitrary despotic and repressive phenomena (e.g. political prisoners and the activities of the KGB). Still, everyday life in an ordinary collective farm in Estonia was most clearly specified by Weber's (1978, 231-232) ideal types of patrimonial bureaucracy and officialdom (Maslowski 1996, 304).

Soviet Community Spirit

Kolkhoz leaders were more a sort of bureaucratic authority, whose personal and *Gemeinschaft*-like (Weber) characteristics formed an additional basis of legitimacy for the *kolkhoz* system. Though their appointment to the position did not occur from bottom to top (even though such a procedure was prescribed in the *kolkhoz* charter), the *kolkhozniks* used to call their leaders 'people's representatives' in varied contexts – sometimes in gratitude, at other times in disapproval. In any case, the parlance was in harmony with the official Soviet ideology.

The Soviet community spirit was not just a fabrication created by Soviet ideologists. Patrimonial bureaucracy, at least in the Estonian countryside, reinforced and reproduced the community spirit: e.g. old people and the underprivileged were well taken care of. The leaders of Soviet farms were not always loved, since many of them were authoritarian or quick-tempered, but they were in general highly respected all the same, and they also felt a sense of responsibility for their employees. However, the nationalist movement managed to shake some of this respect for farm leaders, and in quite a few instances an activist in the nationalist movement challenged the 'communist old guard'. Several collective farm chairmen were overthrown in 'little revolutions', i.e. enterprise-level transfers of power (Alanen 2001a). Although the 'new guard' often accused the 'old guard' of various types of abuse of public funds and resources during the privatisation phase, these appear to have been much more common in farms controlled by the 'new guard' (Alanen 1999). Considering the matter from an ideological perspective, this is not at all surprising. The leaders of the old guard had deeply internalised their patrimonial 'virtues' and most of them were thus unwilling to utilise any (unlawful or immoral) opportunities.

Another element in the community spirit (partly related to the matters above and also reflecting the relative autonomy of ordinary workers in the Soviet mode of production) of Soviet citizens was the ironic expression "the dictatorship of workers and military seamen", which implied the idea that ordinary people are the true holders of power. This was not an entirely fictional fabrication either, since skilled employees in particular (technicians, machine operators, mechanics and milkers) enjoyed a high degree of independence, and often their earnings were higher than those of the better educated 'specialists' (agronomists, zoologists, veterinarians, engineers, economists, etc. with a university or secondary level degree), or even *kolkhoz* chairmen. This is why 'the dictatorship of workers and military seamen' was paraphrased in the countryside to 'the dictatorship of milkers and tractor drivers'.

However, the Soviet Union was not a wonderland of industrial democracy by any means. Although many of the workers expressed their opinions very energetically and were not afraid of clashing with the chairman, if necessary, these conflicts did not shake the hierarchical authority structure personified in the chairman. The Soviet patrimonial system (in its structural context) was not generally conducive to the creation of an active worker, instead, it had the effect of encouraging passivity among the employees since their real chances of influencing their working conditions were very limited (cf. Clarke 1993). As a general rule the common people were always taken care of by someone else, ultimately by the *kolkhoz* chairman. On account of this, the people never learned to look after their own affairs; instead, as one former *kolkhoz* chairman put it from his patrimonial viewpoint, they remained where they were and in a way 'demanded' that someone above them should act on their behalf even during the *kolkhoz* reform. This is a fitting description of the least skilled employees, who were mostly found in the more remote villages.

When the Soviet Union broke up, its political machinery was relatively quickly replaced by parliamentary institutions (parliament, free press, freedom of assembly, political parties, etc.) Meanwhile, establishing the private ownership (including the abolition of price controls, the break up of monopolies, bankruptcy legislation, etc.) characteristic of a capitalist market economy took somewhat longer. Finally, the ideological framework (various types of unofficial norms and institutional arrangements) to which people had recourse during the process and which they used in interpreting the rapid changes in the surrounding society did not change at the same pace. The understanding that any enterprise manager should be a 'representative of the people' endured, and moreover, the privatisation of former *kolkhoz* or *sovkhoz* property, even if the transition was carried out in a legally impeccable and morally faultless manner, was often called 'robbing' by ordinary people. However, it must be pointed out that parallel to legally impeccable privatisation processes there occurred a lot of acts that were either morally reproachable or downright criminal under the current legislation.

Methods of Privatisation and Interest Groups

Two methods dominated the privatisation of collective farms in Estonia. In principle, the primary method should have been the restitution of former ownership relations (e.g. land and livestock) either in kind or through compensation vouchers. The workshare voucher method was usually applied to the greater part of total collective assets, which included cowsheds, piggeries, and other production complexes, such as the sawmill, wood processing workshop, tractor depot, plus offices and other non-production facilities, as well as the majority of the machinery and livestock. These assets were only offered for sale to those entitled to workshare vouchers on each farm. The vouchers were handed out to *kolkhoz* members and employees on the basis of time served in the collective farm or on the basis of wages paid (some farms used a combination of these two methods) – the law allowed for a variety of procedures. People were offered an opportunity to realise their workshares and compensation vouchers by purchasing collective assets at book value or by bidding for them at an auction (or any combination of these two methods). In Estonia, there were huge problems in synchronising the processes of restitution and purchasing with vouchers, which resulted in a great deal of unnecessary material destruction and human suffering (Alanen 1998a and b; Alanen 2001a). The decisive part of the privatisation process was (due to the amount and nature of privatised assets) purchasing with vouchers.

The Board of Management of the Soviet farm usually had a key role in the privatisation process since it prepared action plans for the Reform Committee. For legal reasons, the Reform Committee could not include any members of the collective farm Board, consequently the membership of the Reform Committee lacked expert knowledge on the affairs of the collective farm, and hence they had to turn to the *kolkhoz* leadership for information and advice. In the end, there were only a few rare cases where the *kolkhoz* leadership did not have a central role in carrying out the reform, and Reform Committees were usually only relegated the role of the formal executor of the reform.

The interests of the different groups inside the collective farms varied a lot. So did their ability to recognise these interests and to act accordingly. The real power in *kolkhozes* and *sovkhozes* belonged to (i) the Board of Management: actually the small and exclusive core of it formed by the chairman and his most closed trustees. The other two status classes, much larger in membership than the first one, but of nearly equal size when compared with each other, were (ii) the 'middle class', and (iii) the 'rest'. The struggle for the future of the *kolkhoz* was largely fought between the core (chairman and his trustees) of the old Board of Management and the new nationalist activists arising mainly from the *kolkhoz* 'middle class', with both groups trying to appeal to the ordinary workers, since the 'rest' formed about 50 percent of the total workforce of the *kolkhoz*, and the formal approval of the *kolkhoz* General Assembly was required for all major decisions. The Board of Management had at its disposal all the important data on farms and society at large, including account books, precise information on the continuously changing legislation, and market situation. Parts of this information (bookkeeping) were potentially subject to manipulation by the inner circle, which could also to a varying extent monopolise the information for itself (for more information on interest groups, see Alanen 1998b).

The Importance of the Soviet Community Spirit in the Preservation of Kolkhozes and Sovkhozes and the Formation of the New Large-Scale Production System

The middle class was actually composed of highly specialised employees (agronomists, veterinarians, economists, and engineers) and of the most skilled workers (tractor drivers, truck drivers, and persons in charge of technical matters). Even if this 'middle class' consisted of two subgroups, specialists and skilled workers, the group was rather uniform: for example, many skilled workers had a university degree. There were more highly educated individuals available than corresponding vacancies for specialists. The fact that their work as well as living quarters were concentrated in the *kolkhoz* centre eased the formation of a uniform frame of mind, and at the critical phase it evolved into organised opposition activity. The middle class cultivated private plots less often and with less intensity than 'ordinary workers'; and they were also younger than the average *kolkhoznik.* Most of the activists in the nationalist movement would rise from among this group of people.

The prospect of breaking the *kolkhoz* up into smaller units was unappealing and unpopular among the 'middle class'. The idea of replacing the existing system with a new system of pure family farms threatened their privileged position which depended on the large volume of production and highly specialised division of labour. The basic ideology of their oppositional activity was to prevent the breaking up of the *kolkhoz*, although it was undoubtedly seasoned with suspicion about the self-centred privatisation designs of the *kolkhoz* leadership. In those cases where the *sovkhoz* or *kolkhoz* leadership attempted to preserve the farm as an integral unit in the middle of conflicting interests, active resistance most often failed to arise from among the 'middle class', at least in the sense that a pro-dissolution group would have had its support as a social group. In such cases the criticism usually arose from among the ordinary employees, who were generally more eager than the other status groups to establish small farms of their own. But even the 'rest' as a status group was predominantly disinclined to break up the large production units of *kolkhozes* and *sovkhozes*. However, in the end only four out of the approximately 350 collective farms in Estonia remained undivided and continued with agricultural production. The small number of undivided collective farms was partially due to the anti-collective mentality that dominated the first few governments after the re-establishment of independence. The governments also utilised indirect legislative means in an effort to enforce the division of *kolkhozes* and *sovkhozes* into smaller unit, for example by imposing punitive taxes on collective farms. The small number of surviving collective farms (albeit under a different legal guise) is also the result of the extremely difficult economic conditions during the first years of independence. The ramifications of these problems came to a head in the agricultural sector: producer prices collapsed, at one point markets for farm produce were almost entirely blocked, prices of production inputs skyrocketed, and there was a shortage of many vital raw materials.

With regard to the role of the 'middle class', in an attempt to maintain the unity of the *kolkhoz*, or (as a lesser evil) in the attempt to divide it into technological units as large and viable as possible, their representatives resorted to various tactics which were partly political and partly demagogical (political accusations against the 'red barons', charges of misconduct, etc.). These tactics also served a purpose in the power struggle within the collective farms, but my focus lies on the element of Soviet community spirit in this battle. When the 'middle class' mobilised resistance against the then *kolkhoz* leadership, they was often accused of attempting to 'privatise' some parts of the *kolkhoz* 'for themselves'. From the perspective of Soviet ideology this was tantamount to 'robbing'. The word robbing was indeed often used as a synonym for privatisation, even in such cases where privatisation was carried out according to the regulations prescribed by the law without any sign of foul play or deceit. The ones who claimed that the 'middle class' was robbing pleaded that collective farm units were the result of hard work by *sovkhoz* or *kolkhoz* workers (to a large degree they were), and their underlying conviction was that *kolkhozes* and *sovkhozes* should also under capitalism remain in the collective ownership of working people.

Often the 'middle class' succeeded in organising a 'little revolution' inside the collective farm, replacing the former 'red barons' with representatives of the 'middle class'. Paradoxically, the political line of the new leadership frequently bore a close resemblance to the rightist policy of the government, which at the national level was very hostile towards large-scale agricultural production. In a few cases the 'revolution' may in fact have been a 'coup', a plot by a small group of people, who benefited from the fears many 'middle class' members felt for their future. No matter how sincere were the political programmes and promises set forth in the heat of the battle, after a successful 'revolution' the new guard was offered a wide variety of opportunities to 'privatise for themselves' – such was the moral climate in Estonia at the time.

While the preservation of an entire collective farm was successful in only a handful of cases, there were numerous cases where large-scale production units continued operation on a more modest scale. Usually a new enterprise was established on the basis of one or more of the production units (cowshed, piggeries, etc.) of the former collective farm. The founders of these new enterprises, which were co-operatives in legal terms, appealed to the workers' Soviet community spirit and consequently established a great number of these co-operatives, where the one man, one vote principle prevails. The establishment of such large-scale farms was made possible by a new law formulated by the Estonian government with an altogether different goal in mind. It was the government's intention to create a legal foundation for the recreation of co-operative dairies and other processing and service industries that had proliferated in the 'peasants republic' during Estonia's first independence in the 1920s and 1930s. However, the 'middle class' of the collective farms made use of this law to bring their own ideological project – the preservation of large-scale collective enterprises – to a satisfying conclusion. Many right-wing persons disparagingly call these new co-operative enterprises *kolkhozes*.

Today, the most common legal enterprise form in large-scale production is not a co-operative, but a limited liability company or direct private ownership by one person. In the beginning, limited liability companies shared many characteristics with co-operatives, since these companies were often set up by collecting the amount of vouchers necessary for the purchase of a cowshed, for example, from collective farm members and employees, and the charter of the company prohibited the sale of company stocks to outsiders. In this sense the companies remained in the ownership of the workers. However, in a limited liability company the vote of each owner is based on the number of his stocks, which has made the acquisition of a large number of the stocks or the controlling majority easy for the enterprise directors. All across the country these neocollectivist enterprises are gradually transformed into ordinary, capitalist enterprises. An increasing number of their workers do not own a single share in the company today.

As surprising as it may appear, purely capitalist large-scale enterprises were also founded with the help of the Soviet ideology. The vouchers of individual people or families were not enough to purchase one of the larger *kolkhoz* production units. Therefore, the purchasers had to acquire vouchers from other local people on loan. Specialised credit institutions were often established for this purpose. These Savings and Loan Associations operated like banks, lending money (i.e. the nominal value of unrealised vouchers) deposited on their savings accounts to entrepreneurs, whose own workshare or compensation vouchers did not suffice for the purchase of a required object or complex.

Many private persons (often they were already active businessmen at that time) hoping to purchase an ex-collective production complex went around collecting vouchers from local people (particularly from former employees of the complex and inhabitants at the village where the complex was located) and many businessmen managed to amass large amounts of them. Formally, these vouchers were given on loan to the entrepreneur, but the legal status of many transactions remains in doubt. Some entrepreneurs have said they received the vouchers as presents. Why would that happen? The reason might be that the villagers perceived the role of the prospective buyer in the light of Soviet ideology: They may have believed that the new owner would be the 'representative of the people', i.e. a leader vested with patrimonial power and responsibility for his community members. Naturally, there were also those who believed the invested vouchers would enable them to keep their jobs, but there were many pensioners, who must have been well aware that the enterprise would not be able to offer them employment. It is widely known that many promises were made by businessmen collecting vouchers from villagers, most of which they have not been able or willing to keep.

For example, a typical businessman pledged help to a pensioner in ploughing and cultivating the field plot surrounding his modest hut (the pensioner hoped he would be able to set up a family farm on the land he would receive in restitution). The villagers were in fact expecting even more than this from the businessman – they expected all-round care from their new farm owner. The interviews in Estonia clearly show that people who had handed over their vouchers to the new entrepreneur, plus the majority of other villagers, were totally perplexed when they realised that the new farm owner would no longer cultivate their plots or supply pensioners with firewood for the winter, etc.

Hence, the cultural tradition of Soviet community spirit indirectly contributed to the creation of a large-scale agricultural system based on direct private ownership in Estonia. The moral code originating from the pre-decollectivisation 'communal order' no longer determined the 'moral subject' of all its members. However, in the final part of the chapter I will demonstrate that the community is capable of re-interpreting this cultural tradition in the post-collective context.

Anomie and Protest

Although the newly emerged society is undoubtedly capitalistic, its final form is still imperfect, and the predominantly large-scale agricultural production system is still taking shape in the countryside. The shape of things to come is also influenced by the moral state of the community, which could today be characterised in Durkheimian terms as anomic, but which on the other hand still displays the remains of the community spirit of the Soviet times. The reverse side of anomie is robber capitalism – arrogant and shortsighted pursuit of self-interest. In the following treatment it is linked with Durkheimian thinking, as presented in his works *The Division of Labor* (English translation 1964, Chapter 7) and *Suicide* (English translation 1951), although the term as such is not actually used in these books. According to *The Division of Labour*, social circumstances analogous to robber capitalism are characterised by incompleteness of the moral foundation of society. Although the foundation of new social relations may have already been laid qualitatively, time is also required for the development of a new kind of 'authority' on the basis of 'experience'. Only that will create a moral order as a prerequisite for everyday social exchange between individuals. Hence, we are primarily dealing with a moral principle, although legislation in line with this principle is also being developed.

The moral order is not grounded on explicit interpersonal interests, but on a system, and – in Durkheimian functionalist terms – 'the supra-individual sphere of transcendent values', which is ultimately rooted in every society. By contrast, in robber capitalism people are only connected by immediate interests; moreover, the association is only 'temporary', since ever-changing circumstances divide and pull individual interests in different directions. Thus, in the countryside of the post-decollectivisation era, robber capitalism effectively consisted of everyone at war with everyone and many new businessmen acted without scruples for their own benefit only. Corruption was rampant and even positions of trust were viewed from the standpoint of personal benefit. However, widespread social helplessness (people's inability to defend themselves) has been the reverse side of this selfishness. A large part, perhaps even the majority, of the rural population belongs to this group of powerless, often unemployed and destitute people, who are easily deprived of their petty assets by unscrupulous businessmen. Heavy use of alcohol (heavier than ever before, claimed the interviewees), physical violence and suicides are only some of its manifestations. The former chairman of a *kolkhoz* gave a colourful description of the life of these ordinary people:

> The world of the people has collapsed. They have lost touch . . . From a psychological viewpoint, during the kolkhoz people collaborated with each other, there were possibilities for interaction, it was possible to meet people at the workplace. Now people are lonely, associations are no longer formed. During Soviet times all the needs of the people were satisfied, and in the evening they escaped to the seclusion of their homes. Today, they can't get away from the isolation, you can't get away from the structure . . . The situation is worst in the new villages, kolkhoz villages, those half-villages with blocks of flats . . . Just like in

Russia, they're living from day to day. When they get some money, they drink it away.

Adhering to the terminology used in *Suicide*, these phenomena are manifestations of an *anomic* (normless) social condition. But Durkheim underlines the 'organic' nature of highly developed capitalism, and an Estonian countryside has a long way to go before it reaches a comparable structure of solidarity. All in all, these concepts of classical sociology (organic solidarity, and anomie as a disturbed state of solidarity) portray remarkably well the post-socialist crisis, although they were originally formulated to describe the birth processes of an industrial and partly capitalist society. However, we may presume that the current situation is just a transitional state, which is difficult to avoid because of the nature of the process, but which will eventually produce a normative structure characteristic of a developed capitalist society (cf. Raiser 1997).

Another feature of this state of affairs is that the powerless people described above cannot resort to the formidable support that in western societies arises from various organisational forms of civil society, such as trade unions, political parties, religious communities, or charitable organisations. The powerless also appear to be incapable of organising any collective or public actions, such as a demonstration or a strike. Relatives, particularly the closest relatives, are virtually the only source where people go in search for help and where they are rewarded.

These people who lost out in the privatisation process and are stuck in the countryside are powerless in relation to private enterprises and the formal institutions of society (e.g. municipal administration). Paradoxically these same people still actively influence the way society is being restructured, but in their powerlessness they have to utilise unorthodox means. These means may not be recognised without full understanding of the Soviet ideology, which is still welling up from the former Soviet community spirit, and the re-interpretations of which provide these activities with the necessary moral code and an invisible but tangible social support among the group of powerless who share the same community spirit. One half of these activities is called resistance, while the other half is called protest.

The atmosphere of resistance is reflected, for one thing, in people's language. People felt that the breaking up of the Soviet farms was a socially irrational and unfair act, almost regardless of the other opinions they might have had of the *kolkhoz* system. The people of Estonian countryside described the dissolution process with negatively charged expressions – just as the Bulgarian rural population did in a study carried out on a local community (Creed 1995). In both countries the expressions used originated from the days of forced collectivisation, some fifty years ago. Hence, the break up of a *kolkhoz* is consistently called its 'liquidation', and its last chairman is usually nicknamed the 'liquidator'. The use of the term indicates that ordinary people considered privatisation analogous to the liquidation of 'enemies of the people' in the kulaks.

Resistance is not limited to the verbal level. A woman who had bought a cowshed complex told in an interview that her employees experienced their transformation to the role of wager-workers under her as a shock, which led to a number of unfortunate incidents. At one point some of her employees simply refused to obey her orders, went berry picking and left her all alone with a large herd of cattle. That cowshed owner felt herself completely powerless. In the same manner the employees at the wood processing workshop occasionally walk out and leave work 'to get drunk', as the action is simply described. In both of the cases above the formally non-public collective actions may be interpreted as denial of the legitimacy of the new private ownership – or at least, denial of the legitimacy of the present form of private ownership. However, collective reactions such as the ones above are not typical. A more typical manifestation of resistance might be the following case. An entrepreneur who runs a privatised enterprise asked individually about ten skilled electricians – who had formerly been employed at the *kolkhoz* department now owned by the entrepreneur – whether they would be interested in working under contract for the privatised enterprise, but according to the manager of the enterprise their answer was as if they had spoken with one voice: "A little hunger is better than hard work".

These types of protest could be interpreted simply as negative representations, signs of total denial without any prospects for the future. However, research has shown that it is entirely possible to gain legitimacy in the eyes of the employees with a specific kind of treatment. Instead of brutal robber capitalism it might be possible to discover alternatives that would bridge the gap between the old patrimonial and the modern capitalist industrial relations. The cowshed mentioned above was sold and now the milkers are back at work. As a part of his skilful psychological approach the new owner is offering his employees better wages and long-term employment.

The empirical data gathered in Estonia also included more direct approaches by the powerless, such as anonymous phonecalls to enterprise managers, threats of arson and even murder. These kinds of threats have not remained just talk, since a few cases of murder and arson have been documented. For example, one owner of a cowshed complex decided to move elsewhere, so he sold his livestock and paid his debts. However, right after the cattle had been moved out the cowshed was set on fire. It has never been established who the culprit was, but the general opinion is that the arson was revenge by a former employee or employees. Of course, violence or threats of violence do not always embody a social element. However, it might be that some employees believe that concrete threats of violence will have an effect on the behaviour of the employers. The end result is that matters related to the enterprise are no longer the private affairs of the entrepreneur, although he is not yet facing organised labour and industrial legislation enforced by government officials.

The interpretation presented above is in essence Durkheimian, but my primary source was the theory of the moral economy of peasantry formulated by James C. Scott (1986) for somewhat different circumstances (primarily for the study of peasant communities of the Middle Ages and in developing countries). Resistance and protest of this type is a reaction against proletarisation and marginalisation, and the loss of previous respect – which had been shown towards ordinary *kolkhoz* workers, but especially to such vocational specialists as a milker or a tractor driver in the patrimonial and other Gemeinschaft-like societal structures of the *kolkhozes* (cf. old villages). Scott emphasises the everyday character of resistance. The relations within the peasantry (including relations between classes) and relations between the peasantry and the external elite are to some degree questioned every day, and therefore reproduced as they are or transformed gradually into other forms in the social processes of interaction in everyday life. This resistance is typically both in form and intentional content highly personal. At first sight it is also without any collective elements (e.g. thieving from the rich for the use of the thief's own family, which is also common in the Estonian countryside), but this does not exclude collective cultural conditions and consequences of these actions, nor their culturally determined nature overall.

As with Scott's peasants, the poor and powerless of the Estonian countryside are also characterised by the lack of institutional instruments for organised activity, although its absence has a different explanation. In Estonia former *kolkhozniks* have not yet been able to create such instruments but eventually they will (despite the open hostility displayed by entrepreneurs towards trade unions, etc.), whilst in the developing countries portrayed by Scott, the creation of such instruments may not even be possible in principle, since their development is blocked by the governing elite with violence.

However, both in the Estonian countryside and in Scott's villages each act of resistance and protest is indicative of something that is not anomic in its nature, and in the conditions furnished by Estonia's turbulent transition these acts may contribute to the creation of a bridge between the community spirit of Soviet patrimonialism and organic solidarity characteristic of developed capitalism.

Acknowledgements

This chapter is based on data gathered by interviewing over one hundred former *kolkhoz* and *sovkhoz* leaders and employees in Estonia (and partially also in Latvia and Lithuania). In Estonia interviews were conducted in six municipalities and they covered about ten former Soviet farms, both prosperous and poor ones. Additionally, we interviewed the most important local and county level agricultural officials. Many of the interviews were lengthy and several people were also reinterviewed. A few individuals have been interviewed for a total time of over ten hours (for details on methodology, see Alanen 2001a).

The interview data has been utilised by a multi-national (mostly Finnish and Estonian) group of researchers in two projects. The research projects were 'The Privatization of Agriculture in the Baltic Countries' (1993-1996) and 'The Decollectivisation of Agriculture in the Baltic Countries from a Psychological and Sociological Point of View' (1998-2000). These projects were financed by the Academy of Finland and I served as their responsible leader. The decollectivisation of a single Estonian *kolkhoz* provided us with material for an extensive English language book, which has recently been published (Alanen I, Nikula J, Põder H, and Ruutsoo R. (2001), *Decollectivisation, Destruction and Disillusionment – A Community Study in Southern Estonia*, Ashgate, Aldershot.)

References

Alanen, I. (1995), 'The Family Farm Ideology, the Baltic Countries, and Theories of Development', *Eastern European Countryside*, vol. 1, no. 5, pp. 5-21.

Alanen, I. (1998a), 'Petty Production in Baltic Agriculture: Estonian and Lithuanian Models', in M. Kivinen (ed), *The Kalamari Union – Middle Class in East and West*, Ashgate, Aldershot, pp. 39-70.

Alanen, I. (1998b), 'The Effects of Internal Struggles within Collective Farms on Emerging Enterprise Structures – The Case of Estonia', *Central European Hysteria. (Third) Special English-Language Issue of Replika (Hungarian Social Science Quarterly)*, pp. 145-160.

Alanen, I. (1999), 'Agricultural Policy and the Struggle over the Destiny of Collective Farms in Estonia', *Sociologica Ruralis*, vol. 30, no 3, pp. 431-458.

Alanen, I. (2001a), 'The Dissolution of Kanepi Kolkhoz', in I. Alanen, J. Nikula, H. Põder and R. Ruutsoo (eds), *Decollectivisation, Destruction and Disillusionment – A Community Study in Southern Estonia*, Ashgate, Aldershot, pp. 63-276. Forthcoming.

Alanen, I. (2001b), 'Introduction', in I. Alanen, J. Nikula, Põder, H. and Ruutsoo, R. (eds), *Decollectivisation, Destruction and Disillusionment – A Community Study in Southern Estonia*, Ashgate, Aldershot, pp. 1-14. Forthcoming.

Alanen, I., Nikula, J., Põder, H. and Ruutsoo, R. (2001), *Decollectivisation, Destruction and Disillusionment – A Community Study in Southern Estonia*, Ashgate, Aldershot. Forthcoming.

Clarke, S. (1993), 'The Contradictions of State Socialism', in S. Clarke, P. Fairbrother, M. Burawoy and P. Krotov (eds), *What about the Workers? Workers and the Transition to Capitalism in Russia*, Verso, London.

Creed, G. W. (1995), 'An Old Song in a New Voice: Decollectivization in Bulgaria', in D. A. Kideckel (ed) *East European Communities: The Struggle for Balance in Turbulent Times*, Westview Press, Boulder.

Durkheim, E. (1951), *Suicide*, The Free Press, New York.

Durkheim, E. (1964), *The Division of Labor in Society*, The Free Press, New York.

Kornai, J. (1980), *Economics of Shortage*, North-Holland, Amsterdam.

Lieven, A. (1993), *The Baltic Revolution*, Yale University Press, New Haven.

Maslovski, M. (1996), 'Max Weber's Concept of Patrimonialism and the Soviet System', *The Sociological Review*, vol. 44, no. 2 (May 1996), pp. 294-308.

Palm, T. (1992), 'The Suurupi Program: Estonian Economic Policy for the Transition', *Journal of Baltic Studies*, vol. 23, no. 3, pp. 283-298.

Parsons, T. (1961), 'Some Consideration of the Theory of Social Change', *Rural Sociology*, vol. 26, no. 3.

Parsons, T. (1971), *The System of Modern Societies*, Prentice Hall, Englewood Cliffs.

Raiser, M. (1997), *Informal Institutions, Social Capital and Economic Transition: Reflections on Neglected Dimension*, EBRD Working Paper Series, no. 25.

Raun, T. U. (1989), *Viron historia*, Otava, Keuruu (English language edition: *Estonia and Estonians*, Hoover Institution Press, Stanford, 1987).

Scott, J. C. (1986), 'Everyday Forms of Peasant Resistance', *Journal of Peasant Studies*, vol. 13, no. 1, pp. 5-35.

Smith, G. (1999), *The Post-Soviet States: Mapping the Politics of Transition*, Edward Arnold, London.

Weber, M. (1978), *Economy and Society, Volume 1*, University of California Press, Berkeley.

6 Rural Transitional Problems and Post-Socialist Peasantisation in the Russian Forest Periphery

EIRA VARIS

Introduction

This chapter highlights the various types of rurality in the region of the Karelian Republic in the Russian North, which, unlike the more commonly perceived definitions based on agricultural communities, includes both agrarian and non-agrarian countrysides. The industries found in rural areas of the Russian North are mainly based on forestry because of the natural conditions in the boreal zone and past Soviet policies that prioritised forestry here. Agriculture was marginalised and given only a role in support of forestry. Due to these reasons, rural areas have a specific settlement structure which is linked to the forestry industry and the processing of wood. Additionally, the social system has had its specific influence on rural development.

In the transitional period, i.e. after the collapse of the Soviet Union in 1991, new social structures have emerged and former ones are undergoing restructuring. The ongoing process of development in society is based on changes in the former socialist state structure as it undergoes transformation towards a market economy. Adopting a new social system has brought different problems to rural areas and has forced the accompanying anti-crisis measures to be altered. This chapter presents a new kind of rurality which has emerged during transition. Peasantisation as one possible process in rural transition is discussed. The chapter is based on research undertaken in rural areas of the Karelian Republic throughout the transition period (Varis 1998). Additional materials which include interviews with forest company managers in various types of enterprises across the Karelian Republic were collected in the summer of 1998.

Forestry-Based Karelia

The Karelian Republic covers an area of 170.000 km^2 and has a population of 800.000 inhabitants.[1] The population density of the area is very low: 4.6 inhabitants/km^2. Due to the high level of urbanisation, in fact in marginal areas the population density is only 1-2 inhabitants/km^2. The rural population consists of 24 percent of the population according to Russian definitions (Figure 6.1) (Popova 1998, 22), while if one were to count the population of settlements with less than 500 inhabitants, this figure would only be 5 percent (Varis 1998, 35). The significant feature of the total population is that it is decreasing due to the consequences of the transitional economy.

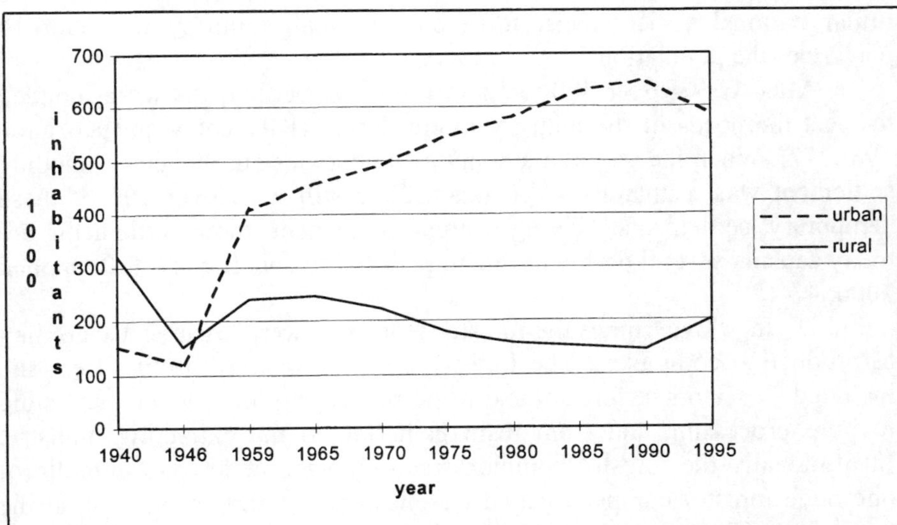

Figure 6.1 Urban and Rural Population of the Karelian Republic in 1940-1995

The present rural settlement structure originates in Soviet state socialism. The rural-urban continuum practically does not exist at all, and instead rural settlements can have very urban features in their physical construction and plans of settlements. The rural population is concentrated in settlement centres, just as the majority of the republic's population is concentrated in town centres. This is due to the dramatic industrialisation and urbanisation processes after WWII and a far-reaching rural policy of centralisation. After the year 1960 more than half of the rural settlements of the Karelian Republic were forcibly abandoned because they were not deemed to have viable development prospects (Varis 1996, 14-15). Overall, a massive out-migration occurred from the countryside and the rural population which remained was ageing.

Forestry is the main source of livelihoods in the entire area (Table 6.1). After WWII the role of rural settlements in the regional division of labour was to provide workers for forestry. Soviet political programmes were committed to this aim. Agricultural collective farms (*kolkhozes*) were suspended in the 1950s and their labour forces were harnessed for forestry. Only *sovkhozes* i.e. state farms and subsidiary farms of the logging companies continued to operate as agricultural units. People from other areas of the Soviet Union were recruited to forestry work, which led to distinct changes in the nationality structure of the population. Today, the titular nationality, Karelians, form only a small minority of about 10 percent of the population.

After WWII over 300 temporary forestry settlements were founded to meet the needs of the industry to cut timber (Klementev and Kozanov 1988, 17). When the required amount of timber was cut, the corresponding settlement was dismantled. Yet today, there still exist over 100 of these 'temporary' settlements. Having become permanent, these settlements are today causing several problems due to poor living conditions (cf. Piipponen 2000, 47-48).

During Soviet times the forestry industries were organised according to production complexes. The forestry complex of the Karelian Republic included all sectors of forestry and wood-processing: from wood harvesting to pulp processing and from resin gathering to the extracting industry. Institutionally the forestry complex was organised under the umbrella of one huge forestry company called Karellesprom, which in practise, at the turn of the 1990s, had itself become the entire forest complex (Myllynen and Saastamoinen 1994, 93-95; Myllynen et al. 1996, 57). The forestry complex accounted for half of all the volume of production of the Karelian Republic, thus it has had a significant role in the development of the entire region (Table 6.1).

Table 6.1 Structure of Industrial Production by Branches in the Karelian Republic in 1970-1999 (percent)

Branch	1970	1980	1990	1995	1999*
Energy production	3	2	3	12	7
Iron and steel industry	1	1	11	13	9
Mechanical engineering	14	22	20	6	7
Forest industry (complex)	**55**	**42**	**37**	**52**	**55**
Construction materials	7	9	7	3	2
Light industry	4	4	4	0.3	0.3
Food industry	13	14	13	8	8
Other	3	6	5	6	2
Total	**100**	**100**	**100**	**100**	**100**

* estimate

Sources: Karel'skaia ASSR v tsifrakh v 1986-1990 godakh: Respublika Kareliia v tsifrakh 1996; and Respublika Kareliia v tsifrakh 2000.

The forestry complex not only took care of production, but also of the social sphere of rural life. Public services were organised through the offices of the forestry complex. The main exception were the schools, which were maintained directly by state organs, in practice by the village soviets i.e. local governments (cf. Varis 1998). Otherwise, the forestry company maintained food shops, canteens, subsidiary farms, public bathhouses, and kindergartens. It also supported cultural clubs and leisure activities. Thus forestry settlements, as well as other rural settlements in the Karelian Republic (cf. Varis 1996), can be characterised as 'one company towns'.

After the collapse of the Soviet Union, Karellesprom was privatised in 1993 and incorporated into joint-stock companies. In 1998 there were 67 private forestry companies functioning in the Karelian Republic (Svanhovd miljösenter 1998). The privatised Karellesprom is now a holding company, which includes 28 joint-stock companies, and of which 20 deal with forest harvesting (Emel'ianov 1998). However, because of the unstable economic situation in Russia, the enterprise structure fluctuates all the time.

Even though the volume of forest production has remained about the same, the real value of total production has decreased dramatically in the transitional period since 1991, as illustrated in Figure 6.2. In 1997 the real value of total production consisted of less than half of the total amount from the beginning of the transition period in 1990.

Forest cuttings have decreased since the prosperous years in the 1960s from 18 million cubic metres to 7 million in 2000, a drop even from 1990 when they were almost 10 million m^3 (Goskomles 1996, Shurupov 2000). In actual fact, Karelian wood-processing enterprises are suffering from the lack of raw materials, because wood logging enterprises are experiencing operating difficulties. In general almost all of the present-day forestry companies have difficulties with their operations. They are experiencing problems in adapting to the rules of the market economy, or as Kosonen puts it (1996, 75): the companies are facing an institutional lag. There seems to be passive resistance to change within the companies and therefore intended reforms cannot be easily carried out (cf. Tykkyläinen and Jussila 1998, 225).

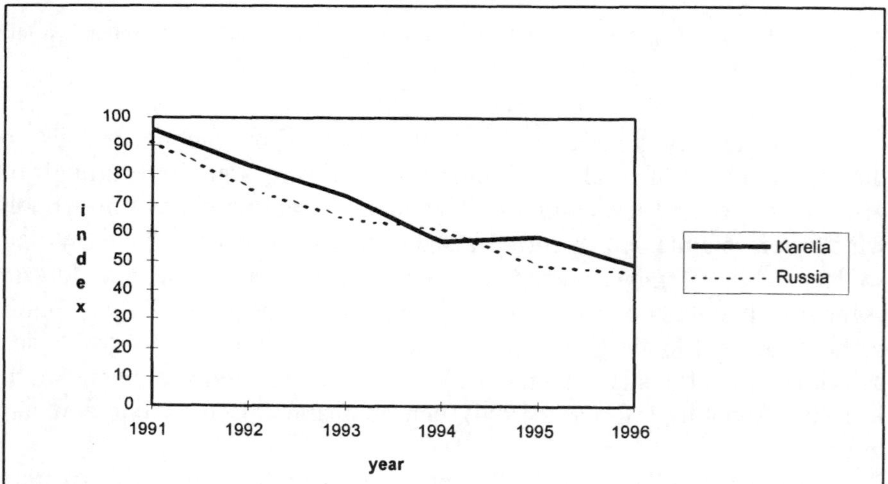

Sources: Based on Sotsial'no-ekonomicheskogo polozhenie Respubliki Kareliia za 1991-1996 and Sotsial'no-ekonomicheskogo polozhenie Respubliki Kareliia za 1997, as calculated by Eira Varis.

Figure 6.2 Index of the Amount of Industrial Production in the Karelian Republic and Russia-wide in the Transitional Period (1990=100)

Rural Problems During the Transition Period

The collapse of the national economy and the restructuring of forest production inevitably were also reflected in rural areas and in their development. Problems, which were unknown during socialist times, emerged. In socialist times, unemployment was a non-existent phenomenon, which only occurred after the privatisation of the forestry complex. New private companies started to rationalise their operations according to the principles of the market economy, which meant that a lot of work places disappeared. By the year 1999 unemployment rose to 15.8 percent in the Karelian Republic (Respublika Kareliia v tsifrakh 2000). However, due to the undeveloped unemployment security system, the number of registered unemployed was only 4.7 percent. Many individuals are not willing to register as unemployed, because unemployment benefits are low and difficult to attain (cf. Klementev et al. 1996, 201). In one of the villages where research was carried out, for example, the distance to the unemployment office is 60 km. Unemployed were expected to register there once a month in order to receive their unemployment benefits. Yet a return bus ticket cost 80 rubbles, which was the same as the monthly unemployment benefit.

The most evident problem of the companies, at least as far as ordinary people are concerned, is a lack of cash money. Due to the bad economic situation across Russia, the majority of domestic trade is barter trade (Smirnov 1998). People receive their salaries in goods if they get them at all. There have emerged salary payments delays for months. This drastically effects people's everyday lives. As less and less cash money is available, the monetary economy has been reduced and people have been forced to resort to a self-sustaining economy. They have to grow their own food and gather the gifts of nature in order to survive in their everyday lives. The situation in rural areas has been made more serious by delays in pension payments. Pensioners do not have sufficient possibilities to maintain their way of life, and are often in dire circumstances. Therefore, the importance of social networks has grown. Adult children have been forced to take responsibility for their elderly parents' livelihoods. In cases where the elderly have no relatives who can acre for them, a lot of human suffering is faced.

As commercial services, such as shops, are today mainly controlled by private retail companies, the price level of consumer goods has risen. Municipal administrations are still maintaining some shops, but they are unable to compete with private ones. This unavoidably leads to shop closures over time and thus to increases in price levels. Since there is hardly any cash present in the villages, these high prices in private shops have limited ordinary people to food purchases only. Table 6.2 indicates the results of household surveys in two forestry settlements in the Karelian Republic. The question asked concerned the adequacy of money in households.

Table 6.2 'What can you buy with your income?' Number of Answers (percent), Koivuselkä n=112 in 1994, n=72 in 1997, Matrosy n=105 in 1997

Income is enough for...	Koivuselkä 1994	Koivuselkä 1997	Matrosy 1997
Everything needed	4	4	6
Food and the most necessary items	22	25	26
Only food	64	29	36
Money is not enough*	-	40	29
Not even food, we are starving	3	1	2
no answer	7	-	1
Total	**100**	**99**	**100**

* This alternative was only in the 1997 survey.

Sources: Household surveys in Koivuselkä in 1994 and 1997 and in Matrosy 1997, in Varis 1998 and Piipponen et al. 1998.

The frequency of answers in the two middle categories indicate that the majority of people cannot afford consumer goods. After the Russian economic crisis of autumn 1998, it can be estimated that the situation is even more critical. On the other hand there were only a few answers falling in the lowest category ('we are starving'). The reason for this is that people are surviving through their self-sufficient home economies and at least are getting their daily meals.

Through the dismantling of the socialist forestry complex its role as maintainer of social services also disappeared. During socialism the typical rural village was a 'one company town'. It meant that the company was not simply an economic institution, but also the primary unit of civil society and the ultimate base of social and political power (cf. Clarke 1992, 7). For example, a forestry company maintained a canteen which offered daily meals to its workers, thus providing adequate nourishment. A company's subsidiary farm produced food products for its workers' household needs and therefore it was not necessary for these workers to maintain their own garden plot. In most cases, they produced some foodstuffs anyway, but not out of necessity, but as a complementary livelihood. The company also used to maintain day-care services for workers' children. After privatising the company the local government became responsible for these social services. With the economic and financial problems of the state, local governments have not received adequate financial transfers for the purposes of maintaining social services. The development of this field is out of the question today.

The downsizing of social services did not happened suddenly in practice. There are still some services left in many forest companies. Housing facilities are the most complicated service to be maintained by state organs. As mentioned above, forestry settlements were originally meant to accommodate forest workers for a limited period only, usually for about 20 years. Some of these settlements became permanent and the original housing, which were mainly barracks and detached houses for several families, are still in use and are in poor condition due to a lack of basic renovations. They lack clean water pipes and sewerage, central heating or washing facilities. For inhabitants' personal hygiene, public bathhouses (the *bania*) were maintained by the forestry company. Now these services, which are essential for everyday life, have been transferred to the district administration's department of municipal housing and service management. One extreme case is where the village's public bathhouse has been closed and those people who need this service, which are usually pensioners, are transported weekly by bus to a bathhouse located in a larger settlement 16 km away. As in the case of social services, local governments do not have adequate finances to maintain housing, thus in some communities companies continue to provide housing, because in practise it is not possible to transfer it to local government. As several people involved commented, any rash housing transfer by companies without heeding the local situation would effectively destroy communities and leave workers inadequately housed (Travkhin 1998).

Cultural services and leisure activities have dissipated in the transitional period. The cultural heart of the forestry villages was the club house. Here were organised hobby clubs for all age-groups, dances, theatrical performances and also official meetings. The club house included also a village library and cinema. These services were financed by the district cultural administration, which employed local workers. The hobby clubs especially were sponsored by the forest company. In the situation of financial crisis the public sector has reduced cultural services and forest companies have cut their support because the market economy no longer makes allowances for such expenses. In many villages the club houses have been closed and the remaining ones are working on minimal budgets. People are not used to pursuing hobbies or initiating clubs on their own, due to the Soviet social environment which organised society's free-time, and has thus replaced the former rich cultural and leisure life with a vacuum, poisoning the social atmosphere.

Emerging unemployment, delay of salaries and pensions, rises in prices and dissipation of services have not meant only financial disaster to household economies, but has also created mental stress and social depression in rural populations. Social problems, such as alcoholism and the diseases and mental illnesses that accompany it, have increased. The occurrence of social problems can already be found in state statistics: the life expectancy of rural people has decreased from 67.3 years in 1991 to 58.1 years in 1994 (Sotsial'no-ekonomicheskogo polozhenie... za 1991-1996). These phenomena may also have unanticipated consequences in the future. Rural problems which have occurred or increased after the collapse of the Soviet Union are summed up by the following:

- unemployment;
- lack of cash;
- disappearance of the social role of the production unit;
- poor housing;
- privatisation of public services → expensive prices;
- dissipation of cultural services;
- social problems.

Rurality of Transition in Karelia

After the structures created by socialist society collapsed and while the new ones are in the process of formation, it has been necessary for individuals to initiate measures to manage. Rural transitional problems have had radical consequences on the population's everyday life. In order to survive people have been forced to solve problems themselves and have adopted their ways of life to become self-sufficient (Varis 1998). Farming on the small plots surrounding individual houses has increased to such an extent that the proportional production from household plots has more than doubled in the transitional period (Figure 6.3). In practice, every family grows potatoes and vegetables for its own needs. Most households keep small livestock, such as sheep, goats and hens. More and more frequently cattle is also being kept and new cowsheds have been built in the villages. This kind of rurality has increased after the reorganisation of the forestry complex, in which subsidiary farms have also mainly been dissolved.

Fish are caught and berries and mushrooms are carefully gathered and used for the family's own purposes. Berries are also gathered for sale, one additional way to obtain a cash income for rural households. For instance, today it is very common to see hordes of berry sellers by the side of the road leading to Finland selling berries and mushrooms to Finnish visitors. And Finns are eager to buy them, since prices are only a fraction of the Finnish product equivalent price.

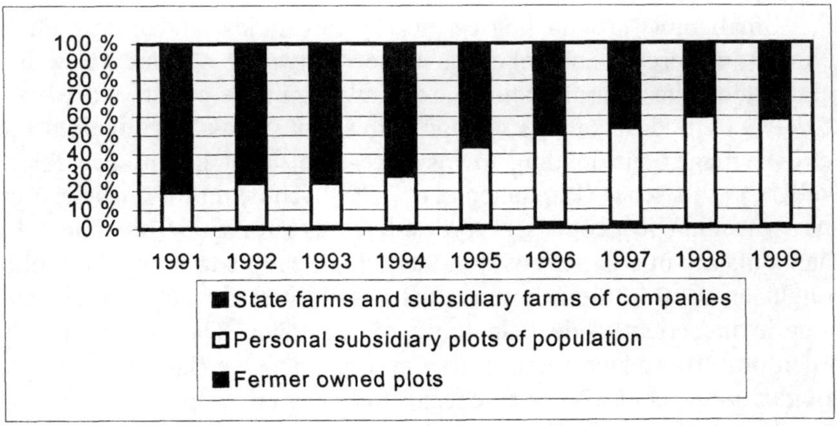

Source: Respublika Kareliia v tsifrakh '97 and 2000.

Figure 6.3 Agricultural Production

This kind of survival strategy can be called one of defensive reactivity (Varis 2000, 132). This suggests an autarkic way of life in which the maintenance of everyday life is based, on the one hand, on the small-scale farming of household plots and on the gathering of nature's products (cf. Ioffe and Nefedova 1997, 177). On the other hand, it is based on strong economically-based social networks. These measures are adopted because the production unit of the resource community can no longer secure the reproduction of the labour force.

While the former rural actors, such as forest companies and local governments, are in the midst of extensive restructuring, new types of actors have replaced them. They are not new as such, but the former role of actors has changed. Growing self-sufficiency needs strong social networks.

Families and kinship relationships are in a vital position in networks. Also, unofficial canals and nets have formed, for instance between berry suppliers and dealers. People exchange not only items that they have produced, but also their private labour services. Social networks are not a phenomenon only of the transitional period – they also existed during socialism – and the family unit has been especially strong in Russia due to cultural reasons. However, the transitional situation has now reformed social networks towards a more economic base. The 'new' rural actors, though playing a major role in rural survival strategies of self-sufficiency, are still only a complementary part of these strategies. The role of local government has been small and it does not seem set to increase in this economic situation. Simultaneously, the traditional roles of the forest companies locally have been fading away.

Small plots around houses, nearby hay fields, and the products of nature from forests and lakes have been given a special value in the countryside. These rural features are available in the countryside, but also the *dacha* (a modest summer cottage with a plot sized a few hundred square metres) villages surrounding towns have significant importance for town dwellers in this sense (Piipponen et al. 1998). Thus, rurality has been given a new meaning in people's everyday lives as rurality of transition. Rural inhabitants are in a better position than urban people to solve the problems brought on by transition, because the possibilities for self-sufficiency are better in the countryside than those of the town. The ability to have a garden plot to produce food, to live in a inexpensive flat (even if it is not provided with modern conveniences), to have access to forests and lakes, are all features of rurality, which combined have drawn some town dwellers to move to countryside. These migrants are mainly people who have been left unemployed in the urban setting. For urban dwellers in Russia today, connections to the countryside, or the possibility to own a *dacha*, is a question of privilege. *Dacha* villages nearby towns have taken on a new vital meaning for urbanites which has replaced the former recreational one.

The question arises if the strategy of defensive reactivity is an effective one. If using economic criteria, then one must conclude that it is not. Rural poverty has grown during the transition period (cf. Granberg and Kovách 1998, 15). Another question is how sustainable it is in ecological terms. Manual labour without excessive farm machinery and an insignificant use of chemical fertilisers has features of sustainable ecological behaviour. The main contradictory issue is that people are forced to do it for their survival, not because of a newly formed environmental consciousness.

In addition to a strategy of self-sufficiency and return-migration to the countryside, the third type of 'new' rurality which has occurred in the transition period in the Karelian Republic is peasantisation. Actually, '*fermer*-isation' would be a more appropriate term in this case. A new group of private farmers have been generated, or *fermers* as they are called in Russian (cf. Nickolsky 1998, 203). At the beginning of 1998 there were 654 registered peasant farms in the Karelian Republic (Sotsial'no-ekonomicheskogo...za 1997). This group of registered farmers has a right to own land through contract and to have title to tracts of forest. In 1995 the average size of private farms in the Karelian Republic was 17 ha, out of which only 1.5 ha was arable land (Izotov 1998). In any case, the proportion of peasant farms involved in agricultural production has been insignificant thus far (Figure 6.3).

The idea of generating this group of farmers has its origins in the Finnish 'smallholder-lumberjack' mode of production in northern and eastern Finland after WWII. In it, the livelihood of a rural household was a combination of private small-scale farming and forest work in state companies' forests (Rannikko 1995). However, Russian farmers own land by contract, and they cannot buy or sell it freely. They also have the right to have title to parts of the forest, but not to own it, i.e. to buy or sell it. Timber from these lands can be used only for the private farm's needs. At the beginning, the new *fermers* were persuaded to start up their farms through advantageous loans offered by the state. Most of the new *fermer*s were previously workers from state farms, i.e. *sovkhozes*.[2] In forestry-based villages, new *fermers* were the exception. In any case, because of high inflation, the attraction of becoming a private farmer quickly disappeared. There also occurred some misuse of loans and many private farms closed before they had properly started up (Sebin 1995).

Post-socialist Peasantisation in the Karelian Republic?

Let us examine the above-mentioned processes of rural transitional problems and the rurality of transition by employing the concept of peasantisation. As presented by the authors in this volume's introductory chapter, the concept of 'peasant' represents a specific set of relations between the cultivators of the land and other groups, and, on the other hand, it refers to a form of production. In the case of the Karelian Republic there are certain specific nuances which must be taken into consideration. Even though the industries of Karelia have been forestry-based, agriculture has also had a certain role during socialism. The area had a supporting agricultural industry, consisting of state farms and the auxiliary farms of companies, even though it was not self-sufficient in its agricultural production. Most food, especially grain, was imported from other regions of the Soviet Union. The role of agriculture changed after the collapse of socialism. When the forestry complex dissipated, the population had to resort to self-sufficiency, and the position of farming in the household economy increased. On the one hand, this was due to employment problems within the newly privatised forest companies. Waged work has become unstable and unemployment has emerged through the rationalisation of the work force. Wage payments have been delayed and much of them have been resolved through a system of bartering. Due to the lack of cash, people cannot buy their food and are forced to grow it themselves.

On the other hand, the meaning of private small-scale farming has increased because of the market-type rationalising by the auxiliary farms of the forest companies. Previously, these farms were responsible for the majority of the food management of rural households by supplying them with basic nourishment, especially with meat. In practice, the dissolution of auxiliary farms has led to a change in the rural diet, consisting more and more of only potatoes and vegetables. The establishment of household animal husbandry takes time, whereas there are few resources to purchase building materials for new cattle houses and other farm buildings. Thus, people forced to adapt a self-sufficient mode of life can be considered as a kind of peasantry, because it has the features of peasantry, such as the farming lifestyle and strong family ties. Yet, essentially, farming is not the primary source of income or living, which remains forestry work.

New *fermers* have, on the other hand, established agriculture as their primary occupation. They have given up work in public companies, which usually were state farms, and have begun to run their own private farms. Or they have been left unemployed by companies. Usually, the 'success stories' among new private farmers are, in fact, among those people who had been in the management of the state farm, so their human capital for small-holding had been high from the start. However, new farms are still very rare and their number is not increasing rapidly as expected, but actually decreasing. The costs met in establishing a new farm are not sufficiently covered by state loans and new farmers have faced many difficulties in the running of farms. One of the main problems is the lack of wholesale trade structures. New farmers have faced so many institutional difficulties that maintaining their small-holding is very difficult work, and has turned private farming into an undesirable profession in the forest periphery.

Another important feature of a peasantry is its relationship to the land. Another distinguishing feature between the peasantry of rural transitional Russia and those in a Western sense is the question of land ownership. In Russia land is still owned by the state (cf. Kozlov 1998). Land privatisation was conceived in such a way that only plots around houses have been privatised. They are, however, very small of about 1500 square metres. Therefore, they are only appropriate for self-sufficient garden vegetable production, but not for corn cultivation or other larger scale agriculture.

Nor do new *fermers* own land in the Western sense. They can own land by contract, once they have registered as a *fermer*. Land is inheritable, but a farmer cannot sell it. Forests are only in the name of a farmer. They can be used for the needs of a farm, such as for firewood or building materials, but not for sale products. Thus, the bonds to the land are weak and a relationship to land does not tally with the Western viewpoint of peasantry.

In addition, individuals are no longer alive from the generation which can still remember private land ownership in the Karelian Republic, i.e. before socialism (cf. Granberg and Kovách 1998, 13). Moreover, the Karelian Republic never has had land ownership in a modern sense at all. Before collectivisation, agriculture was mainly based on the relics of the feudal system (Nikitina 1993), thus an important step in the process of private land-ownership was 'jumped over' in Russia, unlike, for instance, in Hungary (cf. Varis 1998). Land ownership was converted almost directly from feudalism to collectivism.

In addition to its dependence on agriculture as its main industry and a special relationship to the land, peasantry also has a cultural sense. This can be referred to as a peasant mentality, which includes strong family ties and traditions of farming. Strong social networks are an essential part of the self-sufficient mode of life. In this sense it can be considered as a peasantry. The tradition to cultivate the plots around one's house is also long, and during socialism it was also common to have a household garden. Even though the public sector tried to supplement the diets of forest workers with food, it did not, however, succeed in meeting demands very well. Thus, people are used to growing their own food. During the transitional period the meaning of these small gardens and access to products in nature has become vital. Thus, the mind-set to farm is strong, but the self-sufficient mode of life has been limited by the other main features of the peasantry.

New *fermers* also have social networks, but not in a similar sense to the peasantry, in which several generations of farmers have grown up under the same roof. *Fermer*s have no tradition of private farming, and their ancestors have most likely not been peasants. They themselves have been wage workers of state-owned companies, such as state farms or forestry companies. They have no inherited skills for private farming. Even if they have been agricultural workers in state farms, they have only been responsible for a certain part of the production process. They have no comprehensive farmer's skills. Their only experience of the peasantry is from their household plots, and private farming has to be learnt from the beginning. In Table 6.3 the features of peasantry from the self-sufficient and *fermer*-type modes of life are summarised.

Table 6.3 Features of Peasantry from two Rural Modes of Life in the Karelian Republic

	Self-sufficiency	**Fermer**
Position of agriculture in the household economy	supporting, however essential for livelihood during transition	main industry
Relation to land	only a small plot	land is a 'tool' for livelihood, but not possible to sell
Peasant mentality	strong social networks and family relations	no inherited tradition for private farming

This analysis suggests that one can hardly speak about a re-peasantisation (cf. Granberg and Kovách 1998, 13) in the case of the Karelian Republic. Preferably one should speak about post-socialist peasantisation. Rural modes of life have features of peasantry, but they still have their own specific features, which originated in socialism. During the transition period rural modes of life have changed, but they have not turned into peasant ones. The group of individuals most near to the 'Western type' of peasants are the new *fermers*. They have not faced re-peasantisation either, because they have been wage workers and have no inherited traditions of the peasantry. Therefore, the latter could be called post-socialist peasants. On the other hand, the self-sufficient mode of life in the countryside also contains some features of the peasantry. The most important is the significance of private farming to make one's livelihood. Yet this mode of life has not formed as a result of the people's free will from their aspirations, but has been generated necessarily by external conditions. One could even claim that there has occurred an unconscious post-socialist peasantisation process, because people's survival strategy has been one of defensive reactivity.

Concluding Remarks

The restructuring of socialist production has also led the rural areas of the Karelian Republic to social and economic hardship. However, forestry has remained in a central position as the basic rural industry in the transitional period. The forestry sector has lost the majority of its formerly maintained functions in rural settlements, which has accelerated the restructuring of the countryside. Unemployment, lack of cash and the rise in prices, dissipation of services, and increasing social problems are the issues which have arisen during the transition period.

In order to survive the problems of transition, people have been forced to create their own survival strategies. The most important of them is one of defensive reactivity which is based on self-sufficient household economies. The meaning and value of countryside and rurality has increased due to the natural resources available for self-sufficiency. At the beginning of the transitional period a new group of private farmers, *fermers*, a type born of peasantisation, arrived in the Karelian Republic. Their meaning in agricultural production have so far, it seems, been very slight. Self-sufficient and *fermer*-type modes of rural life can be considered as a specific type of peasantry, because of certain features. In any case, they lack the most important feature – the existence of private land-ownership. Therefore, the process which has occurred during the transition period can be referred to as post-socialist peasantisation. The growth of other types of peasantisation, beyond one based on unconscious self-sufficiency, does not seem very presumable at least in the wider sense in the near future, because of still existing socialist structures.

The 1998 economic crisis in Russia once again had negative consequences on rural life. Over the previous few years some signs of economic recovery had appeared which gradually entered people's everyday lives. However, in the present Russian economic and social situation a commitment to a new structural rural policy is unlikely. There are attempts to improve the serious social and economic situation through strong economic policies and massive foreign investments. The question thus arises – is post-socialist peasantisation as a temporary saving measure of the transitional economy, or a first step towards sustainable development in the Green Ring?

Notes

1 The Republic of Karelia is situated in the North-Western part of Russia between Leningrad province and the Kola peninsula. Finland and the Republic of Karelia have 700 kilometres of common border: the natural conditions on both sides are very similar, and forests, marshlands, and lakes predominate. Finland and the Karelian Republic also share much historically and culturally. Karelia is a region which has been divided by a border throughout its history. One part of Karelia has always been in Finland and one part has always belonged to Russia. The border has shifted and changed numerous times. The present border was decided at the post-war 1947 Paris Peace Conference (Eskelinen et al. 1994). As a consequence, the areas which Finland ceded to the Soviet Union were partly incorporated into Leningrad province and partly into the Karelian Republic. The Finnish population fled to Finland and afterwards the area was re-populated by Russians, Belorusians and Ukrainians (Laine 1994). When speaking about Karelia in Russia today it means the area of the Karelian Republic. The multi-phase political history of Karelian nationalism in the 1920s and 1930s and its relationship to the 'Red Finns' is interestingly presented by Markku Kangaspuro in his PhD dissertation (2000, summary in English).

2 Agricultural state farms in the Karelian Republic have also gone through restructuring in recent years, but there still exist about 30 units (cf. Andreoli et al. 1997). Workers of state farms, even though engaging in agriculture as their main activity, do not work for themselves, but can be regarded to be in a similar situation to industrial workers who work eight hours a day for their wages. They are the workers of a public firm, without having the cultural mentality of the peasantry. Even if state farm workers also have private plots, and all of their activities are involved with agriculture, they are still not peasants.

References

Andreoli, M. et al. (1997), 'Food production in transition', in J. Kortelainen (ed), *Crossing the Russian border. Regional development and cross-border cooperation in Karelia*, University of Joensuu, Department of Geography, Joensuu, pp 45-58.

Clarke, S. (1992), 'Privatization and the development of capitalism in Russia', *New Left Review*, No 196. November/ December.

Demograficheskii ezhegodnik Respubliki Karelii. Statisticheskii sbornik (2000), Goskomstat Rossii, Gosudarstvennyi komitet Respubliki Kareliia po statistike, Petrozavodsk.

Emel'ianov, I.G. (1998), Vice Director of Karellesprom, Personal communication, 17.6.1998.

Eskelinen, H., J. Oksa and D. Austin (1994), 'Introduction', in Eskelinen, H., J. Oksa and D. Austin (eds), *Russian Karelia in Search of a New Role*, Karelian Institute, University of Joensuu, Joensuu, pp. 5-12.

Goskomles (1996), 'Rashchetnaia lesoseka i fakticheskii ob'em rubki' (Data from the University of Petrozavodsk).

Granberg, L. and I. Kovách (1998), 'Beyond socialism, beyond the peasant state', in L. Granberg and I. Kovách (eds), *Actors on the changing European countryside*, Institute for Political Science of the Hungarian Academy of Sciences, Budapest, pp. 7-19.

Ioffe, G and T. Nefedova (1997), *Continuity and change in Rural Russia. A geographical perspective*, Westview Press, Colorado and Oxford.

Izotov, A. (1998), 'Maatalouspolitiikan muutokset Karjalassa 1990-luvun alussa', in J. Oksa (ed), *Koivuselkä – metsätyökylä Venäjän Karjalassa*, University of Joensuu, Karelian Institute, Joensuu, pp.42-59.

Kangaspuro, M. (2000), *Neuvosto-Karjalan taistelu itsehallinnosta. Nationalismi ja suomalaiset punaiset Neuvostoliiton vallankäytössä 1920-1939* (Soviet Karelia's Struggle for Autonomy: Nationalism and the Finnish Reds in the Governance of the Soviet Union in 1920-1939, English summary), Bibliotheca Historica 60, SKS, Helsinki.

Karel'skaia ASSR v tsifrakh v 1986-1990 godakh (1991), Goskomstat RSFSR, Karel'skoe respublikanskoe upravlenie statistiki, Kareliia, Petrozavodsk.

Klement'ev E. and A. Kozanov (1988), *Sel'ska'a sreda i naselenie Karelii 1945-1960 gg.* Nauka, Leningrad.

Klementev J. et al. (1996), Local impacts of restructuring – A case study of a forestry settlement, in E. Varis and S. Porter (eds), *Karelia and St. Petersburg. From Lakeland Interior to European Metropolis*, Joensuu University Press, Joensuu.

Kosonen, R. (1996), *Boundary, Institutions and Indigenous Development. The case of Vyborg*, Helsinki School of Economics and Business Administration, Helsinki.

Kozlov, A. (1998), 'Ownership of forests at the period of formation of market relations', in O. Saastamoinen and S. Lipitsäinen (eds), *Social sustainability and the taiga model forest*, University of Joensuu, Faculty of Forestry, Joensuu, pp. 67-74.

Laine, A. (1994), 'Karelia Between Two Socio-Cultural Systems' in Eskelinen, H., J. Oksa and D. Austin (eds), *Russian Karelia in search of a new role*, Karelian Institute, University of Joensuu, Joensuu, pp. 13-25.

Myllynen, A-L. and O. Saastamoinen (1994), *Karjalan tasavallan metsätalous*, University of Joensuu, Faculty of Forestry, Joensuu.

Myllynen, A-L., et al. (1996), *On the ecological and economic impacts of wood harvesting and trade in North-west Russia*, Gummerus, Jyväskylä.

Nikitina, O.A. (1993), *Kollektivizatsiia i raskulachivanie v Karelii* (Avtoreferat dissertatsii na soiskanie uchenoi stepeni kandidata istoricheskikh nauk), Rossiiskaia Akademiia Nauk, Institut rossiiskoi istorii, Moscow.

Nickolsky, S. (1998), 'The treadmill of socialist reforms and the failures of post-communist 'revolutions' in Russian agriculture: is there an alternative?', in I. Szelényi (ed), *Privatizing the Land. Rural political economy in post-communist societies*, Routledge, London, pp. 191-213.

Piipponen, M. (2000), 'Residential issues of two transitional forestry villages in the Karelian republic', in E. Varis (ed), *Transitional survival strategies of peripheral resource communities in Hungary and North-western Russia* (Studies on the Northern Dimension. Vol. 5), Finnish Institute for Russian and East European Studies, Helsinki, pp. 40-55.

Piipponen, M. et al. (1998), *Socio-economic conditions of the population in Matrosy. Report of interviews* (Unpublished report for Taiga Model Forest Project), University of Joensuu, Joensuu.

Popova, M. (1998), 'Maaseutu ja sen asutustyypit Karjalan tasavallassa', in J. Oksa (ed), *Koivuselkä – metsätyökylä Venäjän Karjalassa*, University of Joensuu, Karelian Institute, Joensuu, pp. 19-33.

Rannikko, P. (1995), 'Restructuring of forestry and forest villages in Eastern Finland', in L. Granberg and J. Nikula (eds), *The peasant state. The state and rural questions in 20th century Finland*, University of Lapland, Rovaniemi, pp. 109-120.

Respublika Kareliia v tsifrakh 1996 (1997), Gosudarstvennyi komitet Respubliki Kareliia po statistike, Petrozavodsk.

Respublika Kareliia v tsifrakh '97 (1997), Goskomstat Respubliki Kareliia, Petrozavodsk.

Respublika Kareliia v tsifrakh '00 (2000), Goskomstat Respubliki Kareliia, Petrozavodsk.

Sebin, A. (1995), Vice-Mayor of Sortavala, Personal communication, 7.10.1995.

Shurupov, I. (2000), Minister of Economy of the Karelian Republic, Interview in *Karjalan Sanomat* (newspaper), 30.12.2000.

Smirnov, E.I. (1998), Commercial director of Suojärvi Paperboard factory, Personal communication, 13.8.1998.

Sotsial'no-ekonomicheskogo polozhenie Respubliki Kareliia za 1991-1996 (1997), Goskomstat RK, Petrozavodsk.

Sotsial'no-ekonomicheskogo polozhenie Respubliki Kareliia za 1997 (1998), Goskomstat RK, Petrozavodsk.

Svanhovd miljösenter, Database at http://www.svanhovd.no/ori/orit.asp

Travkhin, V.V. (1998), Director of Lyaskelsky Paper Factory, Personal Communication, 11.8.1998.

Tykkyläinen, M. and Heikki Jussila (1998), 'Potentials for innovative restructuring of industry in Northwestern Russia', *Fennia*, 176:1, pp. 223-245.

Varis, E. (1996), *The restructuring of peripheral villages in North–western Russia*, (Research for Action), The United Nations University, WIDER, Helsinki.

Varis, E. (1998), *Syrjäkylien murros Venäjän Karjalassa ja Unkarissa. Tutkimus postsosialistisen maaseudun restrukturaatiosta ja resurssiyhdyskuntien selviytymisestä*, University of Joensuu, Joensuu.

Varis, E. (2000), 'Rural resource communities and survival strategies during transition', in E. Varis (ed), *Transitional survival strategies of peripheral resource communities in Hungary and North-western Russia*, Studies on the Northern Dimension. Vol. 5, Finnish Institute for Russian and East European Studies, Helsinki.

7 Post-traditional or Post-modern Rurality? Cases from East Germany and Russia

KARL BRUCKMEIER AND MARINA OLEGOWNA KOPYTINA

Introduction

After the end of socialism Eastern Europe became an area of experiments for the building of new rural regimes. The processes under way in rural areas cannot be easily understood since the larger transition processes in the economies and societies of these countries are not finished and new difficulties emerged after some years of reform euphoria in the late 1990s. Perspectives on rural development and post-socialist rural regimes are discussed more at conceptual levels in discourses such as that about post-modern rurality which provides empty theoretical concepts to analyse changes in rural areas. An empirically based analysis and reflection on the ongoing change is much more difficult. Concepts that allow us to interpret East European processes of system transformation do not allow equally well the projection of further courses of rural transition – as can be seen in the example of the ideal types of 'criminal capitalism' and 'democratic capitalism' (Tatur 1998). Such broad framework concepts for system transformation bypass rural changes. Observation of two heterogeneous countries from the former socialist system – East Germany which has been rapidly integrated in the Western system and Russia where the transformation of economy and agriculture is blocked – should allow for some more empirically grounded insights. One of the critical sociological questions for future rural transformation is that of the socio-cultural subjects of rural development in a post-socialist and differentiating rural economy. Contemporary rural development in Central and East Europe is subjected to contradictory challenges of system transformation as well as of the larger globalisation process where rural and peasant traditions are weakened but simultaneously revitalised under the auspices of sustainability.

Post-traditional Rurality – A Framework of Analysis

In this chapter post-traditional rurality is discussed with reference to the present colonisation of rural areas in Western and Eastern Europe by new leisure and producing classes. Although post-traditional is often interpreted to mean 'modern', here we follow a more complex, historically situated notion of traditionality and modernity, which can be shortly formulated as follows: modernism brought for rural and agricultural development different ways of transforming traditions without being able to destroy them, so that every phase of development is characterised by transforming traditions. We refrain from the notion of 'post-modernism' since this is still more unclear, following Tovey (1998), mainly rhetorics. De-traditionalisation is conceived of as a long-term historical process and critically discussed with regard to the prophecies of an end of peasantry that have accompanied the modernisation process. The 20th century, dominated by the industrialisation project, was 'a long 20th century' for Eastern European rural societies that lead from feudalism to post-traditional rurality.[1]

The Transformation of Rural Society in Late Modernity

The notion of 'post-traditional rurality' is at present a vague idea based on the observation that agriculture is no longer the core activity of the rural economy and its institutions no longer guide the use of natural resources in the countryside. When we relate this state of post-traditionality to the agricultural and rural modernisation process in 20th century, we conceive it as a result of modernisation – which reduced the role of agriculture in rural development.[2] During the long modernisation process in urban-industrial society other resource users and social groups articulated their claims to use rural resources and 'colonised' rural areas at several times mainly for the sake of 'consuming nature' or using specific natural resources in the countryside – landscape, soils, forests, water and living resources. What these invaders consumed, in fact, is not nature *per se* but the historical product of traditional rurality, the so-called 'cultural landscape' shaped through hundreds or thousands of years of agricultural and sylvicultural work. The quality of countryside looked for is its advantages in relation to urban life and settlement. One large group of invaders of the countryside in

Western Europe are those who take up residence in the countryside or recreate but do not work there. Among them are people commuting for work to urban areas, urban people with a second and temporary residence in the countryside for recreation purposes, seasonal tourists, and tourists practising traditional resource use such as hunting and fishing in recreational form, and retired people. Ecological innovators or groups making the countryside a project area for experiments in sustainability should be counted among the pioneers of new producer roles in the countryside.[3] Another group of rural colonisers who use rural resources for living and producing can be found in the Eastern European countryside. Not a new leisure class, but part of a new 'awkward class' (Shanin 1972; Stolbov 1995) of producers in rural society is arriving with this new rural poor, the small producers and subsistence producers. The rural poor as a land-less, non-peasant rural population have nearly vanished in the modernisation projects of the welfare state and the 'socialist welfare state' in Europe. After the collapse of socialism in Eastern Europe they are followed to the countryside by urban newcomers who cannot afford to live in the cities, impoverished parts of the rural population under socialism, or part-time and subsistence producers. The arrival of a growing number of rural dwellers and resource users with heterogeneous social status, interests and ambitions, only characterised by the joint property of having no social, cultural and economic roots or basis in the countryside, is the core process of restructuring of rural populations towards a post-traditional rural society.

A post-traditional rural society emerges in a process of 'de-traditionalisation' that refers in unclear ways to modernisation in the dominant urban-industrial society and to the deficient sociological reflection of perceiving traditions as what is to be overcome through modernisation (Giddens 1996, 122). Two notions of de-traditionalisation in rural areas should be differentiated for the sake of analytical clarity: (i) De-traditionalisation at the societal level as a process of rural modernisation that has *longue durée* character as a long-term historical process of institutional and structural change, which, in its late phase, since the end of 20th century, is framed by the transition from industrial to risk society;[4] (ii) De-traditionalisation as a politically guided process of political reform projects which, explicitly or implicitly, aim to modernise, or, in the extreme case, to overcome peasant traditions through supporting entrepreneurial or commercial forms of agriculture. During 'the long 20th century of rural change' the politically framed processes of de-traditionalisation include the transition from the projects of the peasant and welfare state to the internationalised state of the future co-operating as a networked actor in policy arenas with other networked actors, supporting several rural groups and structures ('pluralised rurality').[5]

In most European countries the dominant rural population group during the period called modernity (since 16th century) is the 'awkward class' of peasantry. The awkwardness of peasantry is that it represents partly a class, partly a society, but furthermore, that it survived through adaptive changes in all phases and variants of modernisation, capitalist and socialist.[6] Also the emergence of a post-traditional and post-industrial rurality will not end with complete de-peasantisation, although the fewer remaining peasants may face new societal expectations, other than their traditional roles as producers of food.[7] The new social subject of a post-traditional rural society is a plural one, that arrives historically with the invasion of resource consumers and part-time members of rural society.[8] The emergence of post-traditional rurality differs in Western and Eastern Europe as a consequence of their rural history.

In Western Europe, in the industrial core countries, the emergence of post-traditional rurality is the outcome of the modernisation and marginalisation of agriculture; as a part of the national economy and in terms of social structure agriculture played an ever decreasing role, although not at the political level where agricultural corporatism had a lasting influence on the agricultural policy of the EU (Jollivet 1997). In social-scientific discourse about the dissolution of the traditional values and structures of peasant societies the hypothesis of a 'colonisation of agriculture' has been used to describe the process in which peasant society comes under the influence of other social structures and cultural values, although the peasantry keeps a role and a tradition as part of a changing rural society (Brüggemann and Riehle 1986, Pongratz 1987, 526).

This colonisation process took specific forms in different European countries and regions, and it is only through the late phase of dominance of CAP that it takes on trans-national features.[9] With this colonisation process the peasantry has become economically marginalised and integrated in the modern specialised capitalist economy where production follows market and entrepreneurial standards. Rather late, with the unfolding of CAP as a trans-national political modernisation project, West European peasants have become the unwilling bearers of modernisation in which their socio-cultural identity was challenged economically and redefined politically (after a drastic reduction of the number of agricultural producers through structural change).

Modernisation through CAP led quickly into an ecological crisis of modernised agriculture that has been observed since the 1970s as one of the catalytic processes for sustainable rural development (Jollivet 1997).

Agriculture has recently lost – starting with the 'Declaration of Cork' in 1996 – its political dominance within EU policy in favour of integrated and spatially oriented policies, which can be understood as the beginning of post-traditional rural development (with the European Spatial Development Perspective as a basic framework).

Eastern Europe lies together with Northern and Southern Europe outside the industrial core of Europe; it has a special rural history with the peasant state projects in the first half of 20th century followed by the socialist variants of collectivisation. Socialist agricultural systems with their dissolution transition projects move towards a post-traditional rurality differing from the West European one, which can be partly explained by the semi-peripheral functions of eastern Europe in the globalising world system, although the concrete course of agricultural transformation cannot be derived from world system analysis. Privatisation appears to the political public as the dominant institutional change – as a radical economic and political rupture so that rural development seems to begin anew from a historical zero-point of time.

In a long-term historical view, taking into account the lasting effects of social changes under socialism, such reductionistic views are discredited.[10] Transformation paths are not determined by the mere fact of privatisation and social structures that came with socialist reforms of agriculture continue to exist.[11] After one decade the privatisation of agriculture is unfinished and reveals some striking deficits in most countries of Central and East Europe.[12] Altogether the post-traditional rural society of the future will be shaped under the influence of more and more diverse factors than in Western Europe. The large scale agricultural production units created during socialism, the surviving peasant tradition in the forms of socialist co-operative and small-scale private peasantry (however: in the absence of family farming), the variants of privatisation ('*nomenklatura*-privatisation', 'spontaneous privatisation' or other informal and institutionally deficient forms of privatisation) and its outcomes (*latifundium* structures and commercialisation of agriculture), the redistribution of agricultural land, the influx of new rural population (in contrast to widespread rural exodus in the West), and the development of regional and global agricultural markets, will all contribute to a future rural society the features of which are not easy to predict.

The Transformation of Society at Large

Modernisation as the Core Project of Cultural Change in the West and its Model-character for Eastern Europe

The scientific and political ideology of modernisation accompanying societal change in Western capitalism unfolded fully in 20th century. In this socio-cultural process rural society and the peasantry found no role but tried to keep their values and traditions during modernisation, partly accepting it, partly opposing it. De-peasantisation was the 'major index' of modernisation (Shanin 1987, 468), that is, nothing else was expected of rural change but vanishing of the peasantry with modernisation; this, in fact, never happened. Analysed more theoretically and historically, the course of capitalist modernisation in Western Europe may be characterised since the take-off of industrialisation as a coexistence (seen as 'symbiotic' or 'parasitic' depending on the premises of theoretical analysis) between a modern economic sector (industrial-commercial) for urban and a traditional economic sector (peasant-artisan) for rural society. The agricultural labour force became a cheap reserve of manpower for industrialisation, and rural producers and resources have been deprived of resources for industrialisation. Such a 'dual economy' did not last long in European history (Lutz 1984). It was a component of the rural regime unfolding under the dominance of the welfare state which was historically bound to capitalist industrialisation. It came into crisis with the end of accelerated growth in the industrial core countries since the 1970s. This dual economy was accompanied by the 'historical compromise' negotiated between urban-industrial mainstream society and rural-agricultural society under the prolonged influence of agricultural corporatism to mitigate the rural exodus through a controlled and slow structural change of agriculture, at the price of increasingly politically decided subsidies for agriculture. Such compromises could be realised only in incomplete and contradictory forms – as processes of delayed and politically controlled structural change, but the CAP designed for this became finally the executor of the traditional peasant economy.

In modernisation as a long-term process of social and institutional differentiation (Giddens 1990), which at the macro-societal levels culminates in the project of industrialisation, the exceptional situation of agriculture and rural society is obvious. In a short glance at the European history of modernisation we may note the following points of further influence on the present change of agriculture and rural society.

Traditional forms of agriculture differ – there is no archetype of a traditional agriculture in Europe, neither in property form (different property regimes have co-existed throughout history), nor in the organisation of work (the family farm was not a universal pattern). Only in terms of the relative dominance of the peasantry can we see some similarity in the social structure of rural societies, and can suppose that also in future elements of a changed peasant tradition will endure in co-existence with other components of a new rurality.

The industrialisation process happened rather late in most countries, even in the industrial core – in Germany not before the second half of 19th century, in Russia mainly in 20th century under socialism – and its progress was different. Its main impacts on agriculture have been mechanisation and high-external input production, urbanisation (the drastic reduction of rural population and labour force), its neglect by mainstream industrial society and its development patterns (traditional sector, cheap labour reserve). This is an external, industrialisation-bound view, whereas the impacts from an internal view of rural society (as, for instance, under the hypothesis of colonisation of the peasantry) have less been analysed. They can be understood better from the present changes, when the industrialisation project is turning out to be a dead end of development.

The economic project of agricultural modernisation – to adapt it to competition in capitalist markets – ended in a Pyrrhic victory. Economic success was bought at the price of social erosion of the peasantry and ecological unsustainability. A main cause of the growth of agricultural productivity in 20th century was the ecologically unsustainable external input and use of fossil energy that agriculture borrowed from the industrial mainstream. This energy revolution which made agriculture dependent on the industrial sectors happened in capitalist as well as in socialist modernisation.

As a consequence of its negative impacts, post-traditional agriculture will be challenged by a sustainability revolution the trajectories and duration of which are presently unknown. Post-traditional rural society is the heritage of unsuccessful and misguided agricultural modernisation that will be overburdened with expectations to solve the problems that modernisation has brought. 'Reflexive modernisation', if not debunked as mere sociological rhetoric, is only a formula for the problems faced.

Present Changes – 'Reflexive Modernisation' and the Re-building of Political Capitalism

Globalisation and societal change processes in contemporary Western and Eastern Europe influence rural societies and their populations differently. In theoretical terms both can be reconstructed under the notion of risk society (and the concomitant notion of reflexive modernisation used by Beck and Giddens), however in two different variants. In Western Europe it is the reform project of building a sustainable rural economy, in Eastern Europe the rebuilding of political capitalism, which becomes the core of a political reform project.

In late modernity the industrial system is gradually losing its guiding role for societal development in the West European countries of the industrial core, although the contours of a post-industrial society loom slowly and the notion is still more rhetoric than reality. The ecological crisis which industrialism produces can become a catalyst of change as is theoretically formulated in such preliminary notions as 'risk society', 'reflexive modernity', and 'sustainable development'. Post-industrialism does not mean de-industrialisation (although such projects are under way in some countries) or simply closing down large parts of industry (this can happen in single regions, and happened to a significant degree in East Germany after 1990); its main form is a gradual change in the sense of 'tertiarisation' of national economies in which the heterogeneous 'third sector' of a service-based economy expands in more or less unstable forms. For agriculture the post-industrial change comes with such new roles as that of cultivating regenerative energy sources for society. This is an ambivalent innovation as it can be directed towards more sustainable societal energy use or towards commercialised and destructive use of natural resources in forms of 'monetarisation of nature'.

In Eastern Europe the unfinished modernisation and industrialisation under socialism meant that system transformation started under the aims of a second, competing industrialisation project in the attempt to follow the Western path and to profit from the advantages of the welfare state project. However, this project is contradictory already for the urban-industrial sectors, and it is more unrealistic for the rural economy since at present the resources are not available to build a welfare-state based regime that allows rural areas to develop.[13] The 'third sector' in Eastern European societies takes more the form of the globally expanding informal economy which accompanies the crisis of the Fordist industrial accumulation regime. In terms of political regimes for re-establishing capitalism in Eastern Europe, the future fate and shape of rural society will be influenced by the course of the societal reform process in which two extreme variants of a 'hybrid' political capitalism can be formulated as ideal types for a comparative analysis of further change. The transformation processes oscillate between 'democratic capitalism' (transformation into a democratic state, a productive economy, and a socially integrated society with limited pauperisation) and 'criminal capitalism' (dissolution of the state, a power-based destructive economy, and dissolution of social structures with pauperisation of the largest part of the population). Between these extremes a middle variant of political capitalism can be described that comes close to structures emerging in the present transformation processes (segmented state, extractive economy, pauperisation of the middle classes) but which may change in the longer run in one or the other direction (Tatur 1998, 357, 370).

In most political reform trajectories formulated in the past decade for a transformation of socialist structures the agricultural sector was neglected. Agricultural economy has experienced a withdrawal of the state because of lack of concepts and financial resources, and uncontrolled and politically unguided changes occurred in rural society. With the delayed and unfinished privatisation of land elements of segmented political capitalism and criminal capitalism influence rural society too.

Two Variants of Post-traditional Rurality – East Germany and Russia

East Germany: Up to the End of Socialism – Agrarian Reforms and a Turbulent Rural Society

East Germany may be counted among Eastern European agricultural systems even before the East/West schism of the cold war period. From the 17th century dominance of the Prussian state to the socialist agrarian reforms after 1945 a semi-feudal *latifundist* agriculture persisted in large areas east of the river Elbe (*ostelbische Junker*) as a striking contrast to peasant family farming in the West. After 1945, under socialism, attempts were made first to break the dominance of agrarian *latifundism* and to redistribute land to new settlers, later to collectivise peasant farms and to build two main types of large farms – co-operative farms and state-owned farms. As a result of collectivisation large-scale agricultural holdings re-emerged. The agricultural sector, as in most socialist countries, was of more importance in the national economy than in Western Europe – until the end of socialism about 20 percent of the labour force worked in this sector – less due to technological and economic lags, more because of the different social and economic organisation of the rural economy with agriculture as the core of a territorially organised rural structure.

Post-socialist Transformation – Fragmented Rurality

With the collapse of East German socialism in 1989 a specific variant of transformation began. Within a year the country was formally integrated into the institutional and economic system of the EU through the unification of the two German states. The first, short-term consequence was interruption or cessation of agricultural production in large areas for some years (with a permanent reduction of the agricultural labour force to c.20 percent). Through privatisation of farmland a post-traditional agricultural system is emerging, which in institutional structures and economic functions is dominated by large-scale commercial agriculture. Such agricultural holdings, whatever their legal-institutional form, follow more easily the functional change of agriculture with de-traditionalisation from producer of food into producer of non-food products and every kind of raw material (new energy sources etc.) that the short-term criteria and demands set by EU agricultural policy and the agricultural market allow.

With regard to further change in East Germany, Laschewski has argued that "For the process of agricultural development in the 90s, historical farm structures are not very important, but the mechanisms of social differentiation of socialist society are" (1998, 43). Interpreting this statement in the sense of a path dependency hypothesis means that the components from the changed peasantry tradition (e.g. in co-operative farming) will remain in East Germany, which differs from dominant views of the process that are superficially oriented to the differentiation of legal property forms and the ideologically overburdened case of family farming. In relation to family farming, a territorial cleavage remains in Eastern Germany: family farms re-emerge where they existed before socialism, in the densely populated and industrialised parts of the South, whereas large-scale and commercial farming continues in the agricultural North.

In the first half of the 1990s, parallel to the reduction of agricultural labour force, rural movement projects bloomed in East Germany, including not only the rapidly expanding form of organic agriculture but much more encompassing projects than in West Germany – 'eco-villages', horticultural projects, 'soft tourism' and new forms of nature conservation such as biosphere reserves (Bruckmeier 1994; Bruckmeier and Grund 1994).

The Result of Transformation

The components of a fragmented post-traditional rurality in East Germany that became visible in the 1990s are the following:

At the political-ideological level the integration of East German agriculture into the Western system helped to break the political ideal of the family farm in West Germany and to establish as a new political paradigm the entrepreneurial farm. This can be seen as a major component of de-traditionalisation, the turn from a social-agricultural view of the countryside to a spatial-management view, strategies of nature protection and spatial development being another component. Transformation of East German agriculture became a politically guided experiment with different types of entrepreneurial, commercialised and market-oriented agriculture in which traditional regional products and production forms are given up in favour of reacting to short-run changes in demand and subsidised products (including subsidised non-production on large areas). The new forms of private agriculture have been codified in different legal and property forms and are sometimes understood as a pluralisation of agriculture or a differentiation of agrarian structures, organisation forms and interest groups of the agricultural producers who no longer organise in one dominant union (Clasen 1997). The institutional changes observed in the first years after socialism (Wielgohs and Wiesenthal 1997) supported assumptions that agriculture has been successfully transformed, has found its optimal structures, and is altogether better adapted to future competition than West German agriculture (Clasen 1997, 430). Such an idealised picture has been criticised (Küster 1998) for neglecting social structures and the dominant economic pressure on the new production units, and for its superficial analysis of the differentiation of legal forms that covers other forms of adaptation and differentiation. The wishful thinking of the political reform ideology evident in this picture does not adequately recognise the social structures within the legal ones and does not take account of economic (growth) constraints and ecological impacts of the farms (Küster 1998, 456f).

Structural components of a post-traditional rurality looming in East Germany are: (i) commercialised farming oriented to maximisation of incomes on large areas through compilation of subsidies or easily marketable crops; (ii) the phenomenon of 'absentee farming' where West German or foreign farmers rent large areas in East Germany and farm these 'from a distance', with minimal presence (transporting their machines just for a short time from their main farms to the remote fields); (iii) the post-modern phantom-reality of an agricultural landscape that is preserved in its traditional structure as a 'cultural landscape' in museum form, not farmed, only managed at the landscape level (which can be done by paid landscape managers, an activity which more and more becomes a differentiated role of the agricultural profession). This could be called an extreme variant of the redefinition of rurality from a social to a spatial project (Jollivet 1996).[14]

Beyond such problematic trends supporting social desertification of agricultural areas other forms of post-traditional agriculture emerge from the independent projects and social movements. Among these, one of the important is the breakthrough of organic farming and other forms of extensive, environmentally sound agriculture in East Germany. The 'project-rurality' of manifold local projects and movements, symbolised in the eco-village movement, has included projects for a 'rebirth of non-industrial agriculture', however, they have come under pressure to adapt to EU rural policy (in the forms offered with LEADER; see Bruckmeier 2000). As agriculture increasingly becomes 'part-time farming' the complementary activities and sources of income for the farming population become more and more important and open up side paths to post-traditional rurality: (i) complementary local and regional marketing activities; (ii) the farmer as a landscape manager (separate from his producing activity and his care for his own land); (iii) rural tourism as a supplementary income source (Fink and Siebert 1998); (iv) the new leisure class of urban middle-class people who bought residences in East German villages (many of them people from the former West or West Berlin), contributing to social segregation and conflicts in rural areas.

Transformation toward more sustainable forms of agricultural production and rural resource use has progressed well in East Germany during the post-socialist decade. This can also be seen as a success for the new rural movements; movement based projects (together with the Western political-institutional system that allowed transfer payments to East Germany) helped to mitigate the negative impacts of privatisation of agriculture, the impoverishment and corruption that accompany political and criminal capitalism.[15] The new forms of environmentally sound agriculture (Bruckmeier and Grund 1994) represent a platform of compromise between heterogeneous interest groups for a joint strategy of rural development (the new leisure class, the new social movements, the urban consumers, the conventional agricultural lobby in search of a future for agriculture). This can be seen as one element of post-traditional rurality; another is the 'mainstream project' of a large-size commercial agriculture. Post-traditional rurality is not a homogeneous and standardised system that comes out of present transformation processes and will last forever in a kind of 'end of history' state. In East Germany as in other countries the future rural society is built from a mixture of older and newer structures and bears from its beginning the property of a non-simultaneous historical process. With the integration of East Germany in the 1990s the process of institutional and structural change is proceeding in accelerated and paradigmatic forms, whereas institutional and structural change in West Germany is slower and differs with regard to the absence of *latifundist* structures. East German rural society shows some of the features that West German rural future can adopt in the longer run.

Russia: Until the End of Socialism – Unfinished Agrarian Reforms

Politically, the modernisation of Russian agriculture began with the abolition of feudal serfdom in 1861 which brought into existence a peasant class of agricultural producers, but not however the Western type of family farming.[16] After the forced Stalinist collectivisation of agriculture from 1929 on, two new institutional types of large-scale agricultural enterprises developed with the *sovkhozes* (collective farms) and *kolkhozes* (state farms). Rural life and society were reshaped around the socialist farms. Whereas in Western Europe modernisation meant that agriculture is only functionally integrated in the industrial and market economy (reducing it to the producing function with shrinking manpower and labour input), in Russian socialist modernisation (similar to East Germany) a territorial model of integration dominated with agriculture as the core of the rural economy around which the 'agro-food sector' was built, and rural society had a holistic social organisation of its own. 'The collective-state farms were the centre of local life and the local population, with its pensioners, agrarian specialists, local schools, and much else, all depending on the network of social and economic relations that centred on them. Take away the farm, and the village society and economy was in danger of collapse' (Sakwa 1993, 237).

The period after the second world war brought Russian agriculture several reform programmes (from Chrushtshov, Brezhnev, Gorbatchov) and an enforced outmigration – between 1956 and 1970 about 25 million people left the countryside (Hopkins, Wallerstein et al. 1996, 99). This migration was possible because of strong industrial growth. When growth declined in the 1970s informal economies emerged in the semiperipheral (including Eastern Europe) and peripheral zones of the world system. In Eastern Europe and the USSR this took especially the form of piece-rate work. It coincided with a temporary form of 'second urbanisation' in small towns, with unstable forms of seasonal and temporary work (Haraszti, see Hopkins, Wallerstein et al. 1996, 103, 112).

Widespread neglect of nature and natural resources resulted as a kind of 'negative synergy' from (i) uncompleted agrarian reforms that did not take account of the ecological impacts of natural resource use; (ii) the popular illusion that reserves of land and natural resources are overabundant in the large Russian territory; (iii) the industrial and technological ideology of disregard for nature; and (iv) a scientifically supported struggle against 'the powers that do not serve an ideology', that is, hostile nature in forms of frost and ice, storms, floods, deserts and high mountains (Woinowitsch 1990, 5). Indifference to natural resources is rooted in older socio-psychological orientations among the rural population that had chances to endure during and after socialism – supported by such factors as vastness of the country, great distances between settlements, lack of territorial management beyond the near surrounds of settlements (Sevan 1998, 73). The space around the settlements, however, was not developed according to ecological criteria either – it was devastated through industrialisation programmes before and after the second world war; industrial and military projects shattered the landscape in many regions, agro-industrial settlements led to stereotyped building, and sprang onto illegally occupied land (Sevan 1998, 70-72). Nickolsky describes the Brezhnevian agrarian reform between 1965 and 1985, which tried to solve the problems of agriculture by centralisation and enormous investment to raise productivity and yields, as successful but with large negative external effects: technological intensity and rapidly increasing use of chemical fertilisers and pesticides made this strategy "highly capital intensive and not particularly friendly towards the environment. Growth was achieved at prohibitive financial and environmental costs – this was unsustainable growth indeed" (Nickolsky 1998, 194).

Post-socialist Transformation – History Continues

Post-socialist transformation can be understood as continuing agrarian reform in the tradition of unfinished reforms, now including pre-socialist reform ideas.[17] The first steps of post-socialist agrarian reform in Russia were more towards marketisation than privatisation of agriculture (Sakwa 1993, 236). The early 1990s were characterised by a deficient reform project which included only two aims, the destruction of large collective and state farms and the introduction of market-oriented family farms. This project was not supported by the rural population (Nickolsky 1994, 163). Agricultural production volumes declined throughout the 1990s and the project to establish private family farms failed quickly, among other reasons because of lack of government support for new farmers. The 1990s became a decade of unprecedented decline in Russian agriculture, "a crisis of declining production, productivity, and employment which has been deeper – though not as costly in human life – than the crisis created by Stalinist collectivisation" (Nickolsky 1998, 196). The privatisation strategy chosen under Yeltsin proved to be wrong and agrarian reform failed to happen, one more of the economic, cultural and moral shocks that the peasantry faced during the long 20th century which in Russia comprised ten attempts at agrarian reform and modernisation (Nickolsky 1994, 165; for the environmental situation see Zlotin 1996).

By the end of 1994 the number of individual farms reached 285,000 and these farmers owned or rented 5 percent of agricultural land. In 1993 they produced 2 percent of gross agricultural output. Since 1994, and after the government stopped purchasing agricultural goods from farmers on anti-inflationist grounds, the number of individual farmers and their production has declined (Nickolsky 1998, 204). At the end of the 1990s agriculture is characterised by incomplete institutional reforms (privatisation was more a formal re-registration of agricultural enterprises under new legal forms). Most agricultural enterprises (in 1996 nearly 90 percent) do not make profits but are deeply in debt. The percentage of production of the dominant 'societal enterprises' (former *sovkhozes* and *kolkhozes*) has fallen below 50 percent of national agricultural production and the production of the new farming sector is insignificant, but the production on smallholdings (that is, the unstable, part-time and subsistence production) produces more than 50 percent of national agricultural production (Thede 1997, 279). Part of the crisis and pressure on Russian agriculture derives from increasing imports of food from Western countries in the 1990s. Originally only a phenomenon of consumer testing of new products (*snickerisatsija*) there has developed an enduring change in consumption patterns especially in the large cities, even given limited purchasing powers (Rieger 1999, 44). Agrarian reform has lost momentum in the 1990s; the history of inconsequent and incomplete agrarian reforms in Russia has continued.[18]

In contrast to the dramatic economic failure of privatisation and the impoverishment of large parts of the urban middle class and the rural population, social bonds in the countryside seem still strong. Community attachment as social capital for rural development was studied by Starosta (1998). Based on a survey in two rural communities in the administrative province of Ivanovo (central Russia) and in several Polish and Bulgarian rural communities Starosta reports a high level of social and psychological bonding of rural inhabitants with their communities of residence, even stronger in the Russian than the Polish and Bulgarian communities.

According to this survey, attachment to community of place was stronger than to communities as reference groups. Does this identification with place of residence and attachment to the rural community – given the critical situation of broken infrastructure, unemployment, impoverishment, and the collapse of local agriculture for many rural communities in Russia – indicate an unused capacity for rural development or is it more to be understood as an expression of lack of choice or of chances to leave the countryside? Starosta's conclusion is that there are capacities unused for rural development, and he argues for linking local and supra-local resources to develop rural communities.[19] However, the potential for new social conflicts in the countryside should not be denied – it may come, especially in the vicinity of cities, with buying of rural houses by richer urban people who only spend weekends and vacations here (Sevan 1998, 72).

What is the rural situation at the beginning of the new century? In the years 2000-2005 agricultural decline should stop and growth in the agro-food-sector be stimulated, but without increased governmental spending (Rieger 1999 reports stronger growth of the food sector in 1999, however, this was mainly a short-term consequence of the financial crisis and devaluation of the Ruble in 1998, which led to reduced foreign investments and a stronger demand for food products from within the country). Return of the state to agricultural development after the failure of privatisation seems likely, and some of the features of a rational reform project that Nickolsky outlines may come into effect (Semjon Novoprudsky, Alexej Sinizky in *Iswestija* 01.07.2000).[20] What would be necessary in the longer run to overcome economic and ecological disruption in the Russian countryside? Purely economic reforms are a source of the problems; they need to be supplemented by 'holistic' reform strategies that take into account social and cultural structures as well.[21] Although ecological reform paths are culturally shaped (Bruckmeier and Xu 1995), it is not easy to see how Russia can break through the barriers of blocked development, to open a path towards not only socially integrated but also ecologically sustainable rural development.

The Elements of a Post-traditional Rurality – Heritage of the Past

Post-socialist rural development is blocked in Russia by problems unsolved in the first decade after socialism. The transformation shows fragments of a post-traditional rurality which are quite different from the East German case. (i) The new subsistence producers and the new rural poor are a significant phenomenon which may not dissolve as a short-term phenomenon of the crisis. (ii) The political disintegration of the former Soviet Union into a series of new states has, as a consequence of growing ethnic and social tensions in the newly independent states, caused a return migration of ethnic Russians from the formerly Soviet republics outside of Russia. These return migrants are mostly part of the rural poor – they settle in rural areas because they cannot afford to live in the cities. That they can become a human and cultural capital for further rural development seems wishful thinking. (iii) The progressive devastation of nature, continuing the bad practices of the past, is developing new forms, including a destructive commercialisation of natural capital similar to the one that started earlier in developing countries with projects for the 'valorisation of nature'. The main trend is to use natural resources in unsustainable forms, be it for reasons of subsistence or for commercial resource extraction. (iv) Old social bonds – the human, cultural and social capital that may be mobilised for rural development – cannot be activated effectively as long as the short-term constraints of transformation leave large parts of the population struggling daily for economic survival, while new middle classes hardly emerge in cities or countryside, and *nomenklatura*-privatisation has created extreme inequality and oligarchy.

Comparison

The post-communist countryside does not look much different from the way it looked under communism. Firm size has somewhat reduced, but is still *latifundium*. The equipment used is about the same as it was before. The same people occupy the managerial positions too. The main differences are: changes in property rights – from socialist co-ops, enterprises were turned into capitalist *latifundia*; this happened together with a drastic reduction in the number of employees and the liberation of the *latifundium* from "social costs, which are now transferred to local governments" (Szélenyi 1998, 14). Some of this account of change after the first post-socialist decade is true also for the East German and Russian cases. However, this is largely a 'snapshot' at a certain point of time towards the end of the 1990s, when the agrarian reforms still have not gone far. How the future post-traditional rural society will look cannot be described with these elements only; other questions need to be answered. The comparison of the two cases can be guided by orienting research questions such as those formulated by Granberg and Kovách (1998, 17) in an effort to counteract political speculation and ideological projections onto the rural: What are the processes observed in the society at large? Who are the actors and groups in the countryside, where do they come from, what are their interests in rural development? How is the future of the countryside constructed by people living and staying in the countryside? With such questions one remains within the scope of Mormont's premise that rural society makes sense only as long as rural people make use of the concept.

People and actors follow a West and East European constellation in terms of new colonisers of the countryside. Colonisation of the countryside through new rural resource users in Eastern Europe is quite different from the West European situation of a new leisure class realising their claims of resource use. In East Germany this 'Western path to post-traditional rurality' is starting – with the new leisure class often West German (and former West Berlin) urban middle class people, buying houses in East German villages. In Russia the phenomenon is not unknown, but outweighed by other forms of rural recreation, that is, the traditional form of rural summer houses of urban people (which are usually not located in the villages or rural communities but in separate villages with summerhouses only). The dominant new users and newcomers in the Russian countryside are the new urban people trying to farm, subsistence producers and part-time farmers.

The transformation process in the first post-socialist decade indicates the difficulties of rural development more than the perspectives. East Germany's rapid integration into the western system and the blocked transformation of agriculture in Russia exemplify two different, if equally unfinished, processes of transformation toward a post-traditional rurality. The failure of the hasty introduction of individual and family farming in Russia, and the partial success of family farms in Eastern Germany in regions where they have existed before socialism, can be explained through the different long-term history and traditions in both countries. The persistence of collective types of farming may become a partially successful heritage of socialist agriculture in both countries under which also peasant traditions may be able to renew themselves. The exceptional transformation in East Germany, with the introduction of West German laws and EU policies, should not be counted as complete success for the Western system. Although this institutional framework mitigated the social problems of transformation, it is obvious that the rural population had little influence on this process, and that many former agricultural workers and specialists had no chance to work in the new agriculture.

The experimentation with new ecological variants of agriculture and resource use was open only to small parts of the rural population. In Russia, discourse about sustainable transformation is still absent – the destructive consequences of socialist industrialisation require a period of environmental clean-up before an active strategy of sustainable development can follow.

The patterns of agriculture resulting from transformation show the persistence of older problems of unfinished reforms. The withdrawal of the state from the countryside that happened in Russia is a consequence of misguided privatisation and it has favoured uncontrolled concentration of agricultural capital and resources. The Russian rural crisis provides an example of what Swain (1998, 17) saw as an outcome of East European rural transformation: very large agricultural production units (more and more as private companies, less as co-operatives) that emerge from spontaneous privatisation, coexisting with very small scale agriculture for self supply and subsistence (if possible as part-time farming, supplemented by non-agricultural income) and the relative absence of family farming. However, even with partial re-development of the family farm in East Germany a broader and more equitable distribution of agricultural land has not followed – the new agricultural holdings came immediately under constraints of growth and competition, and the future is not that of the family farm.

There are numerous new problems emerging which may make rural transformation persist for a long time. The absence of the state, the lack of strategies for agricultural policy, the lack of support for individual and family farming, delayed legislative decisions, problems of access to financial resources and credits for agricultural producers, the enlarging of disfavoured areas, the naturalisation of rural economy, the degradation of the socio-cultural and technical infrastructure of the rural society, and the absence of alternatives and choice – such problems seem very much part of the Russian experience. However, they do not only exist there, but also under East German transformation conditions.

Conclusions

"An uncertain future for agriculture everywhere and under very different conditions" seems to be the current outlook not only in East Germany and Russia, but for most Eastern European countries. The problems to be solved seem to block any unfolding of long-term perspectives of development. Future developments can hardly be predicted, however the problems and constraints observed can be discussed with regard to their influence on future changes.

The new actors in post-traditional rural development create a multifaceted rurality. The West and East European constellations of new colonisers of the countryside may not lead to the emergence of geographically separated production and recreation areas with agriculture dominant here and tourism there. A mixture of both, situations of multiple resource use and the co-existence of a plurality of groups with different interests seems more likely. Such a multifaceted future rurality can be paradigmatically seen in European coastal areas where changes in traditional production systems and the advent of the new leisure class happened earlier. Although post-traditional rural society will be composed of institutions and practices from a different historical age, we cannot expect the future to be only a struggle between pre-modern or traditional and modernised styles of life and consumption. The alternatives for traditions in the de-traditionalised society that Giddens (1996, 182f) discusses – 'discursive traditions' and revival of traditions in the form of 'fundamentalism' – do not have to become the fate of rural society – the struggle for peasant traditions does not have to become agrarian fundamentalism. New social conflicts and movements will arrive in the countryside as in the globalised mainstream society and one opportunity for traditions to survive will be their provision of arguments to combat the ecological crisis.

Successful modernisation and involutionary adaptation happen simultaneously.[22] Farmerisation, pauperisation, proletarisation of large parts of the rural population as the components of involution may not simply end in a re-emergence of wealthy/developed and poor/underdeveloped countries along a newly defined boundary. Reproduction of growth and poverty centres within every country may reveal a more splintered pattern of development. Rural growth areas may still concentrate around cities and metropolitan areas, although territorial location becomes less important as an economic criterion for development. Social desertification, presently more a phenomenon of Western Europe, may also happen in East Europe and Russia in remote, ecologically damaged or disfavoured areas. Neither social desertification, however, nor economic and ecological marginalisation, has to be the final state of rural society – they are temporary consequences of the presently unsolved transformation crisis.

The Janus Face of Risk Society Comes to Be Seen in West and East Europe

The affluent societies of Western Europe, with their political and economic control over large parts of global resources, seem able to begin the project of sustainable transformation – 'internalisation' of the negative external impacts of industrialisation and modern intensive agriculture – on their own terms and with some freedom of choice, without pressures for institutional changes and reforms, or without the economic deficits and problems of the East European countries. Reflexive modernisation and sustainable development in East Europe is blocked by the same factors that block economic development. Linking the hypothesis of involutionary adaptation to the concept of risk society allows only a rather negative outlook on development in East European risk societies. West Europe has 'only' to solve the ecological crisis, for which, however, no solutions are yet visible, and which must be seen as a historically unprecedented problem. East Europe however has also to solve the societal transformation crisis for which no previous experience or models exist. The rural is that part of society where the synergetic impacts of both crises take effect, now as economic and ecological marginalisation; it is also that part of society where the instruments and the resources to counteract these effects are mostly lacking.

The Future of Rurality Between Globalisation and Sustainability

That part of the future that is visible in the present allows a limited conclusion: The future of rural areas is not the past but the re-emergence of some elements of the historical past and presence in new contexts. We should not attempt a new oversimplification in the form of diagnosis of another 'rapid de-peasantisation' (Shanin 1966, 357), but a more detailed reconstruction of actual social change. The future of rural society is neither dissolution into the industrial mainstream society nor revitalisation of patterns of traditional agricultural society the moral economy of pre-industrial peasants is not a model for the renewal of total society, but a case from which to learn from the solidarity networks of the moral economy how to solve current problems of rural society (Fafchamps 1992), among others, the present ecological problems. The chances that rural areas can in the long term contribute to a more sustainable development can be formulated theoretically in one summarising premise: progressive ecological scarcity of resources limits the institutionalisation and continuation of the monetary principles of scarcity that drive economic development and globalisation. The globalising system in the long run will be forced to find a balanced development which starts in rural areas and with their natural resources.

After all: in what sense does post-traditional rurality make sense? It makes sense as a hypothetical construction to find out how rural and peasant traditions will change, but not vanish (Shanin 1987, 473). Furthermore, it makes sense to fill the concept of post-traditional rurality with ideas that link social, economic and ecological development – and for such linking, there is not much more than the formula of sustainable development presently available.

Acknowledgements

A prior version of the text was presented at X World Congress of Rural Sociology – Sustainable Rural Livelihoods, 30.7-5.8. 2000 Rio de Janeiro, Brazil, Workshop 26 – Tradition, De-traditionalisation and Peasants.

Notes

1 The 'long 20th century of rural change' is formulated in contrast to the 'short 20th century' (Hobsbawm) relating to the duration of the societal modernisation project of socialism (1917-1989). Theoretically influenced concepts of modernity and modernisation, de-traditionalisation, peasantry and de-peasantisation, rurality and rural society are not discussed anew here, but used as introduced concepts to analyse historical changes.

2 'Rurality' can be used to describe a spatial as well as a social cleavage; it can be linked with agency by the argument "that rural society itself only exists insofar as, or as long as, social movements of rural people see it as useful to mobilise themselves under the category 'rural' " (Tovey 1998, 23, referring to Mormont).

3 Since 19th century, as a concomitant of industrialisation and modernisation of European societies, the countryside was discovered as a recreational area by the new urban bourgeois and middle classes – country summer houses can found especially in Northern and Eastern Europe.

4 World system theory is one theoretical approach to analyse such long-term changes from a global perspective, although it did not focus on agriculture (Hopkins, Wallerstein et al. 1996).

5 The two levels of analysis can be combined with the concept of 'rural regime' (Granberg and Kovách 1998). Rural regimes can be understood as specific constellations of political and societal structures during limited periods of time.

6 Modernisation was characterised by a preoccupation with urban-industrial development and a conscious neglect of peasantry that was mainly understood to be a traditional element vanishing through modernisation. 'De-peasantisation' has been an ideological component of societal modernisation which is discussed critically by rural sociologists (Shanin 1966, 356; Shanin 1987, 468; Araghi 1995; Granberg and Kovách 1998, 13).

7 Jollivet (1996, 24) sees the advent of post-traditional rurality with the redefinition of rurality from a societal to a spatial concept under the influence of such managerial processes as farmland consolidation, landscape management, and the EU regional and spatial policy that led from a social and historical conception of rurality (concretised in agriculture, production and peasantry) to an abstract formal conception of the rural (concretised in activities of managing space, landscape and natural resources). Such shifting from social to spatial thinking, however should not only be reviewed for its problematic effects, but also for its potential of supporting

the continuation and revitalisation of peasant and rural traditions under the approaches of endogenous rural development (see Ray 1998).

8 The unstable and part-time membership of rural society that comes into existence with part-time agriculture and the temporal rural residence of non-agricultural groups resembles a caricature of the communist idyll as characterised by Marx – as the idyll where people spend their time (freed from constraints of alienated work) to fulfil their wants in part-time work in agriculture, fishing, hunting, cultural creativity etc.

9 The metaphoric expression of Habermas, 'colonialisation of lifeworld' is not used here, although it may be used to describe similar processes where traditional, culture based rule systems come under the influence of modernisation-dictated imperatives of large-scale political and economic systems.

10 Criticism is also launched under the complaint that the political and economic discourse outmanoeuvred the sociological one – for the Eastern European countries in general by Swain, see Laschewski (1998, 41), who describes this for East Germany; and Nickolsky (1998) who mobilises the argument for Russia.

11 With this argument the historically oriented theorising as that of path-dependent development (Szélenyi 1998) and regime change (Granberg and Kovách 1998) is taken up. David Stark "calls the change in institutional constellations 'path-dependent development'. This implies that the transformation that East European societies experience will be decisively shaped by the historic 'path' they were on at the time of the break, in 1989 or 1991" (Szélenyi 1998, 12).

12 "Privatisation was not paralleled by a rational loan-system, or by efficient development of infrastructure nor by an establishment of organisations of political representation. ... About one-fourth to one-third of the former agrarian population has fallen to rural poverty. A smaller part of the agrarian population has a potential for family farming in the future. However, their fate is so closely linked to urban society that at present it is difficult to predict their future" (Granberg and Kovách 1998, 15).

13 "The socialist economic redistribution is transformed into a new system, which has aspects of the welfare state without the resources of the other welfare states" (Granberg and Kovách, 1998, 15f). This can be seen in the processes accompanying transformation – "new social segmentation in the form of farmerisation and proletarisation" (Granberg and Kovách 1998, 13).

14 This is not yet reality, but it is not too difficult to see such possibilities as economically feasible and rational when rural development come sunder the dominance of space and landscape management thinking (see the research at the Institute for Research on Agricultural Landscape in Müncheberg, Brandenburg).

15 Assessing the success or failure of agrarian reform with the present privatisation may still be too early and unsecure – Tuma (1965) who has analysed agrarian reforms throughout history did not come to a very encouraging results with regard to successes of such reforms.

16 Chayanovs analysis of the Russian peasant economy as based on family work and absence of wage labour, based on data from pre-revolutionary Russian agriculture, should not be equated with family farms the Western sense.

17 "In contrast with the Baltic republics and the western marches of the Soviet Union, the great mass of Russia had never known a system of consolidated family farming. Before 1929 peasants in effect owned land, but in the form of strips governed by the commune and subject to periodic redistribution. After 1991 the aim was not to reprivatise or reconstitute the family farm system, but to build it for the first time

and thus to complete what Stolypin had tried to achieve when he was Prime Minister between 1906 and his assassination in 1911. The obstacles facing the independent farmer, however, were formidable. Not only was there a lack of tools, seeds, fodder, breeding stock sand so on, but they also faced hostility from the collective farms and in many cases from the villagers themselves" (Sakwa 1993, p. 238).

18 "The Russian peasantry has too long been abused and exploited. No reform scenario has a chance to succeed or to receive the support of the peasantry unless the peasant sees that the plan does address the problem of social justice. Reforms implemented forcefully by the government without the approval of the peasantry can only destroy agriculture and set the country back to subsistence production" (Nickolsky 1998, 191).

19 "The inhabitants' strong psychological bonds with the place of residence can represent a local capital for the development of these localities. It should be underlined, however, that this human capital does not encompass the youngest and the most dynamic individuals. The young and better-educated people are clearly oriented to the escape from the rural and small-town local communities. In the situation, it seems important not only to supply an appropriate knowledge to the local level but also to ensure an initial financial and institutional support; this will facilitate the achievement of success and consolidate a positive inclination to act in support of the local development. Consequently, it is necessary to combine the local and supra-local resources. Relying exclusively on the principle of internal mutual assistance and on the local structures and resources seems to be an insufficient model for the mobilisation of many rural and small-town communities in Eastern Europe" (Starosta 1998, 127).

20 An alternative scenario for 'rational reform' is formulated by Nickolskys as follows: "the sequence of reform measures most appropriate under Russian conditions would begin with prices and harmonization of growth. After this is achieved, a gradual, relatively slow, step-by-step movement toward private ownership is imaginable. Democratization and decentralization can come only last. As such a reform progresses, the role of the state will eventually change. In the first phase – as a country establishes equitable prices and reallocates resources – a strong government is needed. During the phase of property reform, the role of government can be somewhat reduced, and the third phase assumes a rather weak state" (Nickolsky 1998, 192).

21 "Attempts to reconstruct agriculture or industry relying only on economic mechanisms failed not only to reverse, but even to slow down the destructive economic processes taking place both in villages and in town. It becomes evident that the formation of economic links is impossible without involving the cultural dimension (historical reasons for changes, psychological characteristics of population, support of entrepreneurship, etc.). The culture of management and the culture of land-use should be shaped along the lines of contemporary modernization-oriented trends, with a careful application of traditional forms. With such an approach, one can hopefully preserve and judiciously develop a natural landscape environment in Russian settlements" (Sevan 1998, 73).

22 "An alternative hypothesis might be that we see in post-communist Eastern Europe not modernization, but involutionary adaptation. What we see here is not the making of competitive and efficient capitalist agribusinesses, but the growth of post-communist *latifundia*, oriented in the long run to extensive production based on cheap labor, limited domestic markets, and an export orientation (in a world market where there is little market for food). Eastern Europe therefore may not be on an

evolutionary path towards the peak of world economy; on the contrary, it may be sliding down towards the Third World, into the periphery of the world economy" (Szélenyi 1998, 16).

References

Araghi, F. A. (1995), 'Global depeasantization', 1945-1990, *The Sociological Quarterly*, vol. 36, no. 2, pp. 337-368.

Bruckmeier, K. (1994), 'Lokale ökologische Ökonomie – Ökologische Projekte im ländlichen Raum in Ostdeutschland', *GAIA*, vol. 3, no. 6, pp. 319-332.

Bruckmeier, K. (2000), 'LEADER in Germany and the discourse of autonomous regional development', *Sociologia Ruralis*, vol. 40, no.2, pp. 219-227.

Bruckmeier, K. and Grund, H. (1994), 'Perspectives for Environmentally Sound Agriculture in East Germany', in D. Symes, A. J. Jansen (eds), *Agricultural Restructuring and Rural Change in Europe, Wageningen*, Agricultural University Wageningen, pp.180-194.

Bruckmeier, K. and Xu, C. (1995), 'Cultural Specificity of Sustainable Agriculture: A Comparison of China and West Europe', *Zemedelska Ekonomika/Agricultural Economics*, vol. 41, nos. 10-11, pp. 445-455.

Brüggemann, B. and Riehle, R. (1986), *Das Dorf*, Campus, Frankfurt und New York.

Clasen, R. (1997), 'Die Transformation der Landwirtschaft in Ostdeutschland und ihre Folgen für die Agrarpolitik und die berufsständische Interessenvertretung', *Prokla*, vol. 27, no. 3, pp. 407-432.

Fafchamps, M. (1992), 'Solidarity Networks in Preindustrial Societies: Rational Peasants with a Moral Economy', *Economic Development and Cultural Change*, vol. 41, no. 1, pp. 147-174.

Fink, M. and Siebert, R. (1998), 'Tourism – An Alternative for Rural Regions in East Germany?', *Eastern European Countryside*, no., 4, pp. 123-138.

Giddens, A. (1990), 'The Consequences of Modernity', Standford.

Giddens, A. (1996), 'Leben in einer posttraditionalen Gesellschaft', in, U. Beck, A. Giddens, S. Lash (eds), *Reflexive Modernisierung – Eine Kontroverse*, Suhrkamp, Frankfurt am Main, pp. 113-199 (orig. 1994).

Granberg, L. and Kovách, I. (eds, 1998), *Actors on the Changing European Countryside*, Budapest, Institute for Political Science of the Hungarian Academy of Sciences.

Hopkins, T. and Wallerstein, I. (1996), *The Age of Transition – Trajectory of the World-System 1945-2025*, Zed Books, London and New Jersey.

Jollivet, M. (1996), 'Les fondements agraires de l`Europe', in M. Jollivet, N. Eizner (eds), *L´Europe et ses campagnes*, Presses de Sciences Po, Paris, pp. 34.

Jollivet, M. and Eizner, N. (eds, 1996), *L´Europe et ses campagnes*, Presses de Sciences Po, Paris.

Jollivet, M. (ed,1997), *Vers un rural postindustrie*, l´Harmattan, Paris.

Küster, K. (1998), 'Die EU-Agrarpolitik und der Strukturwandel in den neuen Bundesländern', *Prokla*, vol. 28, no. 3, pp. 443-458.

Laschewski, L. (1998), 'Continuity and Change – Agricultural Restructuring in East Germany', *Eastern European Countryside*, no. 4, pp. 37-48.

Lutz, B. (1984), *Der kurze Traum immerwährender Prosperität*, Campus, Frankfurt/Main and New York.

Nickolsky, S. (1994), 'Radical Reform of the Russian Village – A route to the destruction of Russian agriculture?', in D. Symes, A. Jansen, (eds), *Agricultural Restructuring and*

Rural Change in Europe, Agricultural University Wageningen, Wageningen, pp. 162-166.

Nickolsky, S. (1998), 'The treadmill of socialist reforms an d the failures of the post-communist 'revolutions' in Russian agriculture: is there an alternative?', in, J. Szélenyi (ed), *Privatizing the Land*, Routledge, London and New York, pp. 191-213.

Pongratz, H. (1987), 'Bauern - am Rande der Gesellschaft?', *Soziale Welt*, vol. 38, no. 4, pp. 522-544.

Ray, C. (1998), 'Towards a meta-framework of endogenous development: repertoires, paths, democracy and rights', *Sociologia Ruralis*, vol. 39, no., 4, pp. 521-537.

Rieger, F. (1999), 'Russland: Nahrungsgüterindustrie im leichten Aufwind', Ost-West-Contact, no. 12, pp. 44-46.

Sakwa, R. (1993), *Russian Politics and Society*, Routledge, London and New York.

Sevan, O. (1998), 'Impact of Agricultural Reforms in Russia on the Shaping of Rural Settlements', *Eastern European Countryside*, no. 4, pp. 65-73.

Shanin, T. (1966), 'Peasantry in Political Action', *Sociological Review*, vol. 14, pp. 12-27; quoted after T. Shanin, (ed), *Peasants and Peasant Societies*, New Edition, 1988, Penguin Books, Harmondsworth, pp.357-364.

Shanin, T. (1972), 'The Awkward Class. Political Sociology of Peasantry in a Developing Society, Russia 1910-1925', Oxford University Press, London.

Shanin, T. (1987), 'Short Historical Outline of Peasant Studies', in T. Shanin (ed), *1988, Peasants and Peasant Societies*, New Edition, Penguin Books, Harmondsworth, pp. 467-475.

Shanin, T. (ed, 1988), *Peasants and Peasant Societies*, New Edition, Penguin Books (orig. 1987), Harmondsworth.

Sivignon, M. (1996), 'Les systèmes agraires européens: héritages, mutations, frontières', in M. Jollivet, N. Eizner (eds), *L'europe et ses campagnes*, Paris, pp.37-55.

Starosta, P. (1998), 'Inhabitants' Feeling of Social Bonds with Rural and Small Town Communities in Bulgaria, Poland and Russia', in, L. Granberg, I. Kovách (eds), *Actors on the Changing European Countryside*, Institute for Political Science of the Hungarian Academy of Sciences, Budapest, pp. 104-130.

Stolbov, V. (1995), 'New People – The Awkward Class in Russian Agriculture', *Eastern European Countryside*, no. 1.

Swain, N. (1998), 'A Framework for Comparing Change in the Post-Socialist Countryside', *Eastern European Countryside*, no. 4, pp. 5-18.

Symes, D. and Jansen, A.J., eds. (1994), 'Agricultural Restructuring and Rural Change in Europe', Agricultural University Wageningen, Wageningen.

Szelényi, I. (ed. 1998), 'Privatizing the Land', London and New York, Routledge.

Tatur, M. (1998), 'Ökonomische Transformation, Staat und moralische Ressourcen in den post-sozialistischen Gesellschaften', Prokla, vol. 28, no. 3, pp. 339-374.

Thede, S., et al. (1997), *Die Entwicklung der Unternehmensstrukturen im Transformationsprozeß der Russischen Föderation*, A. Spitz, Berlin.

Tovey, H. (1998), 'Rural Actors, Food and the Post-Modern Transition', in, L. Granberg, I. Kovách (eds), *Actors on the Changing European Countryside*, Institute for Political Science of the Hungarian Academy of Sciences, Budapest, pp. 20-43.

Tuma, E. H. (1965), *Twenty-Six Centuries of Agrarian Reform. A Comparative Analysis*, Berkeley and Los Angeles, University of California Press.

Wielgohs, J. and Wiesenthal, H. (1997), *Einheit und Differenz. Die Transformation Ostdeutschlands aus vergleichender Perspektive*, Berlin, Berliner Debatte Wissenschaftsverlag.

Woinowitsch, W. (1990 (1988)), *Die Mütze*, Piper, München and Zürich.

Zlotin, R.I. (1996), 'Contemporary State of the Rural Environment in Russia', in L. Ryskowski, G. Pearson, S. Balazy (eds), *Landscape Diversity: A Chance for the Rural Community to Achieve a Sustainable Future*, Poznan, Research Centre for Agricultural and Forest Environment of the Polish Academy of Sciences, pp. 167-175.

PART II:

MEDITERRANEAN COUNTRIES IN THE GREEN RING

8 The De-agriculturalisation of the Greek Countryside: The Changing Characteristics of an Ongoing Socio-economic Transformation

CHARALAMBOS KASIMIS AND APOSTOLOS G. PAPADOPOULOS

Introduction

Towards a Post-agricultural Countryside?

The Greek countryside has been extensively transformed in the post-war period. The changes that have occurred with respect to the structure of agricultural production have had considerable impact upon its spatial organisation and on the adjustment of local agricultural systems. In other words, due to the modernisation of the agricultural sector the Greek countryside has been significantly de-agriculturalised. Moreover, the Greek countryside has also been an arena for expansion and intensification of processes of both urbanisation and socio-economic integration. These two – actually interrelated – processes have played a major role in internally transforming Greek rural areas.

In this respect, agricultural modernisation played the role of challenging and altering the characteristics of the Greek agricultural production system (e.g. its competitiveness, introduction of an entrepreneurial logic etc.) thus constituting an *externally induced* factor in terms of spatial development. Additionally, urbanisation and socio-economic integration processes operated inside Greek rural areas and, implying closer and multiple connections (e.g. in the sphere of employment, consumption, in culture and social practices) between the countryside and the urban areas and/or the wider system, constituted an *internal* factor to the spatial development.

Early in the history of the modern Greek state, agriculture, and especially family farming, became export-orientated, developing linkages with foreign markets mainly with products like currants and tobacco. The delayed unification of the country among other reasons seriously hampered attempts to find a solution to the land question. It was only in 1917, after years of significant peasant revolts, that the Law of Land Reform was introduced although it was not immediately implemented.

The Asia Minor defeat, however, and the massive inflow of refugees to Greece after 1922 forced a faster implementation of the Law of Land Reform. Thus, land distribution created the small family farm as the dominant unit of production in agriculture. Besides this late solution to the land question, other important cultural factors like dowry and inheritance practices have, for decades, contributed to slowing down land concentration processes, the result of which has been a small, fragmented family farm, the size of which has remained around 4 ha for years.

The domination of the small family farm signalled at the same time the semi-agricultural character of Greek rural household reproduction. Movement between the city and the countryside was widespread and in the inter-war period, a large number of farmers, labourers and petty-commodity producers in general followed a flexible employment pattern characterised by geographical and professional mobility. This led to puzzlement regarding the social structure of the rural areas which seemed to acquire some urban social structural characteristics, maintaining both peasant and urban elements. This puzzlement has been expressed in attempts to define and analyse the transformation tendencies in Greek rural areas.

Karavidas – an agronomist researcher and writer of the inter-war period – has argued that the Greek family farm early combined agriculture with non-agricultural activities to ensure subsistence and/or a better standard of living.[1] Such a combination does not, of course, represent a process of social polarisation. It rather reflects the 'know-how' about survival practices of rural households within a rapidly changing social and economic environment (Pizanias 1993). The social mobility (cyclical, seasonal, or occupational) of farmers, along with the Greek state's involvement in land reform to establish a small family farm agriculture, the tax system and the enlargement of the public sector (Dertilis 1993), promoted an initial de-ruralisation and outward orientation among the rural population. This is normally linked to the so-called petty-bourgeoisification of family farming initiated during the inter-war and intensified during the post-war period (Tsoukalas 1987).

The 'fluid' nature of Greek farmers and the wider polyvalent social and economic structure which prevailed in both rural and urban areas, can be considered the first sign of a process of de-agriculturalisation of the Greek countryside. In fact, farming has never been the exclusive activity of Greek rural households. In this respect, the evolutionary explanatory schemas of 'Modernisation' and 'Marxist-Leninist' writers alike have proved inconsistent with the transformation process in the Greek countryside (Papadopoulos 1994). However, the evolution of certain phenomena such as rural exodus, agricultural development etc. has had a differential impact upon the rural social structure. Their impact has been interpreted in a dual fashion since the 1980s. Some writers stress that a proletarianisation process has been forced on family farming, while others underline the petty-bourgeoisification of the rural social structure in general (Kasimis and Papadopoulos 1997). Despite the fact that both interpretations provide valuable insights for the socio-economic transformation of the countryside, they underestimate the steadily differentiating character of Greek family farms and their changing weight in rural localities.

The extent and size of agricultural land ownership, employment on the farm and primary production are the main components of the conventional definition of 'rurality' in Greece. Admittedly, the process of agricultural modernisation has had significant impacts upon Greek countryside. Moreover, it is a process which is still ongoing and cannot be considered merely a historical phase. However, the 'old' (conventional) perception of 'rurality' has severely constrained the analysis of other phenomena related to the inter-penetration of city and countryside. Despite the fact that certain phenomena were recognised already in the inter-war period – as mentioned earlier – they remained minor issues given the sweeping domination of agricultural modernisation and transformation. In this respect, new 'rurality' is the result of both transformation processes in the Greek countryside and the recognition of certain aspects as now legitimately to be included under the heading of 'rurality'. Now that agricultural policy has been integrated into regional policy Greek 'rurality', once more, needs to be adapted to the new conditions of development of rural areas.

As already argued, the main characteristic of new 'rurality' in Greece, as well as in other Southern European countries, today is the close relationship between city and countryside (Papadopoulos 1999, 248-249). Thus, notions of distance and/or divide between the city and the countryside are frequently conceived in an arbitrary way. Often, only geographically isolated rural areas are conceived as distant and as very different from the city or from 'urbanised' rural areas. But rurality cannot be attributed only to less-favoured or isolated areas where agriculture remains as an activity which cannot anymore adequately sustain the rural household. In Greece it is family farms in less-favoured areas for whom EU subsidies are most important. constituting a large proportion of their admittedly small agricultural income (Papadopoulos and Daouli 1999). How important, therefore, is the distinction between an era of more 'traditional' – agriculture based – rural areas and an era of 'modern' de-agriculturalised rural areas? The de-traditionalisation of Greek countryside did not exactly follow or parallel the process of de-agriculturalisation experienced in other Western European countries (Hoggart et al. 1995, 76-109). A significant inter-regional and intra-regional diversity in Southern European and Greek rurality is evident which cannot be accountedfor by linear evolutionary thinking.

In our view, the passage from an agricultural to a post-agricultural rurality in Greece needs to be explained through analysis of what constituted rurality then and what constitutes rurality today. Since every definition of rurality actually reflects how 'the rural' is conceived by each society, it may be argued that Greek rurality has been modernised and reshaped to the extent that rural areas have been affected by the growth in size of larger farmers and the activities of newcomers, hobby farmers and farmers with average or high status professions such as public employees, doctors, lawyers. This implies that changes in the perception of what farming is or who is a farmer play a vital role in challenging evolutionary thinking about the transition from an agricultural to a less agricultural countryside. In that sense the Greek case can be considered an illustration of a constantly changing countryside. This change is not isotropic but it can be rather conceived as a discontinuous and regressive process. By this we mean that drawbacks and delays are equally important with progressions and/or innovations.

Rural exodus and migration which peaked during the 1960s and early 1970s, agricultural policy, the new survival strategies of rural households, and the dynamics of family farming which emerged in the 1980s and 1990s may be cited as explanatory factors for the social and economic integration of the countryside into the wider economic system.

In this chapter we first discuss arguments which consider the transformation of Greek rural society as an externally induced process, and second, analyse recent work regarding changes in family farm strategies for survival. Each refers to a different aspect of the de-agriculturalisation of the countryside. The first part deals with de-agriculturalisation as a process linked to agricultural modernisation, urbanisation and economic development, leading to a reduction in agricultural employment and rural population. The post-war period up to the mid-1970s is marked by an industrial-modernist stance toward economic development in the country which had significant repercussions on analysis and research into the countryside. The second part deals with de-agriculturalisation as a process of social and economic integration of Greek countryside, in which family farm strategies for survival reflect the dynamics of family farm transformation and of state agricultural policies. It became increasingly apparent in the 1980s and 1990s that older views about the development of the countryside had reached their limits. We suggest a more pragmatic view of the internal dynamics of family farming.

The Transformation of 'Old' Rurality: Rural Exodus and Agricultural Modernisation

Rural exodus has been a historic feature of the process of economic development on a world scale and especially where integration of the agricultural sector to the market and modernisation of agricultural production released significant amounts of labour power. In the course of economic development rural exodus goes through different historical stages. The two most important are, *first*, exodus related to the early stage of industrialisation during which accumulation leads to the concentration of resources and of labour power from the rural periphery to the urban areas and *second*, exodus related to a process in the opposite direction connected with the growing diseconomies of scale resulting from the previous phase of urbanisation and economic concentration. At this stage it is the diffusion of urban based economic activity into the periphery and the development of other new activities which halts the production of a surplus population in rural areas. Rural exodus is now expressed in the partial or complete abandonment of agricultural activity by family members – a phenomenon connected with changing internal and external conditions in the operation of the family farm without, however, a geographical mobility.

Following the Second World War and in the period 1950-70 more than one million people left agriculture and the rural areas for the internal labour market and the markets of Western Europe, North America and Australia. After 1970, however, rural exodus declined significantly as a result of the decline of employment opportunities, especially abroad. In periods of economic development like that of 1950-70 agricultural policy was targeted at facilitating rural exodus whereas in the stagnation period following 1970 it provided incentives to retain marginal producers in the rural areas. An incentives policy for industrial decentralisation in this second phase together with the development of tourism and construction works in rural areas gave large numbers of under-employed farmers with no other alternative employment possibilities in either the internal or the external labour markets an opportunity to remain in the region, very often combining more than one job.

In the immediate post-war period, the conditions of demand and supply of agricultural production did not seriously alter land structure. The position of the family farm as a unit of production was strengthened but some interesting differentiation appeared in terms of mechanisation, labour productivity and types of production. Labour surpluses were very high even when rural exodus had already started.

Agricultural policy which emphasised infrastructural works and agricultural credit contributed to improving efficiency but on the other hand did not force an exodus upon 'non-efficient' producers on small farms, since lack of alternative employment opportunities kept the surplus population in the rural areas. Some authors argue, however, that it was rather the credit policy of the Agricultural Bank of Greece (ABG) which, to a large extent, contributed to keeping farmers in the rural areas. For example, Vergopoulos (1975, 229) writes that, between 1960 and 1963, ABG credit to farmers was three times larger than the actual investments in the sector. It seems therefore that a part of credit was directed towards consumption. Agricultural production and investment showed a slow increase associated with a slow increase in demand. The problem of diversification in production remained as long as the relation of crop to livestock production stayed more or less the same (a three to one relationship). Labour productivity increased significantly as a result of a decline in employment, an improvement in production methods and increased mechanisation.

As far as the agricultural production structure is concerned, two tendencies were noticed in this period: first, the relative socio-economic 'differentiation' of agricultural producers, and second, an inter-regional 'homogenisation'. The evidence suggests that a slow process of concentration and centralisation of agricultural production had been on the way, generating 'hidden' unemployment. This process was now facilitated by the existence of prospects for alternative employment in both the internal and external labour markets (Stathakis 1983).

The migration 'boom' put in motion a cumulative mechanism. 'Hidden' unemployment found a way out in migration and by doing so it favoured the development of 'differentiated' agricultural production, expressed in increased mechanisation and consequently in increasing pressure on the family form of production. This generated new 'hidden' unemployment and remained a consistent pattern in the whole 1961-1971 period despite the fact that after 1971 external migration was replaced by internal. Thus it became clear that in the 1960s, the process of industrialisation of the country was accompanied by serious structural changes in agriculture, which strengthened and even facilitated further the migration tendencies that reduced the agricultural population by one third. It is characteristic that at the peak migration period the economically active population in agriculture fell from 54 percent in 1961 to 41 percent in 1971. Additionally, it has been estimated that almost half of the total number of migrants who went abroad originated from the rural sector. Almost 80 percent of these people belonged to the 15-39 years age group, highlighting the serious demographic and economic implications of the rural exodus for the sector and the countryside in general. In many regions of the country (the islands, Epirus, Peloponese and some parts of Macedonia) the total population was reduced by one fifth and a few hundred villages were deserted.

Almost three quarters of the case studies on migration in this period refer to declining villages (Wagstaff 1982). Rural exodus was connected with the process of social transformation of the countryside and considered as part and parcel of the structural disintegration of rural society. This disintegration meant either the marginalisation of whole rural communities or increasing polarisation within local rural social structures (Damianakos 1987). One way or another, such analyses of rural exodus led to an underestimation of the internal processes of rural social transformation related to the survival strategies of family farms.

A number of empirical studies carried out in the 1970s attempted to shed light upon these internal processes, focusing on three interrelated issues: first, the inequalities of agricultural land ownership, which on its own was not considered adequate for grasping the traditional characteristics of Greek rural society; second, the influence of the city upon the rural economy through the non-agricultural engagements of farmers; and third, the relationship between rural exodus and agricultural structures (Burgel 1977, p. 192). Analysis of studies of concrete, highly diverse, Greek rural localities has given rise to a number of observations, two of which are particularly useful for our analysis here. The first is a recognition that supplementary resources originating from off-farm activities play a significant role in the economy of the farm itself. The second refers to consequences of the exodus for rural localities. Of interest is the fact that migration did not lead to an expected redistribution of the land in the rural localities, while it invariably affected plains and mountainous areas (Burgel 1978).

A number of rural community studies, in the exodus period, are characterised by a dichotomous approach to rural and urban space. Rural localities are handled by certain writers in terms of a 'closed' social and economic system, which operated by reproducing itself in isolation from the rest of the society. According to this approach, post-war urbanisation provided the opportunity for an exodus of rural population from this isolation. The influence of modernisation theory is evident here.

In Greece, however, as mentioned earlier, a large number of rural communities have had a long history of external economic relations. As a result, rural exodus often constituted a response strategy of the family to these external economic pressures (Sutton 1983). It is interesting to note here that what differentiates the farm family from other (more traditional) forms of organisation has been historically its open, extrovert character; in other words its external relations with the surrounding economy and more specifically the market. The other forms of organisation remained 'closed' relying on self-sufficiency and an internal division of labour under a patriarchal structure.[2] Despite its more extensive integration to the market, however, the farm family operated similarly on patriarchal moral codes. It was because of this integration that the family became more and more conscious of its weak position vis-à-vis its external economic relations and bargaining power. Independent producer on the one hand, but a small proprietor on the other, the farm family could not really face the demands of the market economy. Often it exploited its cohesion in a laborious production effort 'facilitated' by the patriarchal family principles. Against the dangers of a possible collapse the family adopted various strategies. In this connection, the migration of some members and the strengthening of kinship relations supported the operation of the family farms and enabled inheriting farmers to fulfil the moral obligations connected with the endowment of their sisters and the education of their younger brothers.

Sutton (1988) considered modern urbanisation an initial phase of a long historical period which does not necessarily imply a break with the past. She writes that any type of migration is a strategic engagement of the rural household with the wider economic system. This 'problematic' differs from that of Vergopoulos (1975) who considered rural exodus to be a consequence of the commercialisation of agricultural production and of the 'unequal exchange' which leads to a rural household budget deficit. For Sutton, therefore, rural exodus is rather a strategy for direct engagement to the 'commercialised' production system dominant in Greece for centuries. This view together with the historical continuity of the phenomenon of migration provides an alternative dimension for understanding the internal dynamics of rural communities.

Karapostolis (1978) similarly explained migration in terms of the social physiognomy of the Greek farm family. For Karapostolis the exodus of some family members is planned, reflecting the reaction of the rural family against the threat of dissolution connected with moneylenders, middlemen and taxation. In other words, it reflects a response to a family deficit which only migrant remittances are in a position to cover. The family calculation and speculation which, in principle, exists behind the exodus of its members complies also with the education and support of its youngest members and their occupational reinstatement. This is

> [B]ecause the successful career of the migrant is expected to determine the social and economic position of the family as a whole. For this reason the preparation and the planning [of migration] concerns everyone in the family and sometimes least of all the migrant himself (Karapostolis 1978, 370).

Migrant members in most cases maintain bonds with their families and seek and devise ways and practices for the social protection and support of their parents and kin. This is imbricated with a moral code considered to be the main characteristic of the traditional peasant society which has survived until recent decades (e.g. the inheritance and dowry system). Despite this moral imbrication one may conceive farm family survival as the main objective, strategically pursued through a complex combination of practices and actions which expanded to both the agricultural and the non-agricultural sectors.

In fact, migration in the 1950s and 1960s contributed immensely to the survival of the family. Until the 1970s, remittances accounted for almost 20 percent of household income, but they had at the same time some negative results because they became a regular income directed mainly towards consumption, contributing therefore to the preservation of marginal rural structures.

The diffusion of urban consumption patterns into the countryside pushed the differentiation from traditional rural behaviour even further. Family labour still constitutes the prime productive force of the farm family unit which is responsible now for securing a constant enjoyment of the 'new' and 'higher' consumption standards.

In the period between the 1970s and accession to the EEC in 1981, Greek agriculture, supported by a declared state policy of self-sufficiency showed positive results without, however, significant structural changes. Productivity increased significantly and new crops, methods of cultivation and technologies were introduced under a state protected system. Support for Greek agriculture was ensured under the price support schemes and CAP subsidies guaranteed a satisfactory income for the majority of the farming population. Directed mainly towards consumption, this income did not lead to any serious structural improvements of family farms. The competitive position of agricultural goods in the domestic and foreign markets worsened and gradually the balance of payments of these products became negative. In other words, these protectionist policies guaranteed incomes which were not indicative of the real productive capacity and efficiency of Greek agriculture (Kasimis and Papadopoulos 1994).

During the 1990s, the reformed CAP's orientation towards unprotected national markets, followed by the GATT agreements and prospects of EU enlargement towards the countries of Central and Eastern Europe, led to a loss of agricultural income despite increased support for structural modernisation and the adoption of policies for environmental protection and regional development. However, agricultural modernisation has proceeded and it cannot be denied that there is significant reinforcement of large farm enterprises despite the persistence of a large number of small-medium farm enterprises whose survival depends on CAP subsidies and/or family labour.

In this context, the structural constraints affecting the large majority of Greek family farms are, first, the small size of farm and its high fragmentation; second, the low level of economic and technical infrastructure; and third, the adverse exchange terms between agricultural products and industrial inputs to agriculture. These constraints have strongly affected the relative agricultural modernisation and adaptation of family farms. Since the bargaining power of the agricultural sector has been weak against the industrial or the urban sector, family farms have increased intensification of the farm labour process combined with a more extensive use of farm technology. In parallel, the search for non-farm sources of income has produced a variety of combinations between agricultural and non-agricultural activities which have 'protected' the family farm from the threat of proletarianisation. Under the new conditions of CAP reform and new market demands for quality products, the Greek farm family has been reorganising its (family and gender) division of labour on and off the farm. Engagement in trade, tourism and industry has become a widespread phenomenon, involving almost 50 percent of farm households. It has become the dominant engagement for the demographically healthy part of the rural population, a kind of guarantee against falling incomes. However, such issues, which express the trends of 'new' rurality, will be dealt with in the next section.

The Trends of 'New' Rurality: Polyvalence of Rural Structure and Wider Economy Linkages

In discussions concerning the de-agriculturalisation of the countryside, arguments frequently stress the declining importance of agricultural employment for the economy and the rural areas, and the decreasing contribution of agriculture to the national and/or regional GDP. However, we want to propose here that the recent trends linked with de-agriculturalisation can be better understood when they are related to family farm strategies for survival.

In Greece, the size of the rural population which is predominantly employed in agriculture has declined over the period 1961-1991. This decline, as shown in the previous section, has been largely the result of rural exodus. Rural exodus has slowed down in the 1980s, having been stabilised in the last years as a result of the elimination of employment opportunities outside rural areas and the development of new activities (in the service sector mainly) which secure complementary employment for the rural labour force. At the same time, maintenance of the population in rural areas results from a process of de-agriculturalisation of their employment, reduction of labour time on the farm, and under-employment which, given appropriate economic conditions in the non-agricultural sector, can lead to the development of pluriactivity.

In the 1990s underemployment (below 140 labour days on the farm) concerned over two thirds of the farm family population. Under-employment is particularly evident in the plains and coastal areas, where modernisation (mechanisation, agricultural investments etc.) had played a significant role in reducing labour on the family farm but where, however, opportunities for off-farm employment were definitely higher making pluriactivity a more widespread phenomenon there. In Greece, pluriactivity has been growing over the past three decades and is mainly confined today to farms under 5 ha. The degree of pluriactivity among family farm heads has decreased to 30 percent approximately, but the percentage of all household members employed off-farm has risen to one fourth. The proportion of spouses who are pluriactive has doubled, reaching 15 percent overall, but with a greater increase recorded on larger farms where the household head tends to remain virtually full-time employed while his spouse may seek off-farm employment. In the mid-1990s for the first time the pluriactivity of other family members exceeded that of the farm head, providing some proof of the strategic adaptation of family farms to the external setting, either by considering the internal labour capacity of the farm and/or by taking advantage of externally available opportunities.

These trends in family labour should be set against an increase in non-family farm employment and an even greater increase in seasonal and foreign labour where an actual decline in the late 1980s was sharply reversed in the 1990s.

In other words, at the level of the farm unit, the process of 'de-agriculturalisation' can be related to the combination of agricultural and non-agricultural employment, often at the expense of the former. This tendency may be related to changing family roles, shifts in the gender division of labour, increasing flexibility in use of family and of cheap, non-family labour, and economic restructuring of the farm holding.

At the macro-level discussion on restructuring and modernisation of agriculture, the professionalisation of farmers, the need for integrated rural development and the reinvigoration of rural areas refers to a conscious transformation leading to an extended change of the rural social structure. It should be said here that despite the operation of market forces and of macro-social changes Greek rural society is characterised by social and economic differentiation and by labour market segmentation which is not easily accounted for in terms of social polarisation and social class crystallisation. This is due to a number of factors, a few of which are mentioned here. First, the consecutive and delayed land reforms which led to an agrarian structure characterised by small farms; second, the continued predominance of small and medium-scale production units in all economic sectors; third, the significance of regional diversity for compartmentalising both the national labour market and national social identity; fourth, the late formation of the modern Greek state; and fifth, the family-centred social and economic organisation of society. These elements provide some indication of an agricultural sector characterised by social and economic diversity, small-scale, delayed structural modernisation and family-centrality.

The treatment of family farming as the basis for analysing the Greek rural economy and society has remained a major issue. Vergopoulos (1975) argued that integration of family farming into the capitalist economy has been achieved through linkages (input and output) developed with different segments of capital rather than through the introduction of capitalist relations of production at the level of the farm production. This analysis de-emphasises the arguments concerning the social and economic integration of family farming at the level of agricultural production relations. In this way, the integration of family farming into the wider economy is conceived and organised outside the immediate agricultural production sphere and increasingly refers to the exercise of control over the agricultural labour process. In this respect, the lever through which the family farm unit of production is integrated into the wider economy relates to the non-agricultural elements of Greek family farming. Recent research on the survival and/or modernisation practices of Greek family farming shows that both agricultural modernisation and off-farm employment have played a significant role in integrating family farms into the economy (Papadopoulos 1994).

In this context and since there is considerable diversity of approaches to the significance and the continued existence of family farming, we argue that the concept of *strategy* is vital for considering how farmers and, more specifically, rural households are in a position to sustain their livelihoods and utilise their resources in the Greek countryside (Kasimis et al. 1998, Long and van der Ploeg 1994, 79). The concept of strategy is not specific to the agricultural sector but it signifies the combination of agricultural and non-agricultural activities to sustain the existence of family farms. The embeddedness of such a concept in the Greek socio-economic formation entails characteristics that cannot be considered as specifically agricultural.

In our view, the concept of family farm strategies for survival and/or expansion includes by definition a significantly de-agriculturalised segment of family farming. In this respect, regional, spatial or local practices/activities appear to complement the agricultural activity of the farm family members. Moreover, the agricultural sector seems to follow tendencies which predominate in the wider economic system. Tsoukalas, for example, has argued that modern Greek society has shown significant 'resistance' to adaptations which lead to structural and organised forms of 'rational' development. In short, "according to the criteria of a rational liberal notion Greek society appears, perhaps not anomic, but certainly without any norms" (Tsoukalas 1993, 23). More specifically, the lack of crystallised social collectivities leaves significant room for 'resistance' to externally induced changes or integration movements. In this way, even apparent convergence to modernisation patterns may be linked to a resurgence of 'informal' phenomena which express some form of 'social resistance' to the formal rules of the division of labour within a modern society. In the Greek context, such 'de-normative' behaviour encourages a generalised activity of 'free riders' that aim at securing their own individual living. This 'resistance potential' or 'norm blindness' of family farms constitutes and is reconstituted by a set of practices, which sometimes formulate strategies outside the agricultural sector. As a result, the adaptability of family farms, which may be considered a form of 'reflexive responsiveness' against the calls of state agricultural policy and the ensuing European integration process, is both cause and effect of an inherent de-agriculturalisation of the Greek countryside. 'De-agriculturalisation' is the result and not the cause of the transformation process of Greek countryside.

In terms of farmers' practices and strategies it may be argued that Greek farmers have a long established history of strategic adaptability against state policies and market calls. This strategic adaptability is both inter-sectoral and cross-sectoral, the result of significant 'room for manoeuvre' on the part of small family farms reflecting their flexibility for a differentiated utilisation of family and farm resources (labour, land, machines, gender division of labour etc.) (Papadopoulos 1998).

We argue that the continued adaptation of Greek family farming can be attributed to this 'strategy' of reflexive responsiveness. This strategic responsiveness can be analysed more particularly if placed within three tendencies that characterise the countryside today. The *first* refers to the *increasing professionalisation* of Greek agriculture which is partly due to the measures exercised by the state, in conjunction with CAP measures; the *second* relates to the *feminisation of agricultural activity* in Greece, particularly in the smaller farm units; and the *third* is connected with the *increased marginalisation* of mountainous and less favoured rural areas which depend mainly upon agriculture for their economic survival. All three tendencies signify not only increasing social and economic differentiation and polarisation of the rural social structure but also increasing similarities in the economic and social development of rural and urban areas. The evolution of the countryside more and more approximates that of the urban areas while spatial characteristics emerge as more important for determining socio-economic differences within both rural and urban areas. Consequently, de-agriculturalisation goes hand in hand with an increasing social and economic diversity of rural areas.

With respect to the *first trend*, a recent report notes that what can be considered as 'large-scale professional agriculture' (i.e. farms over 8 or 10 ha) covered nearly half of the agricultural land in 1993 while representing only around one tenth of total farms in the country. This professional agriculture seems to be close to the average of the EU in terms of labour intensity; it includes only a small number of aged farmers and contains mainly full-time farmers (CSLEP 1998, 21-22). At the other end of the pole there is a 'small-scale agriculture' or 'residual agriculture' (of less than 2 ha) which covers only 9 percent of the land and contains 44 percent of farms. This residual agriculture is located mainly in the mountainous and less favoured areas of the country. In the middle of these two 'agricultures' lies the largest proportion of Greek agriculture (47 percent of farms) consisting of small and medium sized farm holdings which occupy 57 percent of the agricultural land. This intermediate stratum of farms contains farms that lean either towards the 'professional' or the 'residual' agriculture.

Farming families depend increasingly on non-agricultural income sources for their survival and/or expansion. The incomes of farm and non-farm households converged in the 1980s and the increasing importance of the latter is based upon two factors: the significant support (subsidies, interventions etc.) provided by the CAP to Greek farmers, which today amounts to almost half of agricultural income; and the significant degree of pluriactivity of farm household members, which has expanded to spouses and family members other than the household head, as mentioned earlier (Kasimis et al. 1998). Both factors can be considered external infusions of income into the rural household triggering farm differentiation trajectories, which cannot be confined within the agricultural sector. In this connection, farmers cannot any more be considered less privileged in comparison to other social strata. The increasing dependence of farmers upon off-farm incomes has considerably decreased their proportion among the lower, poor household strata; despite the fact that the proportion of the poor among farmers still remains large when compared to the country average (Sarris 1997).

As far as the *second trend* is considered, women's participation in the labour force is directly linked to their agricultural employment. Today, 22 percent of all women and 17 percent of all men are economically active in agriculture. It should be noted that rural women's participation in the labour force is still lower than that of rural men. Analysis of women's employment in agriculture shows that there is a significant tendency towards an increase of women farmers. First, the proportion of women farm heads has increased from 12 percent in 1971 to 17 percent in 1991. Second, two thirds of the women employed in the agricultural sector are 'unpaid helpers in the family unit' despite the fact that the number of women employed as unpaid helpers has dropped significantly during the last decade. The number of 'self-employed' women in agriculture has increased significantly, reaching 29 percent of employed women by 1997. This indicates a small but steady tendency for women to change their status from unpaid helpers in the family enterprise to self-employed farmers and employers. This tendency, however, does not seem to be homogeneous in the entire country since a number of historical, local, cultural and regional factors exert a significant influence on it.

The intensity of women's labour in the agricultural sector is considered lower than that provided by men (44 hours for men against 36 hours for women), while the number of women working less than 25 hours on the farm has significantly increased. This agrees with the generally shared view that men have to secure full-time employment, either within the farm or – if there is no possibility of this – outside the farm. Women's employment in agriculture is flexible and their integration is mediated through their role as spouse and mother rather than as farmer (Gasson 1992, Reimer 1986). This shows that women's agricultural employment tends towards higher engagement of women in farm management, depending on the extent of male farmers' involvement in agriculture; and towards the use of female flexible employment in response to the farm family's needs and resources.

Both tendencies are interconnected and provide a strong indication of the transformed character of women's agricultural employment, which is today significantly de-traditionalised and, finally, de-agriculturalised. Today women's role in family agriculture increasingly resembles their role in other employment sectors, while in the past women's position in the farm was considered as pertaining to a number of patriarchal codes of honour which correlated to a traditional rural social structure. Moreover, the contemporary 'invisible' character of women's employment cannot be considered specific to the agricultural sector but rather, and more importantly, to the domestic social and economic sphere. Women's changing roles in family farming and farm management can be seen as a sign of higher adaptation of women into the wider economic environment and also of the inter-penetration of agricultural and non-agricultural activities and practices.

Finally, with respect to the *third trend* it has been observed that both the distribution of the CAP support and the opportunities for additional or supplementary off-farm work directly influence and differentiate the income of rural households in Greek countryside. There are areas that are predominantly based upon agriculture enjoying a high proportion of subsidies and a per capita household income close to the country's average. In some other areas extensively subsidised agricultural production, dynamic cash crop cultivation and incomes from tourism and industrial employment have created incomes exceeding the national average. Finally, there are other rural areas which share a small proportion of CAP subsidies and reduced opportunities for off-farm employment, whose average income falls significantly short of the national average (CSLEP 1998, 3). This unequal distribution of resources and of agricultural dynamism shows the differential spatial de-agriculturalisation of Greek countryside.

In fact, it is common place now that CAP has favoured the most productive and most prosperous rural regions of the country while it has provided only supplementary support for the marginal rural areas whose rural households depend mainly upon agriculture for their subsistence and/or economic survival. Consequently, either through higher support for agricultural modernisation or through provision of off-farm employment, the countryside is led to de-agriculturalisation. This is due to the fact that agricultural subsidies are linked to the expansion of agricultural production for modernised farms and the maintenance of agricultural employment. However, for less modernised family farms and/or farms which are not adaptive to the changing conditions of agricultural production, non-agricultural sources of income and employment are necessary. More and more family farms need non-agricultural resources for their reproduction and/or expansion. This 'externalisation' of farm family reproduction conditions leads to a higher dependence of Greek family farms upon the wider economy. Thus, the farm family unit of production needs closer linkages with the external economy and/or seeks higher integration into the market economy.

To sum up, these three trends provide the basis for a reconsideration of Greek agricultural policy which, despite changes in the agricultural structure and the employment of rural household members on and off the farm, basically remains rigidly agricultural and slow to reform. From another angle, it may be argued that these trends provide support for the differentiated persistence of Greek family farming through their long-lived survival practices and strategies which, however, appear to be highly diversified and heterogeneous when considered regionally and/or locally (i.e. within a specific socio-economic context). Thus, the argument that Greek agriculture is becoming professionalised, albeit increasingly or demonstratively feminised and regionally diversified, endorses the view that family farms have persisted either by adjusting their strategies and adopting new practices for their survival or by resisting official agricultural policy which has marginalised a good proportion of them. Family farms aim at achieving survival or expansion by either becoming professional, or transferring the farm title to the woman when the farmer seeks full-time employment outside the farm; by turning to pluriactivity, where there are off-farm opportunities, or by retreating to subsistence farming when commercialisation and farm modernisation proves to be ineffective. These strategies are closely related to de-agriculturalisation which, in turn, constitutes the changing character of Greek (or Southern European) 'rurality'.

Additionally, one may note today the development of a complementary trend. This refers to an increase in agricultural activity in urban and peri-urban areas. Along with the diffusion of pluriactivity among other farm family members, the increasing employment opportunities in other economic sectors in rural areas and the changes in employment patterns and gender roles, one may note a corresponding expansion of agricultural activity in urban areas. In fact this is not a new trend; Karavidas described it in the inter-war period. Thus agriculture ceases to be an activity which solely characterises the rural population, but has started to be relevant for urban and/or peri-urban populations. According to official statistics 23 percent of those who work in the primary sector do not live in rural areas. A recent analysis of 'peri-urban' rural areas has noted the transitional and dynamic character of agriculture in those areas (Moissidis 1994, Moissidis and Duquenne 1997). The centripetal characteristic of peri-urban agriculture provides a contemporary model of an 'open'-extrovert family farming reliant upon urban demands, tendencies and short-term needs for agricultural production. Small urban or peri-urban agricultural activity is linked to the operational-integrated character of family farming which strictly follows market demands (Goussios 1999). Recognition of the differentiated functionality of rural areas and the need for spatial planning of production and consumption activities in the countryside are put forward as the main domains for intervention and/or regulation (Louloudis and Arahoviti 1997). The location of de-agriculturalised farm household operations within highly urbanised areas encourages the neglect of significant forces favouring social and economic integration of the remotest rural areas. In this way, planning for integration of rural areas frequently neglects those areas that cannot be easily accessed by the external economy. In other words, the remote rural areas (or 'less-favoured areas' in EU jargon) are purposefully de-agriculturalised (from above) through planned interventions aiming at rural household income differentiation and support. Less effort is expended on sustaining agricultural development and more on reducing the dependence of remote rural households' income upon agricultural production.

Undoubtedly, the emphasis on 'urbanised' agricultural activity signifies changes in the research questions posed by the expansion and intensification of the socio-economic integration of contemporary rural areas into the wider economy, highlighting the multi-functionality and inter-sectoral determination of the countryside. However, once again this change of emphasis is primarily dictated by policy questions posed by external forces, defined by EU regional policy and the reformed CAP. New research questions and tools are needed to examine the complex process of *'re-urbanisation' of the countryside* which – as stressed earlier – is not a new phenomenon. The revisiting of 'rurality' remains basic to the reconstruction and the specialisation of agricultural policy. Resistance to the marginalisation of rural areas does not have to be reinvented in the Greek or Southern European countryside (Bazin and Roux 1995), but has rather to be interpreted and adapted to the new conditions determined by European integration and the globalisation process (Papadopoulos 1999).

Conclusion

De-agriculturalisation has been the result of the transformation of the Greek countryside. However, it cannot be seen merely as incidental to the wider transformation process. Rather it constitutes a multifaceted process which has significantly evolved as part and parcel of the differentiation and restructuring of rural areas. Rural exodus and migration have proved effective mechanisms for reducing the agriculturally employed rural population and the expansion of agricultural employment in wider areas. The socio-economic transformation of rural areas has signified the de-traditionalisation of Greek society (e.g. erosion of the domination of the patriarchal code of behaviour) and the inter-penetration of 'rurality' and 'urbanity'. This inter-penetration has been evident in the persistence or even intensification of the linkages between rural and urban areas.

In the past, the migrant maintained the relationship with his/her farm family and pursued his/her family's upward social mobility. Later, the combination of farm and non-farm activities/incomes was related to the polyvalence of rural households. Currently, the frequent geographical mobility between the city and the countryside and the horizontal and vertical networks across social and economic space have interconnected localities and production units. In the changing conditions of the countryside the family farm still holds the key role showing some resistance against market rationalisation. The family farm's capacity to resist has relied upon the differential use of practices and construction of adaptive and/or responsive strategies pursuing survival and an improvement of household consumption levels. New types of family farm differentiation can be described on the basis of their capacity for expanded consumption (dignified survival) or conspicuous consumption (prosperous survival) (Tsoukalas 1991).

During recent years a number of trends reveal the de-agriculturalised future of the countryside: the growth and diffusion of pluriactivity to other than the head family members, the increase in the flexibility of labour and especially of female labour within the family farm, the increasing inter-relationship between urban and rural labour markets and the changing gender roles within the family farm are only some of the trends. They are significantly reinforced by the extensive use of foreign immigrant labour in the 1990s. Foreign labour contributed positively to overcoming a 'disguised' crisis of Greek agriculture by reducing labour costs of production and satisfying labour needs in peak seasons. Additionally, it gave the family an opportunity to 'redesign' its allocation of family labour on and off the farm in the context of a strategy for either survival or expansion of the farm. This sharpened the survival strategies of the farm families and contributed to a (post-)modern transformation of the countryside.

In sum, the adaptive capacities of Greek family farms have played a significant role in adjusting their behaviour to the pressures and regulations of official agricultural policy, by altering their farm size (through renting or sharecropping), by modifying the gender division of labour within the farm family, by allowing the succession of their younger members in farming and by exploiting agricultural subsidies. It is our contention that the survival strategies and the flexible coping devices of Greek farm households, elaborated under new circumstances, may secure the future of rural areas, if they are not sacrificed on the altar of normative behaviour and operational rationalisation largely imposed by external forces. One way of approaching contemporary agricultural and regional policy measures is to turn responsive de-agriculturalisation into a structural factor and vital force for the sustainable development of Greek countryside.

Notes

1 In the inter-war period, Karavidas analysed family farming in Greece in terms of two basically different types. The first type he called 'nuclear peasant family of the free smallholder' characterised by over-employment of its family members, production for subsistence, dependence upon the market and the state for its reproduction. This small peasant family is gradually led to incorporation within the capitalist economy, while it also develops a *bourgeois-peasant* (*astochorike*) identity. He called the second type the 'mixed urban-peasant family of small landowners' consisting of those professionals who lived in the town but still cultivated their agricultural land with the help of their relatives or other villagers (see Karavidas [1931]1978, Karavidas [1930/1936]1981). Of interest is a recent publication which stresses the content and importance of similar phenomena in contemporary rural society (Zakopoulou 1999).

2 See the analysis of Karavidas ([1931]1978) for a number of agricultural formations (e.g. *zadruga*, *chiflik* etc.) which coexisted with the 'small peasant family unit' but could not survive expansion of the commodification of agricultural production and penetration by a capitalist economy.

References

Bazin, G. and B. Roux (1995), 'Resistance to Marginalisation in Mediterranean Rural Regions', *Sociologia Ruralis*, Vol. 35, No 3/4, pp. 335-347.

Burgel, G. (1977), 'Rural Research in Greece: Part I', *The Greek Review of Social Research*, No 30/31, pp. 190-213 (in Greek).

Burgel, G. (1978), 'Rural Research in Greece: Part II', *The Greek Review of Social Research*, No 33/34, pp. 190-213 (in Greek).

Committee for the Study of Long-term Economic Policy (CSLEP) (1998), *Competitive Agriculture and Rural Development: Current Structural Problems and their Solutions*, National Bank of Greece, Athens (in Greek).

Damianakos, S. (ed, 1987), *Processes of Social Transformation in Rural Greece*, EKKE, Athens (in Greek).

Dertilis, G.V. (1993), *Ineffective or Effective? Taxes and Power in the Modern Greek State*, Alexandria, Athens (in Greek).

Gasson, R. (1992), 'Farmers Wives – Their Contribution to the Farm Business', *Journal of Agricultural Economics*, Vol. 43, No 1, pp. 74-87.

Goussios, D. (1999), 'Countryside, Rural Space and Small Town: From Ruralisation to Local Development', in D. Economou and G. Petrakos (eds), *The Development of Greek Cities*, University Press of Thessaly/Gutenberg, Athens, pp. 157-207 (in Greek).

Hoggart, K., H. Buller and R. Black (1995), *Rural Europe: Identity and Change*, Arnold, London.

Karapostolis, V. (1978), 'Economic Rationality and the Family Farm', *Greek Review of Social Research*, No 36/37, pp. 365-371 (in Greek).

Karavidas, K.D. [1931] (1978), *Agrotika: Study of Social and Economic Morphology in Greece and in Slavic Countries: A Comparative Study*, Papazisis, Athens (in Greek).

Karavidas, K.D. [1930/1936] (1981), *The Issue of Autonomy: Socialism and Communitarianism*, Papazisis, Athens (in Greek).

Kasimis, C. and A.G. Papadopoulos (1994), 'The Heterogeneity of Greek Family Farming: Emerging Policy Principles', *Sociologia Ruralis*, Vol. 34, No 2/3, pp. 206-228.

Kasimis, C. and A.G. Papadopoulos (1997), 'Family Farming and Capitalist Development in Greek Agriculture: A Critical Review of the Literature', *Sociologia Ruralis*, Vol. 37, No 2, pp. 209-227.

Kasimis, C., A.G. Papadopoulos and S. Stravoravdis (1998), 'Farm Family Labour and Pluriactivity: Rural Labour Markets and the New Physiognomy of Greek Family Farming', paper presented in the 14th World Congress of Sociology, International Sociological Association (ISA) University of Montreal.

Long, N. and J.D. van der Ploeg (1994), 'Heterogeneity, Actor and Structure: Towards a Reconstitution of the Concept of Structure', in D. Booth (ed), *Rethinking Social Development*, Longman Group Ltd, Harlow.

Louloudis, L. and E. Arahoviti (1997), 'Unprotected Land: Greek Land-Use Policies in a Period of Farming Contraction', in H. de Haan, B. Kasimis and M. Redclift (eds), *Sustainable Rural Development*, Ashgate, Aldershot.

Moissidis, A. (1994), 'Peri-Urban Farming: Some Theoretical and Methodological Considerations', *Greek Review of Agrarian Studies*, No 12, pp. 22-36 (in Greek).

Moissidis, A. and M-N. Duquenne (1997), 'Peri-Urban Rural Areas in Greece: The Case of Attica', *Sociologia Ruralis*, Vol. 37, No 2, pp. 228-239.

Papadopoulos, A.G. (1994), *Class and Social Stratification in Contemporary Rural Greece:*

A Comparative Study of Three Communities, Unpublished D.Phil Thesis, University of Sussex, Brighton.

Papadopoulos, A.G. (1998), 'Flexible Agriculture: Survival Strategy and Resistance to Marginalisation or Form of Integration into the Agro-Food System', in C. Zioganas (ed), *Competitiveness and Integrated Development of the Rural Sector: New Challenges for Greece*, ETAGRO, Thessaloniki, pp. 255-268 (in Greek).

Papadopoulos, A.G. (1999), 'Revisiting the Rural: A Southern Response to European Integration and Globalisation', in C. Kasimis and A.G. Papadopoulos (eds), *Local Responses to Global Integration*, Ashgate, Aldershot, pp. 245-270.

Papadopoulos, A.G. and J. Daouli (1999), 'The Impact of EU Policy Measures upon Areas Lagging Behind in Development and the Survival of Family Farming in Greece', in A. Nemenyi (ed), *Rurality in Europe at the Threshold of Centuries*, Editura Alsand, Cluj, pp. 18-33.

Pizanias, P. (1993), *The Poor of the Cities: The Know-how of Survival in Greece in the Inter-war Period*, Themelio, Athens (in Greek).

Reimer, B. (1986), 'Women as Farm Labour', *Rural Sociology*, Vol. 51, No 2, pp. 143-155.

Sarris, A. (1997), 'Myths, Realities and Prospects for Greek Agriculture', paper presented in the Economic Congress of the Economic University of Athens, Athens (in Greek).

Stathakis, G. (1983), *Industrial Development and the Regional Problem: The Case of Greece*, Unpublished PhD Thesis, University of Newcastle Upon Tyne, Newcastle Upon Tyne.

Sutton, S.B. (1983), 'Rural-Urban Migration in Greece', in M. Kenney and D. Kertzer (eds), *Urban Life in Mediterranean Europe*, University of Illinois Press, Urbana.

Sutton, S.B. (1988), 'What is a Village in a Nation of Migrants?', *Journal of Modern Greek Studies*, Vol. 6, No 2.

Tsoukalas, K. (1987), *State, Society, Employment in Post-war Greece*, Themelio, Athens (in Greek).

Tsoukalas, K. (1991), *Images of Civilisation: Freedom, Equality and Fraternity in Modern Society*, Themelio, Athens (in Greek).

Tsoukalas, C. (1993), 'Free Riders in the Wonderland', *Greek Political Science Review*, No 1, pp. 9-52 (in Greek).

Vergopoulos, C. (1975), *The Agrarian Question in Greece: The Social Integration of Agriculture*, Exantas, Athens (in Greek).

Wagstaff, M. (1982), *The Development of Rural Settlements*, Avebury, London.

Zakopoulou, E. (1999), 'Pluriactive Population and Agriculture: Towards a New Tracing of a Multifarious Phenomenon', in C. Kasimis and L. Louloudis (eds), *Countryside in Greece: Rural Society in the End of the 20th Century*, Plethron/EKKE, Athens (in Greek).

9 Shifting Rurality: The Spanish Countryside after De-peasantisation and De-agrarianisation

JESÚS OLIVA AND LUIS A. CAMARERO

Introduction

This chapter deals with recent changes affecting rurality in a Mediterranean context and the consequences of the continuous decline of farm predominance in the rural world. First, we examine the role of Spanish agriculture as part of a European 'Green Ring'. Next, we set out the key role that the *agrarian* and *peasant question* has played in modern Spanish social research, and the main traits that have characterised Spanish agrarian society historically are summarised in broad outline. Thirdly, the meaning of the castilian word *campesino* (peasant) is considered, and other equivalent terms used in other languages of the Spanish state for defining people working in agriculture and living in the country in different regions. Next, we define and differentiate between the *de-peasantisation* and *de-agrarianisation* of rurality, exploring the significance of both processes and the way they affect the theoretical and methodological assumptions that have predominated in rural studies until now. The suitability of expanding our view of *the rural* is proposed, to analyse it as a social representation that fits with our everyday social practices, discourse and lifestyles.

Following this, we describe in more detail the changes that Spanish rurality has undergone since the end of Franco's autarchy period in 1959 (de-peasantisation, de-agrarianisation, changes in migratory patterns, residential and labour strategies, etc.). We examine the new cultural and ideological context in which these recent restructuring processes have developed and have taken place. In addition, we identify and analyse the changing discourses and social representations about the *modern* and the *rural* that legitimate these processes, as well as the new social actors involved in them. We suggest that this new emergent context could be studied as a shifting rurality. Finally we explore what we call a cultural *re-peasantisation*, the social construction of *the rural* on the basis of the traditional peasant references but in a new post-productivist context.

Spain in the Green Ring

The lateness of the Spanish experience of modernisation compared with central Europe, and its geographic situation in the south of the south of Europe, have come to shape its position in what has been called here Europe's 'Green Ring'. However, this backwardness of social and economic structure was overcome by an accelerated and dramatic process of capitalist modernisation. In 1950, the agricultural labour force was nearly half the total active population (48.8 percent) while industry employed just under 20 percent. Nowadays, only 8 percent of the workforce is employed in the primary sector while industry still employs under 20 percent. The figures illustrate the rapid transition that has taken place in Spain from an agrarian economy to one based on services and tourism, without the consolidation of an industrial structure or the paradigmatic type of 'urban-industrial' worker as the key social actor of modernisation. While, in central European countries, modernisation is evident at the time industrialisation takes place, in the Spanish case this process is completed in a post-industrial stage. So, the industrial growth of the 1960s became an obvious trend towards de-industrialisation fifteen years later. In this way, the Spanish process of modernisation (which coincides with the political democratisation in its final phase) differs from the core European country model. In it, the pattern of the ex-farmer turned into an industrial worker coexists with other paths of transition.

Three processes are particularly significant with respect to Spanish de-peasantisation. First, the increasing participation of women in the workforce (mostly employed in service and administrative jobs) that favoured female emigration to urban areas and an increasing male surplus in rural ones. Second, farm modernisation brought about a significant unemployed labour force that can only remain living in their villages through welfare assistance programs. In the south of Spain these programs have succeeded in restraining rural depopulation, and on the other hand they have made possible a wide range of informal and casual working practices. Third, the intensity of the migration flow to central European cities at the beginning of the process of capitalist modernisation. These three processes, together with a highly concentrated and located industrialisation, gave rise to an accelerated transformation of the Spanish economy, from one based on a domestic subsistence farming to one based on wage labour. All these processes favoured change in rural families, from a mono-agrarian type to a pluriactive and multi-strategic one (Sampedro 1996). In this way, the pluriactive rural family became an alternative to the male-wage-based industrial pattern.

The peripheral role of Spain in the European division of labour turned it into a labour supplier country for advanced European economies (Germany, France, Switzerland). In this context, the south of Mediterranean Europe became a tourism-specialised area that depends on the core countries' demand. The expansion of tourism and construction sectors gave rise to an internal seasonal labour market within which the rural workforce was involved on the basis of nomadic and itinerant strategies. More recently some of the southern regions have also become highly specialised in fruit and vegetable farming, oriented to the central and north European countries and extremely dependent on external inputs (seeds, pesticide, technology). This is an agriculture that benefits from the condition of being the first European gateway for African immigration into these areas, since this process provides suitable workers for its highly casual and flexible labour market. The success of this export-oriented agriculture is also based on recruitment of young and female rural workers, favouring a wide variety of mixed strategies of domestic exploitation and wage labour. A recent study (Pedreño 2000) shows that the earlier pattern of Fordist industrialisation of this agriculture is being replaced by another pattern, based on a widespread informalisation of labour relations.

Rurality, Farming and Social Research

The 'Agrarian' and 'Peasant Question' in Spanish Rural Studies

The *peasant* and *agrarian question* has been a key and predominant issue in Spanish social research from the 17th century. The ways it has been considered and focused on in these works were influenced by the different property, tenure and crop systems predominant in each region. Historically there have been significant differences between regions. In the northern part (Basque Country, Navarre, Castile, Galicia, Catalonia) a farm household system that employed family labour as well as a family-oriented subsistence production prevailed. A large number of aged farmers still remain in some of these regions. However, in the southern part (mostly

Extremadura and Andalusia) the *latifundio* system of landlords and large property based on extensive cropping was very frequent, as well as a large number of workers employed on a day wage (*jornaleros*), and a market-oriented form of production. We can still see these differences in the farm workforce figures today (see below). The *agrarian question* in the northern regions is focused on the permanent crisis of traditional family farming and its necessary modernisation. However, when attention is paid to the southern ones, the focus is on the unfair property system and deep social inequality that characterised life in the populous *agrotowns* of *latifundist* regions and led to the outbreak of continuous peasant revolts that were crushed by the armed forces.

In this way, in the context of a pre-industrial agrarian society the *social question* is identified as the *agrarian question*. Unequal distribution of land, absentee landlordism, the secular crisis of traditional farming, rural caciquism as well as periodical peasant riots (particularly in Andalusia) gave rise to works by social reformers at the end of the 19th and start of the 20th century (Caballero 1864, Costa 1901, López-Tuero 1905, Diez del Moral 1929, Carrión 1932). A public research institution, the Social Reform Intitute (*Instituto de Reformas Sociales*), founded at the start of the 20th century, focused on this issue in one of its early research programmes (I.R.S. 1919).

Agrarian reform became a historic demand and the Republican government of 1931 approved a determined plan that was shattered when the civil war broke out (1936-9) and Franco's final victory restored the former state of affairs. The flagrantly unfair structure of property, which was a reflection of the social gap, was reinforced after the armed conflict during Franco's dictatorship. So, in the context of the *political transition* at the end of the 1970s and later on when the Socialist Party won the elections in 1982, the claim for an *agrarian reform* was reactivated as an important symbolic reference. For these reasons, the autonomous parliaments of Andalusia and Extremadura (the historical *latifundist* regions) approved different agrarian reform plans (albeit adapted to a new context in which social pressure progressively loses importance) as well as special farm unemployment benefits.

However, when the autarchy period of dictatorship petered out and liberalisation of the country favoured economic development from the end of the 1950s, an age of profound changes started (see below). The process of de-peasantisation and the rural-urban exodus resulted in a global economic restructuring as well as a large-scale redistribution of population in the territory. Sociological and economic studies that the dictatorship had cut off were resumed at this time, stimulated by these processes of change.

In the postwar period, the repression practised in the academic world restricted rural sociology to a *catholic sociology*, a moral, charitable one, carried out by priests and civil servants of Franco's regime that praised the moral virtues of the rural life (Newby and Sevilla-Guzman 1984). However from the end of the 1950s arose (both in Spain and in exile) an important and plentiful scientific literature that was inspired by the deep transformations that were taking place (De Miguel and Moyer 1979). Economists, sociologists, anthropologists and historians analysed the de-peasantisation process, the consequences of rural-urban exodus, the disintegration of traditional peasant cultures, etc. (FOESSA 1966, Perez Díaz 1972, Aceves y Douglas 1976, Sevilla-Guzman 1979, Barón 1971, Naredo 1971, Giner 1977, Brandes 1975, Etxezarreta 1977, Castillo 1979). Most of the work at that time was based on the theoretical and methodological assumptions of the *peasant studies* perspective that dominated European and Latin-American studies, and constituted a young Spanish political economy and sociology that dealt with the evolution of agriculture and the peasantry in capitalist society (see Ortí 1992).

Campesino: 'One Who Works and Lives in the Country'

The Castilian words *campesino* (peasant) and *agricultor* (farmer) have been the terms most in use for referring to the rural agrarian population. The term *agricultor* alludes to the profession and evokes professional connotations. The word *campesino* (that comes from *campo* = country) alludes to a social condition ('the one who works and lives in the country') and evokes more general meanings in its uses as an adjective that refers to 'what comes from the country or belongs to it' (R.A.L. 1992). Finally, *campesino* has also been used in everyday life for alluding to the *wild*, the *spontaneous* or the *uneducated*. Some of these last meanings can be used contemptuously, as the opposite of the *modern*, the *polite* or the *urban*.

However, both in scientific literature and in daily language, other words from different languages of the Spanish state have been employed for designating more specifically the agrarian groups in each region. Thus, in the Basque Country and Navarre, the word *baserri*, that comes from the basque language (*euskara*), is used and alludes to the peasant household. It indicates the importance of the household in the peasant basque culture in which the peasant himself takes his surname from the *baserri* where he was born or where he lives. In Catalonian the word *payes* is used in a similar way and also the term *rabassaire* (this latter refers to sharecroppers and tenants). The words *labrador* (in Castilian) and *labriego* (in Galician), that derive from the verb *labrar* (to carve, to plough), have been widely used for referring to small landowner farmers (who plough and work their own land). Finally, in the *latifundist* regions, while different local terms have been used for designating different social conditions derived from the structure of property relations, the word *jornalero* (daily-paid-worker that comes from *jornal* = daily wage) is the most widely used to allude to those farm workers, usually without land or any other means, that were employed in the *latifundios*.

De-agriculturalisation, Mobility and Rural Research

Rural, *farm* and *peasantry* were synonymous references in the past but the crisis and change of the society that they represented (agrarian and pre-industrial) have differentiated and distinguished their meanings. Terms that were once interchangeable have been ordered by a hierarchy of meanings in which *rural* occupies the more generic place for designating *what is not urban*. Meanwhile, *farm* has changed for referring to that part of the rural world linked to the farm activity and *peasant* for evoking the pre-modern past when rurality and farming coincided. *Farm* and *peasantry* diverge in their meanings as far as *de-agrarianisation* and *de-peasantisation* refer to different processes (even though they were frequently mixed in the past). *Peasant* refers to a self-sufficient family-oriented farming. Patriarchy and land ownership are characteristic institutions of this familiar socio-productive organisation. In a contemporary way, we could define it as a family involved exclusively in farming on the basis of a self-sufficiency-oriented multi-crop system. The *de-peasantisation* process describes the transformation of this peasant farm into a modern family farm holding; that is, the conversion of the multi-crop family system into a mono-crop market-oriented one, the integration of farm activity into capitalist forms of production (commercialisation) and the emergence of family pluriactive strategies. *De-peasantisation* deals with the end of the feudal regime and with farm modernisation.

De-agrarianisation constitutes a step beyond *de-peasantisation* and describes the process of diminution of the former central place that farming had occupied in rural settlements. That is, *de-peasantisation* deals with the modernisation of farming and *de-agrarianisation* deals with the process of disassociation of farm activity from rural residence. If the process of *de-peasantisation* is seen to start from the end of the 19th century in the most industrialised regions (Basque Country and Catalonia), its acceleration and accentuation took place in the two first decades of the second half of the 20th century. However, the process of *de-agrarianisation* developed at the same time, and is definitively evidenced since the 1980s. Since then some processes not predicted by any social discipline started to be identified and came to call in question some of the theoretical assumptions about the rural world that predominated during earlier decades (EC 1988, 1995). These were unexpected processes (Saraceno 1994) for *agri-centered* research perspectives that, as far as they identified rural with farm, could not deal with the new processes of change that rurality was undergoing – the arrival of newcomers, the spread of second homes, the impact of increasing mobility of capital and labour, the new consumption-oriented roles conferred on the countryside (recreation, leisure, conservation, etc.), changing social representations relating to rurality.

The high mobility of people, capital and labour facilitated by improvements in transportation and telecomunications have accelerated the crisis of the concentrated (Fordist) urban-industrial model of social and economic modernisation, giving way to new de-concentrating and restructuring processes (Lash and Urry 1987). Counter-urbanisation, industrial de-centralisation, long-distance commuting and weekly flows to second homes, holiday trips, etc. turn the cities into deserts during some periods of the year (summer time, weekends) at the same time as many small villages double their numbers of residents and even become part-time cities (holiday villages, coastline and mountain areas). Some urban people (but not all social groups) go to rural areas in search of better places to live out of the cities, and some firms (but not every branch of industry) look for new profitable rural and peripheral locations (areas with low rates of unionisation and high rates of unemployment). The rural restructuring perspective (Bradley and Lowe 1984, Marsden et al. 1990) emphasises the way some rural areas become integrated into broader economic frameworks and are increasingly affected by the vagaries of international competition and trade.

On one hand, in this context it is more and more erroneous to identify the employment structure of an area with the employment of its residents. On the other hand, the farm population become a progressively smaller proportion of the rural workforce in the industrialised countries and fewer and fewer rural residents depend on farming, even in traditional agrarian regions (Oliva 1995). Finally, farming has become highly industrialised and the agricultural economic sector strongly integrated in a vertical food chain in which the farmer is frequently the weakest link. As Commins has noted, "advanced technologies tend to dissolve the distinction between 'agriculture' and 'industry'" (1990, 55). The necessity of integrating a wide variety of external inputs into the process of production (even capital and labour), very often provided by multinational firms (seed, pesticides, fertilisers) as well as the rules imposed by processor industries and delivery and financial firms have changed the farmer into a businessman. Terms such as *agribusiness*, *agri-industrial complex*, *international food chain* have been coined to describe the new forms of the agricultural system. In addition, scientific improvements make farming more and more independent of local resources (land, water, local knowledge, etc.) and increasingly subordinated to the international food market and biotechnology industry.

However, de-agriculturisation implies not just a decrease in the farm workforce and traditional farming but the fading away of the historical representations of land, work, family and community that predominated in rural society. As these processes take place, local economies have been differently reshaped and former farm homogeneity has given way to a complex rural mosaic. As a result, new representations of the countryside arise which are different from representations of farming. In the following section we attempt to reflect on how all these processes erode the classical conceptions, operative definitions and social representations of the rural.

De-peasantisation and Rural Restructuring

As a Mediterranean economy, Spain contrasts with the north of Europe in terms of its high rates of unemployment and higher farm population rates. Underdeveloped forms of welfare as well as the recent flexibility of labour markets, a *post-agrarian flexibility* as Peck (1994) has noted, have facilitated an increase in temporary jobs and widely exploitative conditions. Together with Italy, Ireland, Portugal and Greece, Spain has a higher than average farm workforce for the EU-12. More than 1 million people were registered as engaged in the farm sector in the second quarter of 1998. However, three autonomous regions accounted for almost half of these (Andalusia 22 percent, Galicia 17 percent and Castile-León 10 percent), while 6 of the 17 regions accounted for 68 percent. We can still see today the differences between north and south in their historic land tenure systems and crops in the diverse agricultural workforce figures. More than 232,000 people are engaged in agriculture in Andalusia (62 percent of the paid labour force) in contrast to Galicia that has more than 183,000 people in this sector but this constitutes only 14 percent of the salaried workforce. For this reason, special systems of unemployment benefits have been applied in some Southern regions (Andalucía and Extremadura) since 1984.

De-peasantisation in Spain has accelerated dramatically since 1960, as Franco's autarky (a self-imposed economic and political isolation during the 20 years after the civil war) petered out, and development of urban-industrial areas became the key strategy of modernisation. Foreign investment and transfer of American technology, the arrival of North European tourists, the periodical savings remitted by emigrants facilitated the integration of Spain into the international capitalist economy as well as rapid (but dramatically disorganised) economic growth. Rural-urban exodus encouraged by economic planning and precipitated by the industrialisation of agriculture, the expansion of industry, construction and tourism gave rise to a radical redistribution of population within the country. In 1950 more than 50 percent of the workforce was involved in agriculture and 63 percent of the population lived in rural areas of fewer than 10,000 inhabitants. By 1976 more than 5 million people had migrated from rural villages to major cities and capitals, and more than 2 million Spaniards went to the most industrialised countries of North and Central Europe. The aim of technocratic planners was to concentrate capital, labour and the means of production in selected cities, to maximise internal economies of scale and external economies of agglomeration.

Many rural areas (mostly mountain and inner regions) underwent an irreversible depopulation and rurality as a whole became subjected to a circular process of social and economic deprivation (decline of agricultural employment, depopulation, deterioration of the demographic structure in relation to age and sex) new socio-political values adopted by emigrants in Europe, the arrival of tourists, the introduction of television (1956) and the universalisation of the automobile contributed to spreading a new modern way of life bringing wide cultural and social changes. It was the crisis not only of traditional farming but of a complete way of life.

De-peasantisation continued in the 1970s and 1980s. Employment in agriculture retreated to 1.7 million between 1976 and 1998 and the primary sector of the workforce declined from 22.3 to 7.8 percent. Figure 9.1 shows the evolution of the rural and economically active population in agriculture from 1940 to 1998.

Agrarian modernisation, as a public administration strategy of rural development, failed to offer an appealing role to women and young people in family households (Sampedro 1996, Vicente-Mazariegos et al. 1991, 1992, 1994). Consequently, both rural women and young people continued migrating from rural areas (Camarero et al. 1991). It is not surprising that all the regions lost, between 1976 and 1998, more than 60 percent of their 'non-salaried' farm workforce (mostly small farmers and other household members).

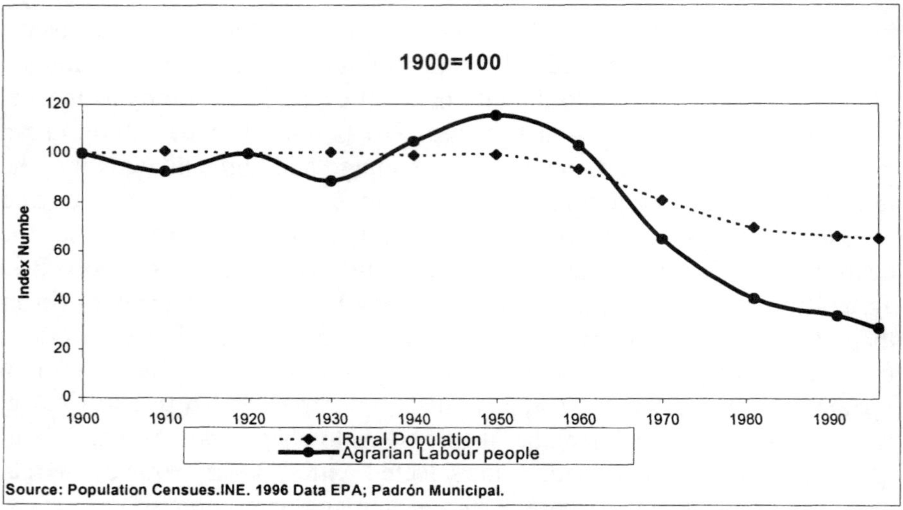

Figure 9.1 Rural and Agrarian Population Trends

By the 1980s, the employed population of rural areas became more significantly *post-agrarian.* In 1981, only 4 out of 10 economically active rural residents were involved in agriculture. Moreover, an increasing number of farm households complemented their earnings with sources of incomes outside farming. *Worker-peasants*, *part-time farming*, and finally *pluriactivity*, as social research categories, evidence this process. Map 9.1 shows the result of these processes.

In addition, the pattern of migration changed at the end of the 1970s. On one hand, the impact of economic stagnation throughout the capitalist world, the congestion of industrial cities, increasing unemployment, the rise in urban house prices, converged to mitigate rural-urban exodus. In 1975 only 4 out of 10 migrants came from rural areas (less than 10 thousand) and the early spread of the urban form (counter-urbanisation) made the large cities and capitals no longer predominant as destinations.

By 1995 the proportion of people who were migrants from rural areas was hardly 3 out of 10. In that year, more than 265,000 people migrated to rural villages (places with fewer than 10,000 inhabitants) against 245,000 that chose capitals and large cities (places with more than 100,000 inhabitants). Rural destinations have almost doubled between 1975 and 1995, reaching more than 30 percent. In fact, 1 out of every 3 people that moved from capitals or large cities in 1995 went to a rural destination (more than 160,000 people).

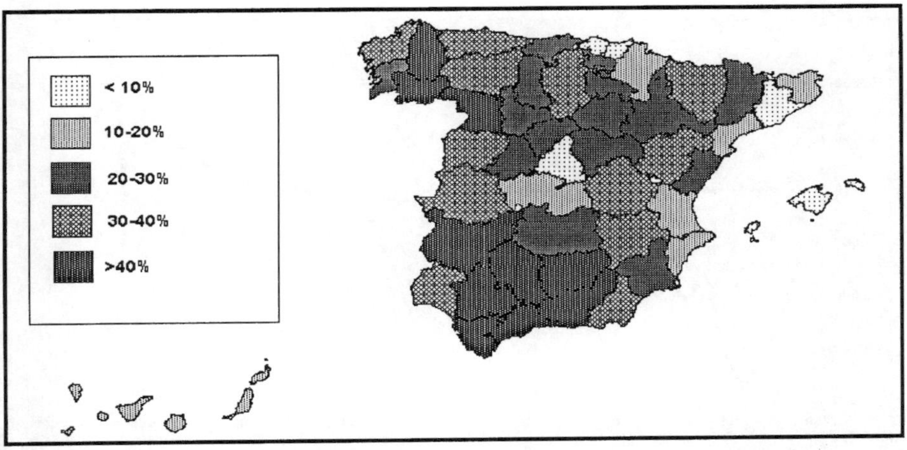

Map 9.1 Results of Rural De-agrarianisation in Spain, 1991 (Proportion of Agricultural Population of Economically Active Population in Settlements with less than 10,000 inhabitants)

As we can see, there has been a change from the rural-urban exodus in 1960-70s to a new situation of equilibrated rural-urban exchange (Camarero 1991). Figure 9.2 shows the changing migratory balance in rural villages. Throughout the 1980s and early 1990s migrant social groups changed too. Most rural-to-urban migrants in the 1960s-70s were ex-peasants and young people who were searching for a job and wanted to integrate themselves into the modern way of life in the city that both the mass media and dominant ideology promoted as the opposite to the peasant world and rurality (with its deprivation, family patriarchalism, isolation and rusticity). Even nowadays young people continue to migrate to cities to study or live, but in many rural areas unemployment rates are lower than in urban ones. However, unexpected urban-to-rural moves of retired workers, middle classes, and neo-rurals have been increasing steadily during the last two decades. Their lifestyles are very diverse and predominantly they do not migrate in search of a job but rather of a paradise out of the congested cities. For some of them it could be the village where they come from. Other groups look for a special idyllic residence in an exclusive area on the coastline or in the mountains. Other people may prefer a safe quiet place located in a convenient crossroads near the city. Some retired workers have converted their second homes into primary residences. Finally, there are people who seek a different way of life in the country.

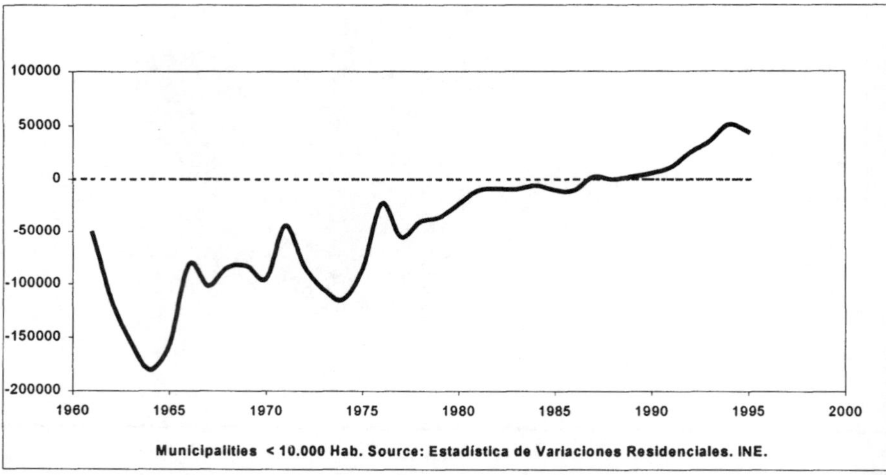

Figure 9.2 Rural Migratory Balance in Spain, 1960-1995

Shifting Rurality

The profound change in residential patterns, labour strategies and restructuring of production, the consequences of changing leisure and travel practices etc., brings forth new social representations about *the rural* and, after a decade of researches about it, we can identify much better these processes and theorise their social significance. Moreover, these changes do not take place in a ideological vacuum; on the contrary, they require a new cultural-ideological configuration on the basis of which they are socially legitimated in everyday life. If the industrial city and the *modern* urban way of life occupied a central place in the ideological configuration of *modernity*, so *the rural* as a social representation (and its associated symbolic references to nature, health, tradition, etc.) plays a key role in the *postmodern* ideological configuration and the fields in which it is deployed (consumption, leisure, social identities, etc.). Morin (1973) has used the term *neo-archaism* for referring to the way *the rustic* became fashionable after the 1960s and how it is integrated symbolically in a wide range of products and consumption patterns (rustic furniture, natural and home-made food, country clothing etc.).

Industrial cities (as a final product of the old Enlightenment project) have become a 'pathology of modernity'. As Vicente-Mazariegos has pointed out "that planned Jerusalem that claimed to be the city produced by the industrial bourgeoisie is metamorphosed in a muddled and despotic

Babylonia that is gradually and quietly abandoned by its residents" (1993, 5). But the crisis is extended also to other aims of the program of modernisation developed by the bourgeoisie. Thus, in opposition to universal homogenisation as citizens and the nation-state identities formed during the last three centuries, 'others' emerged, through the discovery of multiculturalism, the revaluation of difference, and local identities. The earlier picture of the countryside and the peasant world searching for anonymous integration into the city and the market, is followed by the exploration of their own identity. As peasant references (land, work, family and community) and the correspondence between urban and industrial faded away, rurality became a shifting society. The past cannot help us to manage our present, neither is the road to the future clearly outlined.

As long as agriculture dominated rural life and industrialisation dominated urban, this favoured the construction of a rural-urban dichotomy through which country and city were imagined as different and as themselves homogeneous. But nowadays we find that rural and farm (as well as urban and industrial) no longer coincide in meaning. There is no longer something like a durable, natural 'urban' or 'rural' object but social categories constructed and defined in a specific historical context. We have to explore not only what rural means nowadays but who defines rural and how they do it. In other words, we have to use a reflexive approach that attempts to explain why these social categories are constructed and used in a concrete way. It is no surprise that recent representations of rurality became independent of both the space and the social reality that they claim to represent (Mormont 1987) and that rural became a key element in the dominant representations of postmodern life.

On one hand, the countryside appeared as a segmented world of diverse local economies linked to urban and international processes (tourism and migration flows, environmental conservation, productive restructuring etc.). On the other hand, rural settlements are progressively made up of heterogeneous social groups and contrasting lifestyles. Obviously, the traditional agri-centered perspective had ignored many 'other' rurals (such as people involved in fishing, forestry or tourism). However, a number of miscellaneous new social actors have been identified in the post-agrarian rurality (temporary residents, local holiday makers, farmers, newcomers, returned migrants, non-farm workers, etc.) (Kayser 1991, Marsden et al. 1993, Hoggart et al. 1995), including many people whose occupations, consumption patterns and social values are definitely non-farm. Even more, it includes people whose outlooks, perceptions and attitudes about locality may be completely contrasting. Much of the rural conflict described by writers, usually presented as symbolic battles with no neatly drawn lines of engagement, stems from confrontations between newcomers and oldtimers, farm and non-farm interests, locals and foreign-born, locally employed and commuters, etc. As a result, localities are no longer the placid peasant community frequently described in the past but a controversial rurality. Figure 9.3 summarises some of the processes that are shaping rurality as a new space that is socially heterogeneous, economically diverse and potentially conflictual. It outlines new settings in which several actors who experience rurality differently come together on the basis of peculiar patterns of time-space consumption, social identities and lifestyles.

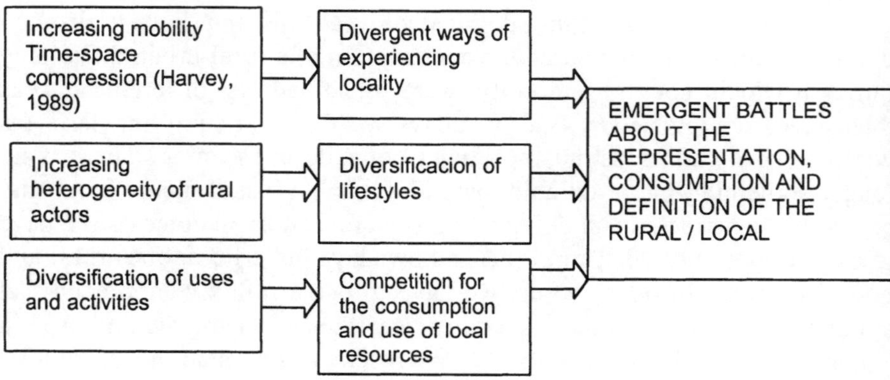

Figure 9.3 Conflicting Rurality

Finally, new representations of rurality emerged which are separate from farming and relate to nature, recreation, leisure, and local identities. City people frequently come to rural localities in search of residence and recreation, bringing with them an idyllic vision of rural life, a perception that stimulates their projects and interventions in the locality. Moreover, as Cloke and Thrift (1987) found in their work on service class ruralisation in England, such people are attracted to rural areas only to the extent that these conform to this idealised re-creation of the countryside. Moreover in both local and supra-local arenas, diverse actors claim to represent rurality as a whole, fighting symbolic battles in which are put forward a wide variety of social representations of the countryside. These involve a wide variety of actors who produce and manipulate a range of rural imagery: mass media, public administration, farmers associations, ecological movements, developers (Granberg and Kovách (eds.) 1998). While disputes in a specific locality frequently stem from economic diversification and social hetereogenity, the symbolic conflicts which occur in national and international settings also embrace identities and political topics. As a result, rurality acquires an increasing ideological significance since social actors compete to define and appropriate it: as a sign of a superior class lifestyle (middle-class), as a word that covers up corporatist interests (farmer associations), as a definition that legitimates specific policies (planning), as an image that validates the authenticity and natural origin of a wide variety of products in mass media advertising (food), as a source of social identities (locals vs. foreigners), etc.

Farmers in the Shifting Rurality

The Spanish peasantry formed a heterogeneous social group composed of day-wage labourers, small tenants, and owners, who had in common agricultural activity as a specific way of living as well as a similar position in the rural social structure. While it brought together people in different conditions in regard to landed property, all of them could be characterised as living from an insufficient agriculture. The Spanish position in the Green Ring as well as in the European division of labour has determined the transition from peasant to farmer. In effect, at the time as agrarian modernisation took place, the industrialisation of farming became polarised. On one hand, intensive farming (greenhouses, hydroponic crops) dependent on the international agri-industrial complex and oriented to the world food market is located mostly in the south and mediterranean coastal regions. On the other hand, cereal and fodder farms (most located in the inland regions), as well as the small cattle breeding farms in the mountain areas, survive because of state protectionist policies.

In this context, the former peasant masses have given way to a new agricultural workforce involved in agri-industrial complexes and small family farming. At the time as Spanish rural society is changing from an agrarian rurality to a shifting one, these new peasants became the remnants of the earlier peasantry as an historical subject. First of all, family farming has undergone a wide process of de-agrarianisation (Hervieu 1996, González y Gómez 1997). The former mono-agrarian family has been replaced by the individual farmer, in a new context of widespread pluriactive family strategies in rural areas. This is illustrated by the proportion of farms which employ only one family member, which reached 42 percent in 1998 (Gómez et al. 1999). The same authors show that only 52 percent of households live essentially from farming. However, two other processes have become relevant in relation to this fact. On one hand, the crisis in family farm succession (65 percent of farmers declare that they have no heir) has brought about, in inland regions, a structure of elderly farmers and an agriculture which is conceived of as a source of complementary income, and which deviates from market guidelines. On the other hand, as a result of the spread of pluriactivity, an increasing urban from of settlement of farmers is evident (20 percent of active farmers lived in towns of more than 20,000 inhabitants in 1991 [Camarero 1996]).

In regard to intensive farming as agri-industry, the nature of its labour market favoured the recruitment of a female rural workforce (in handing, packing, production) and immigrant workers (mainly as day-wage labourers and pickers). In this way a new social division of labour on the basis of ethnic and gender is developing in a highly de-regulated and casualised labour market (Berlan 1987), in which the former day-labourer (*jornalero*) is being replaced by the new immigrant worker.

However, while older features of the peasantry are dissolved in these itinerant and pluriactive practices, the cultural references of peasant society remain in the villages. Thus, local folk and ritual events (dances, pilgrimages, feasts, festivals) as well as sociability practices are reinforced even in those localities in which agriculture has become a secondary activity. At the time when the countryside has lost its agrarian centrality, peasant cultural signs are increasingly revitalized. Thus, in spite of the dramatic rural exodus of some decades earlier, people have been able to continue to live in Spanish rural settlements in a context of accelerated de-agrarianisation. Rural residents have become involved in a wide variety of commuting and pluriactive strategies and, many areas have benefited from the arrival of retired migrants and seasonal settlers. Inasmuch as peasant cultural institutions are appreaciated as sources of collective identity, the countryside is evaluated by new standards now. In this sense, 'the rural' becomes a sign for general consumption and the re-interpretation of peasant culture a key issue in the production of rurality. Thus, peasant culture could be assumed by rural people as a matter of distinction, as signifying (re)construction of local identity and as a resource demanded by postmaterial consumers.

Re-peasantisation

Debate on re-peasantisation and on initiatives for rural development is of particular interest in the Spanish case. On one hand, the revalorisation of rural areas provides new opportunities for the new farmer and small scale producer linked to local resources and identities. Undoubtedly, these farmers will be a key element in processes of production and consumption of rurality. It would not be surprising to see a process of re-ruralisation of agriculture oriented to these issues in coming years – that is, a return to farming as a form of compromise with rural places and populations.

Intensive agriculture (as a world-market-oriented activity more and more uprooted from local places and peoples) follows a different path, and has a much wider social and economic impact. Here, immigrants replace the deprived local peasantry in harvest labour markets. This workforce (politically 'invisible') is part of the modernisation of Mediterranean agriculture but they do not perceive themselves as future peasants since very often they come from urban areas and consider their job as a temporary strategy.

References

Aceves, J. and W. Douglass (eds, 1974), *The Changing Faces of Rural Spain*, John Wiley, New York.

Baron, E. (1971), *El final del campesinado*, Zero, Madrid.

Berlan, J. (1987), 'La agricultura 'mediterránea' y el mercado de trabajo: ¿una California para Europa?', *Agricultura y Sociedad*, 42, pp. 233-245.

Bradley, T. and P. Lowe (1984), *Locality and Rurality*, London, Norwich.

Brandes, S. (1975), *Migration, Kinship and Community: tradition and transition in a Spanish village*, Academic Press, New York.

Caballero, F. (1864), *Fomento de la población rural*, Imprenta Nacional, Madrid.

Camarero, L. (1992), *Del éxodo rural al éxodo urbano: ocaso y renacimiento de los asentamientos rurales en España*, Ministerio de Agricultura, Pesca y Alimentación, Madrid.

Camarero, L. (1996), '*El mundo rural en la era del ciberespacio: apuntes de Sociología Rural*', in García de León, M. (ed), El campo y la ciudad, Ministerio de Agricultura, Pesca y Alimentación, Madrid, pp. 125-151.

Camarero, L. F. Rodriguez and J.I. Vicente-Mazariegos (1993), 'Los campos de la conflictividad en la España rural', *Documentación Social*, vol. 90, pp. 181-95.

Camarrero, L., M.R. Sampedro and J.I. Vicente-Mazariegos (1991), *Mujer y ruralidad. El círculo quebrado*, Ministerio de Asuntos Sociales, Madrid.

Carrión, P. (1932), *Los latifundios en España, Gráficas Reunidas*, Madrid.

Castillo, J.J. (1979), *Propietarios muy pobres, Ministerio de Agricultura*, Pesca y Alimentación, Madrid.

Cloke, P. and N. Thrift (1987), 'Intra-Class Conflict in Rural Areas', *Journal of Rural Studies*, vol. 3 (4), pp. 321-33.

Commins, P. (1990), 'Restructuring Agriculture in Advanced Societies: Transformation, Crisis and Responses', in T. Marsden et al. (eds), *Rural Restructuring*, David Fulton, London, pp. 45-61.

Costa, J. (1901), *Oligarquía y caciquismo*, Fortanet, Madrid.

De Miguel, J. and M. Moyer (1979), Sociology in Spain, *Current Sociology*, vol. 27 (1).

Diaz del Moral, J. (1929), *Historia de las agitaciones campesinas andaluzas*, Gráfica Universal, Madrid.

Etxezarreta, M. (1977), *El caserío vasco*, Elexpuru, Bilbao.

Etxezarreta, M. (1992), 'Transformation of the labour system and work processes in a rapidly modernizing agriculture: the evolving case of Spain', in T. Marsden and S. Whatmore (eds), *Labour and Locality*, David Fulton, London, pp. 44-67.

European Commision (1988), *The future of rural society*, Office for Official Publications of the European Community, Luxembourg.

European Commision (1995), *Europa 2000+. Cooperación para la ordenación del territorio europeo*, Office for Official Publications of the European Community, Luxembourg.

FOESSA (1966), *Informe sociológico sobre la situación social de España*, 1970, Euramérica, Madrid.

Florez, A. (1839), *La cuestión social. Origen, latitud y efectos del derecho de propiedad*, Miguel de Burgos, Madrid.

Giner, S. (1977), 'The Latifundio as a Local Mode of Class Domination: The Spanish Case', *Iberian Studies*, vol. 1 (2), pp. 47-58.

Gómez, C., J.J. González and R. Sancho (1999), *Identidad y profesión en la agricultura familiar*, Centro de Investigaciones Sociológicas, Madrid.

González, J.J. and C. Gómez (1997), 'Clases agrarias, estrategias familiares y mercado de trabajo', in Gómez, C. and J.J. González (eds), *Agriculura y sociedad en la España contemporánea*, Centro de Investigaciones Sociológicas-MAPA, Madrid.

Granberg, L. and I. Kovách (eds. 1998), *Actors on the changing European Countryside*, Institute for Political Science, Budapest.

Harvey, D. (1990), *The condition of postmodernity*, Blackwell, Oxford.

Hervieu, B. (1994), *The champs du future*, Boulin, Paris.

Hoggart, K., H. Buller and R. Black (1995), *Rural Europe. Identity and change*, Arnold, London.

Instituto de Reformas Sociales (1919), *Información sobre el problema agrario en la provincia de Córdoba*, Minuesa de los Ríos, Madrid.

Jollivet, M and N. Mathieu (1989), *Du rural à l'environnement*, ARF Editions-L'Harmattan, Paris.

Kayser, B. (1990), *La renaissance rurale. Sociologie des campagnes du monde occidental*, Armand Colin, Paris.

Lash, S. and Urry, J. (1987), *The End of Organized Capitalism*, Polity Press, Cambridge.

Lash, S. and Urry, J. (1994), *Economies of signs and space*, Routledge, London.

López-Tuero, F. (1905), *Sociología agrícola*, Ricardo Fe, Madrid.

Marsden, T., P. Lowe and S. Whatmore (eds, 1990), *Rural Restructuring*, David Fulton, London.

Marsden, T., J. Murdoch, Ph. Lowe, R. Munton and A. Flynn (1993), *Constructig the countryside*, University College London Press London.

Morin, E. (1973), 'L'Ecologie de la civilisation technicienne', in *Une nouvelle civilisation, Hommage à Georges Friedman*, Gallimard, Paris.

Mormont, M. (1983), 'The emergence of rural struggles and their ideological effects', International Journal of Urban and Regional Research, vol. 7, pp. 559-575.

Mormont, M. (1991), 'Who is Rural? or, How to be Rural: Towards a Sociology of the Rural', in T. Marsden et al. (eds), *Rural Reestructuring*, David Fulton, London, pp. 21-44.

Naredo, M. (1971), *La evolución de la agricultura en España*, Estela, Barcelona.

Newby, H. (1980), 'Rural Sociology', *Current Sociology*, vol. 28 (1).

Newby, H. and Sevilla-Guzman, E. (1983), *Introducción a la sociología rural*, Alianza, Madrid.

Oliva, J. (1995), *Mercados de trabajo y reestructuración rural*, Ministerio de Agricultura, Pesca y Alimentación, Madrid.

Orti, A. (1992), 'Una visión histórica generalista de la sociología agraria en España: las tres modernizaciones del desarrollo capitalista', *Revista de Estudios Agro-Sociales*, vol. 16, pp. 231-280.

Peck, J. (1994), 'Regulating Labour: The Social Regulation and reproduction of Local Labour-Markets', in A. Amin and N. Thrift (eds), *Globalization, institutions, and regional development in Europe*, Oxford University Press, Oxford, pp. 147-77.

Pedreño, A. (1999), *Del Jornalero Agrícola al Obrero de las Factorías Vegetales*, Ministerio de Agricultura, Pesca y Alimentación, Madrid.

Perez Diaz, V. (1972), *Estructura social del campo y éxodo rural*, Técnos, Madrid.

R.A.L. (1992), *Diccionario de lengua española*, Real Academia Española-Espasa, Madrid.

Sampedro, M.R. (1996), *Género y ruralidad. Las mujeres ante el reto de la desagrarización*, Ministerio de Trabajo y Asuntos Sociales, Madrid.

Saraceno, E. (1994), 'Recent Trends in Rural Development and their conceptualisation', *Journal of Rural Studies*, vol. 10 (4), pp. 321-330.

Sevilla-Guzman, E. (1979), *La evolución del campesinado en España, Península*, Barcelona.

Urry, J. (1984), 'Capitalist Reestructuring, Recomposition and the Regions', in Bradley and P. Lowe (eds), *Locality and Rurality*, London, Norwich, pp. 45-64.

Vicente-Mazariegos, J. I. (1993), *'Prólogo'*, in L. Camarero, Del éxodo rural y del éxodo urbano, Ministerio de Agricultura, Pesca y Alimentación, Madrid, pp. 3-11.

Vicente-Mazariegos, J. I. (1991), 'Presentación', *Política y Sociedad*, vol. 8-9 (monographic issues 'Las trayectorias de la ruralidad en la sociedad itinerante').

Vicente-Mazariegos, J.I., L.A. Camarero, M.R. Sampedro and F. Porto (1991-1994), *Situación socioprofesional de la mujer en la agricultura*, IV vols., Ministerio de Agricultura, Pesca y Alimentación, Madrid.

10 Portugal: The Emergence of the 'Rural Question'

ISABEL RODRIGO AND MANUEL BELO MOREIRA

Introduction

Before approaching the issues of agriculture and rural life it is necessary, particularly for the reader less familiar with Portuguese society, to point out a number of aspects concerning the agrarian structure and the overall socio-economic and political context.

Portuguese agriculture was already quite opened up to foreign markets in the 18th century, when agricultural commodities, like wine, cattle, fruits and cork were exported on a permanent basis to the UK, and other northern European countries (Vaquinhas and Neto 1993 and Pereira 1971).

This commercialised and internationalised agriculture was based on agrarian structures marked by a strong structural dualism. In the northern part of the country, and in the Algarve (south), a system based on tenants and sharecroppers prevailed, even if some larger landowners directly cultivated their most fertile land, employing 'free' day workers. This structure was complemented by tiny landowners producing on the basis of the household labour force and essentially for subsistence. The small size of their holding forced a great number of these farm operators to rent more land or to sell their labour to the larger landowners. In the rest of the south (Alentejo) the *latifundia* system predominated, using hired labour, mostly seasonal workers employed on a daily wage basis, and practising quite extensive market-oriented agricultural activities. In this *latifundia* landscape some landowners found it more profitable to lease some plots to small tenants or sharecroppers, while keeping the prevalence of the very large farms.

This kind of structural dualism survived until today, even if there were important changes in the ways landowners related to tenants, sharecroppers and workers, in the areas of both the family farm and the *latifundia*. During the last century the social and structural basis of family farming in the northern part and in Algarve was consolidated. Land rent and sharecropping arrangements, at first based on payments in kind, were progressively monetarised. In Alentejo, after very tough struggles, hired workers obtained better working conditions, and the importance of the seasonal labour force dropped drastically with farm mechanisation and an increased agricultural exodus.

In the beginning of the 1960s massive rural exodus accelerated the changes that led to the diminishing of the former central place that farming held in rural settlements. The consequences of this de-agrarisation process (Oliva and Camarero in this publication) are quite evident, marking the distinctive types of Portuguese rurality. In areas where family farming predominates, the traditionally very complex mixed crop farming systems became simplified in a movement towards specialisation, even if fully specialised farms are still in a minority. This slow process towards farming system specialisation was accompanied by an increase in mechanisation and deepening integration into markets. Some family holdings with enough land adopted the modernisation process as full-time farmers, while others engaged in diversified off-farm strategies.

This process of modernisation of the family farms, identified in this publication as de-peasantisation, is distinctive for its heterogeneous character, as will be described later on. Concerning the issue of de-peasantisation, it is worth noting that the debate among Portuguese historians and social scientists about the existence of peasants in Portugal is so controversial that the word is not listed by the renowned Portuguese History Dictionary (Serrão 1971). In fact, in Portugal the word peasant (*camponês*) is used essentially in two senses: in the social sciences it is used as equivalent to family farmer; in common speech and writing it is used in a broader sense, as one who works and lives in the countryside. In both meanings peasant refers to a social category characterised by a particular way of life based on performing small-scale agriculture and using traditional agricultural skills and knowledge. Nowadays, Portuguese family farmer operators do not use the word peasant to refer to themselves. Rather they call themselves peasant farmer (*lavrador*) – specially the oldest stratum of farmholders; farmer or agricultural producer. In ordinary speech, the word peasant is used as a synonym for traditional rural life style, and in political discourse it was banished in favour of farmer.

In this text we use the concept of family farm to refer to farms operated exclusively or essentially by family labour regardless the ownership of land. Thus, when we refer to de-peasantisation we consider it as the equivalent of the structural transformation of family farms towards modernisation.

In addition to these brief comments, there are other features relevant to understanding the general tendencies of change in Portuguese agriculture and rural life. These features are described in the next section and concern the most important changes in the political situation. From 1926 until the revolution on the 25th of April 1974, Portugal was under the rule of a fascist-corporatist dictatorship. The period immediately after was characterised by a great political turmoil, which began to fade by 1976. The period from 1976 to 1986 can be designated as the period of democratic transition, which finished when Portugal entered the EEC. From then to the present, the political situation has stabilised opening up a phase that we can describe as the integration period.

The chapter starts with a brief overview of this changing socio-economic context. The next section focuses on the generic trends of the evolution of Portuguese agriculture, setting out the main features of the de-peasantisation process and the emergence of the rural issue. As will be shown, in Portuguese society this issue is emerging separately from the agrarian question. Finally, a brief conclusion will summarise the relevant trends and features of Portuguese agriculture and rural society.

The Changing Socio-Economic Context

The Period of Dictatorship

From a political and economic point of view, this period (1926-1974) can be characterised by a number of features. For 48 years Portugal lived under a strong dictatorship. The regime had 'preserved' the country from economic, political and cultural external influence, having been very successful in preventing any attempt at democratisation. Internally, 'social peace' was maintained through the skilful use of measured repression, involving not only brutal force, but also a surgical selective repression, and tight control of the media.[1] It was also notable for its ability to manage contradictory economic interests so as to prevent overall business opposition. Portuguese society was also strongly influenced by the political and ideological hegemony exercised by the 'lords of the land' of the southern *latifundiary* regions, located in Alentejo, along with the very conservative Catholic Church, and local clusters of larger landowner 'notables' in the north.

The Portuguese corporative regime was characterised by a strong nationalism. The wish to develop the Portuguese economy without any outside influence was expressed in refusal to adhere to the Marshall Plan after the Second World War. In fact, at the economic level Portugal lived in deep economic isolation until the 1960s. Portuguese economic isolation only started to open up when Portugal joined the European Free Trade Association (EFTA) following the UK leadership of this international organisation. However, until the end of the regime (1974) Portugal lived under a situation that could best be described as a corporative economy where the market was strictly controlled by the political level, that is, the corporative economy over-regulated the market, not only protecting it from international competition, but also avoiding internal competition amongst Portuguese capitalists. On the other hand, capitalist accumulation was facilitated by the abundance of cheap labour, since independent labour unions and labour movements were strongly repressed. Labour costs only started to rise, especially for some categories of workers, towards the end of the regime. This was due to shortages provoked by the military mobilisation, that put almost half a million men in arms in the colonial wars (in Angola, Mozambique and Guiné Bissau), and particularly by the emigration flow from rural villages to the most industrialised countries in northern and central Europe, mainly Germany and France.

Politically, when the allied and democratic forces won the Second World War, the dictatorship took advantage of its strong anti-communism and, despite its lack of democracy, escaped a full political isolation, participating in NATO, and allowing a permanent USA aerial base in the Azores. However, increasing independence movements in Africa and the stubborn refusal of the regime to consider the possibility of granting independence to the Portuguese colonies caused the political isolation of the regime. This political attitude was responsible for not only a number of United Nations resolutions condemning Portugal, but also led to the colonial wars which lasted for 13 years. These colonial wars were the basis of the military unrest that led to the bloodless revolution known as the 'Carnation Revolution', started on 25th April 1974.

The Revolutionary Period

This period (1974-1976), characterised by enormous social, political and economic turmoil, paved the way for a number of very relevant changes that should be underlined. A very short description focusing on rural issues must include the following aspects. At the broad political level the democratisation of social life must be stressed, i.e. freedom of association and of speech, the end of censorship of the media, and in particular the ending of state repression of labour movements and popular initiatives. This gave rise not only to the creation of free labour unions and several political parties, but also to new forms of democratic participation by popular movements gathering together those eager to take advantage of the new freedom and democracy to participate in political experiences, and in various forms of more spontaneous citizenship mobilisation.

From a more strictly economic point of view, we underline the dismantling of some of the main institutions of the corporatist regime, and the emergence of new forms of state intervention in the management and/or nationalisation of key economic sectors like banks, insurance companies, and major industries. Finally, the changes that occurred in agriculture and rural life mirrored Portuguese agrarian structural dualism.

In the northern part of the country, dominated by small holdings operated by a household labour force and still relying upon traditional low external input farming systems, the agrarian structure kept its shape intact. Although the larger landowners' economic power was not affected, their political power and influence began to be eroded as new actors gained ascendency in local institutions.

In the Alentejo *latifundia* the rise and fall of the agrarian reform must be emphasised. During this revolutionary period about 1,100,000 hectares of land were occupied by landless day labourers and seasonal workers as well as *seareiros* (a social group who have contracts with landowners for the use of land on an annual basis, land use being paid for either through rent or on a sharecropping basis). However, the dissolution of *latifundia* ownership did not give way to a 'peasantisation process'. Rather, it gave rise to collective forms of production and social organisation, ruled and operated mainly by landless labourers and households for whom to be sure of a permanent income source was more important than to farm a piece of land which they owned. In order to better understand the outcome of the Alentejo land reform, which mirrored the goals of the landless households, one must remember that, firstly, the latifundia system had structured the economy of the region since the previous century and, secondly, that this agrarian economy based upon extensive farming systems was generating diminishing labour force demands.

This last aspect becomes particularly relevant when one takes into account the overall context of economic crisis, either external or internal. More precisely, at the external level the economic recession after the 1973 oil crisis had important consequences for Portugal; it contributed to a decline in exports, aggravating the commercial imbalance, and not only stopped traditional outflows of emigration but also put pressure on the emigrants to return to Portugal. At the internal level, the industrial recession reduced demand for industrial labour, which, along with the return of colonial settlers and soldiers from the colonial wars, greatly increased the unemployed labour force, thus creating strong social pressure for the appropriation of extensive or almost abandoned *latifundia* (Moreira 1989 and 1991).

With the decline of revolutionary political power a counter agrarian reform took place and almost all occupied *latifundia* were returned to their original owners. Nowadays only some very few examples of collective farms survive, in cases were the land was acquired by the collective of workers or where they succeeded in making long term rental contracts.

The Transition Period: 1976-1986

This period can be characterised as a period of change and continuity. Continuity because if at the political level the corporatist regime has been dismantled, at the economic level the country continued the slow pace to a complete opening up. In fact the role of the state was probably even greater than before. It maintained a tight import-export trade control, allowing outward investment only on a case-by-case assessment. State monopolies or state-controlled monopolies were maintained in certain commercial networks and economic sectors.

But this period also experienced a number of elements of change. The society calmed down after the turmoil of the revolutionary period, and a new civil society began to emerge and to become able to express its presence.[2] Even if the political situation continued to show great instability, with governments succeeding one after the other at a very fast rate, the democratic institutions were functioning.[3] And the economy, in part as the result of the usual impositions of the International Monetary Fund (IMF), was under tight monetary control, preparing to move toward liberalisation and the achievement of a complete opening up.

It was in this period that Portugal submitted her bid to join the EEC. After a period of several years of negotiations Portugal was finally accepted as a full member in January 1986. Thus, a number of changes in the institutional structures were made during this period in preparation for accession, and Portugal started to receive pre-accession funds.

The Integration Period

From EEC accession up to 1998, when Portugal concentrated its efforts to be included in the countries of the Euro Club of the European Union (EU), the following features stand out.[4] Politically, after 1986 the situation stabilised, starting a long period under the majority rule of one political party, the PSD (Social Democratic Party that, despite its name, is not a member of the socialist or social democratic group in the European Parliament, but is included instead in the Popular group). In 1995 the PSD was defeated, and the Socialist Party (in fact the Portuguese social democrats) won the general election without a majority, but with a sufficient percentage of votes to rule with stability.

At the more strictly economic level the movement towards liberalisation and privatisation went into overdrive, and the overall economy benefited largely from EEC integration. A large amount of EEC structural funds were put at the disposition of the Portuguese authorities to invest in infrastructures, in the modernisation of the productive sectors, and in technical education and training of 'human capital'.

It is not the aim of this chapter to evaluate the strengths and weaknesses of the way that the government and the economy took advantage of these huge incentives. But it is certainly necessary to stress that a lot changed during this period, and that the Portuguese situation in 1999 is substantially different from that in 1986.

The Decline of Agriculture

Let us now draw a brief outline of the main features of the last 25 years of Portuguese agriculture. Agriculture lost a large proportion of its total active population, which dropped from 28.6 percent in 1980 to 13.3 percent in 1997 (ECC 1999:T/116).[5] This demographic decline, which started at the end of the fifties with the rise of emigration, brought about substantial changes in farming systems, which had to adapt to the decrease in agricultural labour force supply.

The adaptation of the existing farming systems was achieved following two distinct patterns: an intensification process which replaced labour by capital – more machinery and more intermediate consumption – and extensification of many areas, that is, practising farming systems which need less labour and capital, increasing the pace of replacement of cultivated land by forestry, and in many cases a clear abandonment of productive activities. The overall result of this contradictory movement was stagnation of agricultural value added, that lasted for a large period of time (Moreira 1986).

Another important feature of the agricultural sector is the ageing of the farm population. In 1997, only 8 percent of farmers were less than 40 years old, whilst the oldest category (> = 65 years) constituted almost 38 percent. But the problems with 'human capital' go beyond age. In fact, in 1997 the rate of illiteracy of the agricultural labour force was still almost 19 percent, a figure that rises to 35.6 percent if we include the people who did not complete primary schooling (INE 1995[a]). On the other hand, despite the increase in agricultural training courses, which are intended to support modernisation and professionalisation processes and whose number grew by 108 percent between 1989 and 1993, only 12 percent of farmers under 40 years old attended such training programmes.[6] Since the ageing of the agricultural labour force and its lack of educational and technical skills reduce the opportunities to implement and take advantage of technological improvements, Portuguese agriculture, particularly when compared with the industrialised countries, appears technologically backward. In addition, it is characterised by a long tradition of a strongly unbalanced productive structure. As already mentioned, structural dualism and a bifurcated land ownership pattern still persist. Structural dualism can be illustrated by the following facts:

- In the southern regions of Ribatejo and Alentejo, along with capitalist farms, modern and capitalised, it is still possible to find very large estates using extensive farming systems framed by a typical *latifundist* economic rationality.[7]
- On the other hand, in the northern areas and in the Algarve, intensive farming systems practised by small family farms predominate. The situation is known as *minifundia* but in many cases is more adequately described as *microfundia*. The income generated by this small-scale agriculture is not sufficient to make a living, thus requiring off-farm income – either from off-farm labour (industrial or service work or participating in temporary emigration flows abroad), or from social security pensions, or based in past savings. In 1997, only 17.8 percent of farmers worked full-time at their farms and only 9.4 percent could live exclusively from farm income (INE 1999[a]).

It is very important to underline the crucial importance of the family *minifundia* and *microfundia*. In reality, besides their social importance illustrated by the farm population, those small landholdings are also crucial from an economic point of view, as recent statistical data indicate. In 1979 family farming represented about 50 percent of the UAA (Utilised Agriculture Area) and 67 percent of the Standard Gross Margin (SGM) (Baptista 1993, 111). Data from 1993 suggest that its importance is increasing in the Portuguese Mainland.[8] Today, family farms account for 96 percent of the total number of farm holdings and 97 percent of the farm population. They use 60 percent of the total UAA and contribute 73 percent to the SGM (Table 10.1).

Coexisting with the structural dualism mentioned above is a strong regional imbalance, illustrated by the importance of mountainous and less favoured areas which represent 75 percent of the total Mainland area. In contrast to the richest coastal area, with its larger amounts of fertile and irrigated soils, the hilly or flat interior landscape lacks fertile agricultural land, and dry farming prevails. In addition, communication and transport networks are worse here, consumer centres are far away, and economic life is clearly less dynamic.

Great hopes for change in the structural and regional imbalances were pinned on the EEC aids for modernisation and restructuring of Portuguese agriculture and to help mountainous and less favoured areas. However, in spite of all the funds directed at Portuguese agriculture, before and after Portugal's accession, through both EEC general regulations and through the PEDAP (Specific Programme for Portuguese Agriculture), their investment did not show any significant structural improvement.

Furthermore, investments made under EEC Reg° 797/85 and 2328/91 particularly benefited larger farms, those with an average size (66.6 ha) seven times higher than the Portuguese Mainland average size (9.4 ha). A more detailed analysis shows a diversified situation. On one hand, considerable amounts were invested in certain areas, and in certain types of farms and production systems (mainly viticulture, dairy, some irrigated productions, and forestry) but, on the other hand, this was counteracted by a decrease or even a complete absence of investment in other areas, farms and lines of production, as well as a substantial increase of abandoned land. This explains the stagnation of investment outcomes overall and indicates that, in many cases, significant amounts of Community funds channelled into the modernisation and restructuring of Portuguese agriculture substituted for national investment instead of supplementing it (Moreira 1996). In order to explain these trends one must look at diverse aspects, among them the following: (i) the already mentioned rentist and risk minimisation character of a number of latifundists; (ii) the poor results of previous attempts to modernise by some family farmers facing poor structural conditions; (iii) lack of technical guidance to select the most appropriate investments, and last but not least, (iv) the macro economic policies implemented by the state at the beginning of the nineties, specially the over-valuation of the Portuguese currency, that have strongly affected farmers' incomes (Avillez 1994).

Moreover, during the first transitional phase of EEC accession (1986-1991) much investment was aimed at developing intensive technological models, similar to those promoted by the Common Agricultural Policy (CAP) in other European countries. In many cases these intensive models are not suited to Portuguese ecological and structural conditions, so their adoption has raised doubts about its perspective on competitiveness.

In fact, under climatic and geological conditions that put pressure on energy needs, mainly for irrigation purposes, the adoption of these kinds of technological patterns in a country with poor energy resources, and with domestic energy costs higher than in the most European countries, will lead to a failure of competitiveness. Thus, in the new context of greater liberalisation and more competitive pressure, resulting from CAP reform and GATT agreements, many of these intensive models emerge as inadequate. Then, it is no surprise that many farmers, in particular those who have recently taken over their farms and whose investments are at the point of reaching maturity, face a terrible dilemma. They are forced to recognise that they do not have a future with the adopted technological model, and they are not emotionally or technically equipped to look for alternative strategies.

Finally, we should note that the most dynamic sector of Portuguese agriculture, i.e. animal production, in particular poultry, dairy and pork, is heavily dependent on feed imports.[9] This indicates that important sectors of Portuguese agriculture are following a path of de-territorialisation, since they do not depend on the Portuguese territorial base to continue with production. This phenomenon has a strong impact on rural life since the growing importance of these most intensive and larger production units contributes to the abandonment of production by the small, less intensive ones.

It is not surprising, when we consider this pessimistic productive situation, and given the growth of food consumption resulting from the increased purchasing power of the Portuguese population and the growth of tourism, that food dependency is increasing. In fact, up to 1986 Portuguese agriculture lived under the protection of barriers to foreign competition, producing mainly for the domestic market; but with European integration, and particularly after the end of the first transition period in 1991, the impact of competition is clear.

Between 1980 and 1985, domestic production already fell short of total consumption in grain cereals and oil fruits and seeds (average percentages below 40 percent), but domestic production in potatoes (95 percent), pork, poultry, milk and dairy products showed only a slight shortfall (percentage values around 100 percent) (INE 1994). The latest figures for the period 1995-1997 indicate that Portugal remains strongly in deficit in grain cereals and oil fruits and seeds and that some other products have also fallen off considerably, with the exception of milk and dairy products and poultry (INE 1999).[10]

In short, Portuguese agriculture, despite the already mentioned technological changes, and the increase in the use of largely imported intermediate consumption,[11] is still low-productive when compared with other EU countries.[12] Portuguese agriculture is not able to satisfy domestic needs,[13] and does not show any exporting capacity, with the exception of sub-sectors producing traditional exports like wine, vegetables, and forestry products such as cork and cellulose. Nevertheless it is a sector with very great social importance when one includes, along with the active farm population, the pluriactive population and that part of the non-agricultural active population that lives on farms while working off-farm. Its articulation with industrial activities like the clothing and footwear sectors is particularly important, and has made it traditionally a shock-absorber for socio-economic conjunctural crises.

The most striking conclusion from the above analysis is that, in spite of the dynamism and competitiveness of a few isolated farmers and agro-food industries, it does not sufficiently cover domestic consumption needs, thus facilitating penetration of the domestic market by imports. Even more worrying is that it is not easy to detect signs that the dynamics of Portuguese agriculture are changing substantially. This fragility of domestic production is not restricted to agriculture but is spreading to agro-industry and marketing agents.

This continuous crisis of Portuguese agriculture is connected, on one hand, to the absence of realistic technical models. On the other hand, one must stress the lack of clear strategies for the sector defined by the state,[14] as well as the lack of dynamism among producers, which is sufficient proof of the weakness of agricultural associations.

Portuguese Agriculture and Rural Society: An Overview

As mentioned above Portuguese agriculture is largely a matter of family farms. However, even if we focus on the family farming sector it is necessary to briefly characterise the other types of farms. Capitalist farming can be divided into different categories: societies or corporations, and individual capitalist farmers.

In 1993, individual capitalist farms accounted for 3 percent of the farm population, used 25 percent of the total UAA, contributed with 16 percent to the SGM and accounted for 3 percent of the total number of farm holdings. Other types of farms (i.e., not individually owned or managed, like societies or corporations, co-operatives, and commons) accounted for 1 percent of the total number of farm holdings, used 15 percent of the total UAA and contributed 11 percent to SGM (Baptista 1997).[15]

Individual capitalists show different dynamics. On one hand, there are those who think of themselves as modern entrepreneurs, look for investment opportunities, and do not avoid taking risks. On the other hand are those who try to avoid risk, relying on rentist behaviour. Rentist behaviour refers to situations where landowners take advantage of their large estates, renting part of the land to small operators, essentially on a seasonal basis, and managing some very extensive production by themselves, especially those that are supported by Common Agricultural Policy subsidies. The rents and the income from agricultural subsidies are complemented by the incomes obtained from strategies of production at minimum cost, which provide relatively low yields per hectare, but multiplied by a great number of hectares. Thus their scale allows them to obtain important annual incomes that are enough to justify their rentist economic rationality. This rationality is one of the main causes of the relatively low productivity levels on capitalist farms: as already mentioned, although using a quarter of the total UAA they only contribute 16 percent to overall SGM.

Finally there are also a small number of co-operatives that constitute a relict of the agrarian reform and resist change with difficulty; a small number of state properties, some waiting to be privatised; and the commons, mainly located in the North and used for forestry production.

Returning to the crucial family farming sector, we wish to stress that in order to understand the social and economic dynamics of Portuguese rural society the family farming sector and its internal differences must be taken into account. Amongst family farmers, those who depend mostly on income derived from outside farming are particularly important. In the next section the heterogeneous character of Portuguese family farming will be analysed from both economic and social points of view. In the ensuing discussion of the main characteristics of the rural society, it will be argued that rural Portugal is highly diversified.

The Family Farming Sector

The heterogeneous character of Portuguese family farming is identified here on the basis of the main household income source. Following this criterion, three types of family farmers are identified: (i) those for whom farming is the principal source of family income; (ii) those who depend mainly on income derived from wages earned by household members, mostly in the industrial and service sectors; (iii) those who depend mainly on off-farm income which does not originate in employment, such as: social security payments and/or emigrants remittances (Table 10.1).

This criterion highlights different kinds of relationship between family farming and the overall society. More precisely, accordingly to the above family farming typology the main links between family farm households and society are: agricultural markets; non-agricultural labour markets, and welfare income sources. The main economic and social features of each type of family farming will be discussed in the next section.

Farmers Living from Farming

Family farmers living from income exclusively or principally generated on their farms account for one third of the total number of holdings and use 38 percent of the total UAA. These farmers cultivate the largest landholdings among Portuguese family farms (an average UAA per holding is 9.2 ha, which is quite small in the European context) (Table 10.1). Since Portugal joined the EEC in 1986, there have been particular changes and an economic and social renewal process in this type of family farming. In order to take advantage of Community Regulations agreed between Portugal and the EEC for the transition period (1986-1991), in particular those oriented to financial support for modernisation of agricultural structures through an increased installation of young farmers, some family farm operators retired in favour of their sons.

Farm succession processes have been studied particularly in the largest family-owned landholdings, located in the northern coastal area. These cases present a productive structure that had already benefited from capital investments made by the retired farm managers who also had already adopted, although not excessively, the agricultural modernisation process, through farming system specialisation (Rodrigo 1992 and Santos 1996). The requirements demanded by the potential heirs for a successful farm succession to take place strongly restricted social and economic renewal of this type of family farming. However, when succession is successful, most of the young farmers have been able to make significant investments mostly channelled to intensive production, allowing them to increase their revenues from farming activities.

Table 10.1 Portuguese Agriculture: Family Farm Types

Principal source of family income	UAA per holding (ha)	No of holdings (%)	UAA (%)	SGM (%)	FP (%)
Farming activity	9.2	34	38	44	36
Off-farm (Total)	2.8	(62)	(22)	(29)	(61)
Wages earned in the agricultural employment sector	2.7	5	2	2	5
Wages earned in the industrial and services employment sectors	2.6	23	7	12	28
Welfare (pensions and retirement)	2.7	25	9	9	18
Entrepreneurial activities	4.6	4	2	3	5
Other off-farm income sources	2.9	5	2	3	5

Source: Adapted from Baptista (1997). The Table is based on statistical data from 1993.

Abbreviations: UAA = Utilised Agricultural Area; SGM = Standard Gross Margin; FP = Farm population, i.e. population living in farms.

These social and economic changes have been accompanied by and are interlinked with cultural transformations, leading to the emergence of a new social profile and identity shared by the young family farmers. By referring to themselves as *professional farmers*, the young farmers benefited from a series of conditions which have facilitated them in achieving their economic and social expectations, and have given material reality to their social identity. Thus, besides the internal conditions of the farm operations described above, there have also been external conditions that have favoured the process of cultural change.

Among these external conditions, the economic incentives and other facilities provided by the EEC funds, and the training activities provided by the national agricultural policy financed by the European Social Fund that provided young farmers with a professional qualification, stand out. In order to gain access to economic and financial EEC aids, applicant farm operators had to attend a specific professional training course, which would entitle them to a certificate of qualification. These training activities provided young farmers with a professional qualification that not only facilitated but also legitimised their social distinctiveness in comparison to other family farm operators. It also allowed young producers to feel fully integrated into the society's dominant values, norms and ideology. The qualification has an important classifier function, as it affects the hierarchies of prestige, as well as income.

The cultural model shared by *professional farmers* clearly marginalises traditional knowledge and practices, and land capital is subordinated to operating capital. Land is thought of as a rational instrument, i.e., an instrument with an economic production utility. To sum up, farming is conceived and lived by *professional farmers* as an entrepreneurial activity, i.e., a way of earning money (Rodrigo 1998).

It must be emphasised that the emergence of these professional farmers is not only a matter of will, know-how and training. Successful installations depend on sufficient owned land. Given the traditional land structure and the absence of land policy in Portugal, we can infer that successful installations of younger farmers are still far from general.

The social, economic and cultural changes we have described are identified by classic sociological literature as processes of agricultural modernisation and professionalisation. On the other hand, the change from farming as a way of life to farming as an way of earning money, which is part of agricultural modernisation, has been described in some sociological studies as 'de-peasantisation' or 'farmerisation', as some contributions in this volume illustrate.

In order to clarify the extent of these processes within Portuguese agriculture and its consequences for rural communities' dynamics we must clarify two points. Firstly, these social, economic and cultural processes of change have occurred in Portuguese agriculture only recently, compared to some other EU countries that had adopted the CAP agricultural model for decades. Secondly, those processes have had a reduced impact, in the agricultural sector as a whole and in the structuring and dynamics of rural society. In fact, one should note that apart from the relatively reduced importance of family farmers whose household source of income is exclusively or principally derived from farming, a restricted subset of these have experienced social, economic and cultural renewal processes.

Thus, notwithstanding efforts at agricultural specialisation and professionalisation, family farmers who depend on income derived from farming continue to practice mainly mixed farming systems, and to run their farms using techniques mainly based on the knowledge and experience of past generations. For the oldest farm holders (aged over 55), who still identify themselves as being *peasant farmers*, farming still means a way of life, in contrast to the young entrepreneurial family farmers.

The *peasant farmers*' resistance to integration into the agricultural modernisation process, i.e., to being involved in the economic and technical dependence that it requires, has prevented them from fulfilling the requirements for successful farm succession. Thus, the cultural pattern of *peasant farmers*, reinforcing the effect of the poor structural conditions, further helps to explain why a great part of them do not have any farm successor.

Farmers with Diversified Sources of Income

Many family farmers depend on income derived from outside farming. This type of family farming accounts for 62 percent of the total, uses 22 percent of the total UAA and contributes 29 percent to SGM. It amounts to 61 percent of the farm population, indicating the social importance of this type of family farming (Table 10.1).

Most of this group (33 percent) work off-farm, 5 percent working on other farms and 28 percent working in other sectors of activity mostly in industry and services. Meanwhile 5 percent have other entrepreneurial activities. A significant proportion receive off-farm income from welfare sources (18 percent), whilst the others (5 percent) have various sources of income (emigrants' remittances and, to a lesser extent, bank interests and rents are the main off-farm household income sources).

It is important to note the range of roles played by these farms in their respective households. They are simultaneously the household's home; a source of monetary income, as these farmers usually sell a significant part of their productions; and a source of food, provided by self-consumed agricultural products. All these farm functions are of the utmost importance in the economy of these households, given the low wage levels in most manufacturing and services activities, particularly to the unskilled labour force, as well as the low rates of retirement and welfare pensions (Barros 1992 and Fragata 1992).

Three aspects of the family farming type dependent on off-farm income sources need further clarification. Firstly, the high rate of off-farm activity results in a significant degree of articulation between agriculture and the other sectors of the economy. In fact, 33 percent of the total farm population is directly involved with other sectors of activity, 28 percent through employment, and 5 percent through entrepreneurial activities. The 23 percent who depend on pensions (18 percent), and other income sources (5 percent) are only indirectly articulated with other sectors of activity; however, the existence of these sources of income certainly contributes to economic activity in rural areas, mainly in services.

Secondly, that articulation is rooted in a complex economic and social network that allows the reproduction of these family farms and, at the same time retains in the countryside important segments of the active labour force.

The third aspect to be mentioned concerns the dominant articulation model through which farming households and the economic system are related. In fact, the Portuguese agricultural model, like that of other southern European countries – see, for instance, Arnalte (1997) for Spain; Fonte (this publication) for Italy – is quite different from that prevailing in most other EU countries. In Portugal, family farmers who mainly rely on off-farm income derived from wages have taken advantage of the diffused spread of industries and services. In addition, the household members (father, sons and daughters) are involved full-time in employment and, by and large, use their free time to farm. Finally, the decision by household members to work off-farm is largely determined by characteristics of the landholding structure, such as low levels of farm capital investment and farm income returns (these farmers do not perform capital intensive farming systems), and a surplus of available family labour *versus* a reduced amount of available cultivated land (the average UAA of these farm holdings is 2.6 ha to 2.7 ha (Table 10.1).

All these characteristics illustrate that Portuguese agriculture sharply contrasts with most other European agricultural models, particularly those whose main trends show that the farm unit is changing from a family type to an individual operator. In the latter, full-time farming tends to be based upon an individual farmer operator, generally male, who practices intensive and specialised farming, and other household members' decisions to look for an off-farm economic activity is determined by their individual career preferences, regardless of farm expectations (Baptista 1997).

To sum up, three major characteristics of Portuguese agriculture are the dominance of family farming, the importance of the agricultural sector in terms of the population that by their work or by their living conditions are connected with farms, and the high rate of off-farm activity. These features, although shared by other southern European agricultures, contrast with many northern European ones, as Eurostat data (1992-93) illustrates.

Using the ratio 'total population who works on farms (including full-time and part-time work)/total population employed in agriculture' as an indicator to express the importance of part-time farming, the Portuguese ratio value is 2.44; 2.21 in Greece, 2.87 in Italy and 2.05 in Spain, while in Holland the ratio value is 1.17; 1.24 in France; 1.14 in the United Kingdom and 1.08 in Germany. Similar discrepancies can be noted when one assesses the importance of the agricultural sector in terms of population. Using the percentage of 'total population employed in agriculture/total population with employment' to express the importance of the farming population, the Portuguese value is 11.5 percent; 21.8 percent in Greece, 7.9 percent in Italy and 10.1 percent in Spain, but only 3.7 percent in Holland; 5.9 percent in France; 2.2 percent in the United Kingdom and 3.7 percent in Germany (Baptista 1997).

The Rural Diversified Society

Simultaneously with the internal differentiation process within family farming, the values and networks of rural society were slowly substituted by new cultural and political configurations (Almeida 1986, Ribeiro 1997, Silva 1998 and Wall 1998). Emigration, the spread of the market economy, and the political democratisation determined the changes.

Emigration had different kinds of impacts. A large part of emigrants' savings were used for land acquisition, or to build a new house with at least a small garden plot. This allowed the more needy and land-dependent domestic groups, who constituted the bulk of emigrants, to rise in the local social scale, reaching the status of landowners. This social mobility process encouraged the disintegration of traditional rural structures that were closely tied to landownership. In addition, the presence of return migrants with their money, professional skills, consumption styles and social success transformed the customs and ways of life of other rural and agricultural actors, and reshaped rural social life and values (Almeida 1986, Ribeiro 1997 and Silva, 1998). Reconfiguration of the cultural system and worldviews was reinforced by the progressive penetration of market economy into rural society, along with popularisation of the new mass media, in particular television.

Finally, the political democratisation process after the 25th April 1974 contributed to progressive disintegration of the corporative organisational forms and the traditional political power (*caciquism*) in rural villages, and introduced a new dynamic into local politics. At the same time that the social mobility process described above occurred, the prestige and power of landlords started to be eroded.

Before April 1974, political power was concentrated in the hands of a few 'notables', who were, by and large, the richest landlords, parish priests and some other individuals benefiting from a social status given by cultural capital, such as teachers and physicians; many of these were also landowners. These individuals, constituting the local political elite, filled the gap between villagers and the State and decoded the messages from one to another. The boundaried society of rural villages, the social status rigidity and very restricted social mobility, and the limited sources of power (economic, political and cultural) open to the villagers also explain the role played by the Catholic Church in imposing a normative world on rural people and controlling their customs.

Dismantling of corporative rule, democratisation and the emergence of the multi-party system gave room to manoeuvre to new actors who replaced the traditional local elite in local political networks. The hegemony of ecclesiastics and landlords weakened at local, regional and national levels. Simultaneously, new alternative channels of power have appeared with the development of farmers associations, co-operatives and trade unions (Silva 1998).

Rural and agricultural exodus, the diversification of non-agricultural off-farm opportunities, the different types of pluriactivity strategies, and the different social opportunities offered by rural areas contributed to the differentiation of rural society. To describe the main characteristics of current types of rurality, we use a descriptive definition of rural, based on population census variables.[16] Rural is here identified with the places where the population live dispersed or in communities of less than 2000 inhabitants.

Acording to the last population census (1991), 89 percent of the active agricultural population lived in rural areas, and rural society accounted for 52 percent of the population residing in the Portuguese Mainland (Baptista 1995). By and large, it is possible to identify two types of rurality, which mirror the importance of family farming within Portuguese agriculture and its internal differences as well. Geographically these ruralities largely coincide with the northern coastal and inland areas and the Alentejo region, in the south.

In the northern coastal area rural dynamics are mainly rooted in the inter-sectorial links between agriculture and the industrial and service sectors which, along with the broadening of transport networks and infrastructures, advances in electrification, and establishment of schools and health centres in small towns and villages have encouraged revitalisation and restructuring of rural society. These structural changes have facilitated rural people's contacts with other social worlds. They have also been responsible for changes in consumption patterns and living standards, ways of life and value systems among the agricultural and rural population, bringing them closer to urban cultural patterns.

Nevertheless, agricultural activity remains important in shaping inhabitants' identification with their place of residence, and this illustrates the relatively recent character of the structural changes which have accompanied diffused industrialisation (mainly since the 1970s) and urbanisation. Identification by inhabitants with their rural community is particularly strong among part-time farmers and older full-time farmers. Part-time farmers do not define themselves as farmers nor as agricultural producers, but as workers; however the waged labour force without links to farming classify farm operators with off-farm activity in the farmers' social group, while full-time farmers refuse to recognise them as farmers (Rodrigo 1992). These contradictions in social identity and identification of family farmers who rely mainly on off-farm income illustrate new features of the rural question in Portugal.

In contrast to the rural vitality of the northern coastal region, the northern inland and the Alentejo region in the south have not developed strong inter-sectorial linkages, which might support rural restructuring. The rural locations in these two areas are marked by a much lower population density than, either, the Mainland average (105 inhabitants/km2 in comparison to 20 in the Alentejo and 52 in the northern interior), or the northern coastal area average (254 inhabitants/km2), according to the last population census (Baptista 1995).

Agriculture still remains the dominant economic activity in these rural spaces, which are affected by agricultural and rural exodus and lack financial, human and knowledge resources and many social services, such as health, education, and transport. Agricultural activity is mainly performed by older farmers and is marginal in terms of the income that it generates. As Fonte (in this publication) describes for the south of Italy, in these Portuguese rural communities also it is not counter-urbanisation but 'return-migration' and the links that emigrants keep with their land of origin, in addition to welfare payments, that offer a glimpse of revitalisation.

The future of these rural territories might be sought in measures to conserve and valorise local techniques and varieties, trying to preserve the existing resources and to improve the population's living conditions, as Fonte (this publication) suggests in relation to the south of Italy. However, in some rural communities implementation of such practices appears difficult. This is particularly clear in rural territories located far from those small villages which have some economic and social vitality, with very low population densities and lacking capital, institutional and human resources. Small rural communities with these types of characteristics generally exhibit ongoing de-population.

Conclusion

We start this brief conclusion with a theoretical remark about the de-peasantisation process and will end it by underlining the most important findings on recent changes in the Portuguese agricultural economy and rural society. The process of de-peasantisation must be analysed taking into account two main criteria. The first is the degree of market integration, including not only the restricted market for agricultural products and inputs, but also the labour market; the second concerns the cultural and identity features of household farms. Looking at the first criterion it is important to know how farmers relate to agricultural products and inputs markets. This issue must not be addressed only in quantitative terms, but it is essential to identify how far the farmer's economic logic is driven mainly by family reproductive needs and goals or whether submission to market relations begins to by-pass the household's logic.

Concerning cultural and social identity dimensions, the emergence of professional farmers, and the multiple social identities of part-time farmers, are both significant. These two different social groups no longer share peasant values and cultural identity. To the first, land is no longer the ultimate condition of family reproduction, or 'a way of life', but tends to be equivalent to any other capital. That is, the economic functioning of these farms has to follow the rational capitalist approach. Part-time family farmers' social identity and cultural values separate them from the typical peasant. Accepting this perspective, de-peasantisation as a process of family farm modernisation can apply either to the typical peasant of the North or East of Europe or to the typical family farm of the South.

In relation to the main features of the recent evolution of Portuguese agrarian economy and rural society, we have noted that during recent decades Portuguese agriculture has been losing its importance as an economic activity. This may be expressed in terms either of the reduction of the active agricultural population or the decrease of the agricultural contribution to GDP. Until very recently, the process accorded with state policies which prevented most agriculture activities from being fully integrated into a typical market economy. Thus, many processes recorded in the most industrialised countries, like the penetration of capital into agriculture and into the agro-industrial complex, and concentration in land use, have only recently begun to be observed in Portugal.

The dominance of family farming, the relatively high importance of the agricultural sector in terms of farm population, and the high rate of off-farm activity are three main characteristics of Portuguese agriculture. Neither full-time nor part-time family farming models share the main characteristics of those prevailing in most of the countries marked by the productivist model promoted by the CAP.

In relation to rural society, agricultural activity and the farming population still play important roles in maintaining territorial equilibrium and the economic and social dynamics of many rural spaces. This appears particularly clearly in the northern coastal area, where rural communities have been experiencing a revitalisation of their social, economic and cultural structures, especially since the last two decades. However, the rural society located in the northern inland and in Alentejo, in the south, has not been experiencing such changes. Lacking capital, institutional and human resources, rural communities located in those areas face strong difficulties in reversing the current trends of depopulation and desertification.

This scenario is the more serious if one takes into account that the Portuguese urban population's demands for non-productive agricultural services from rural communities are restricted to a very narrow social segment, in contrast with in wealthier European societies. That reduced demand is explained by economic and cultural factors. In addition to lower living standards and purchasing power, the urban population still has close links with rural society, through either a rural childhood or family ties with relatives living in rural areas and/or performing agricultural activities, reflecting the recent character of industrialisation and urbanisation processes. This also has important consequences in terms of the social representation of the Portuguese population, and the political discourses and projects addressed to the rural society.

In Portugal, the 'rural' is thus evaluated and valued within symbolic and cultural frameworks which are different from those prevailing in wealthy and oldest industrialised European societies, and from those which underlie the EU rural development policy model. As we have described, Portuguese agricultural and rural society models have their own specificity, which is historically and culturally rooted.

By and large, EU agricultural and rural development policies mirror and are addressed to northern social realities. Taking into account the Portuguese case, as well as the contributions from the southern European countries included in this publication, one may conclude by emphasising the importance to these countries of searching for a development pattern based on the specific nature of the South, instead of an adjustment to northern European models.

Notes

1 It is sufficient to know that the media was subjected to a rigorous form of previous censorship (to every media and cultural expression) ranging from periodic and non-periodical publications, to the radio, TV, cinema, theatre, cultural associations and any other form of public expression of citizens' associations.

2 That is a number of social groups gained voice and intervention power in political, social and economic issues.

3 Functioning does not mean that the democratic institutions were more or less effective.

4 That is one of the countries that now is under the monetary discipline of the Euro.

5 The absolute values show a decline from 1122 to 601 thousands from 1980 to 1997. It is possible that some of the most recent statistics are underestimated since, on the one hand the statistical definition of farm has changed (Baptista 1993, 8-9), and on the other hand, it lacks visibility to the female farming labour force (Rodrigo 1986).

6 The data used come from the comparison between the Recenseamento Geral da Agricultura de 1989 and the Inquérito à Estrutura das Explorações Agrícolas de 1993 (IEADR 1995[a]).

7 Such economic rationality or logic searches the minimum risk and minimum investment, and takes advantage of the large area to obtain a final result that is considered sufficient, knowing the land price is increasing.

8 We say suggests because in order to fulfil EEC statistical requirements, Portuguese statistical definitions and criteria changed, namely the definition of farm, which prevent us to establish exact comparisons between 1979 and 1993 data.

9 Overall average feed balance ratio (that is internal production/net import) is 65.3 for the period 1994 to 1997 (INE 1999).

10 During this period Portugal was only self sufficient in vegetables, poultry and eggs, wine and dairy products (INE 1999).

11 It must be stressed the importance of intermediate consumption that, in 1992, was about 52.4 percent, one of the most higher values in EU, fourth place after Belgium, UK, and Denmark (ECC 1995:T/29).

12 It is enough to give a look at the European Commission Annual Reports on the situation of EU agriculture, to confirm this general statement.

13 Agriculture Commercial Balance kept the structural deficit in spite of the higher export propensity of the forestry sector. The rate exports/imports has been inferior to 70 percent in the last two years (IEADR 1995).

14 And this in a country that one decade ago was characterised by a very strong dependence of the state, a kind of economic tradition and mentality that is very difficult to lose.

15 The quantitative data concerns the Portuguese Mainland. This note is also valid for Table 10.1.

16 As the current debate taking place in the rural literature concerning the fundamental issue of the rural spaces' definitions discussion, there are several approaches to define and categorise the rural (Halfacree 1993, Hoggart et al. 1995, Blanc, 1997 and Mormont 1996). However, the limitations and advantages of the descriptive approach, will not be discussed here, since this issue is clearly out of the aims of this paper.

References

Almeida, J. (1986), *Classes sociais nos campos*, ICS, Lisboa.

Arnalte, E. (1997), 'Formas de producción y tipos de explotaciones en la agricultura espanhõla: viejas y nuevas líneas de diferenciación', in Gómez Benito, C. and Gonzalez Rodriguez, J. (eds), *Agricultura y sociedad en la España contemporánea*, Ministerio de Agricultura Pesca y Alimentacion/Centro de Investigaciones Sociologicas, Madrid, pp. 501-532.

Avillez, F. (ed.) (1994), *Análise da situação actual e das perspectivas de desenvolvimento da agricultura das zonas transfronteiriças do Minho e da Galiza*, IDARN, Porto.

Baptista, F. O. (1993), *Agricultura, espaço e sociedade rural*, Fora do Texto, Coimbra.

Baptista, F. O. (1995), 'Agriculture, rural society and the land question in Portugal', in *Sociologia Ruralis*, Vol XXXV, Nº 3/4, pp. 309-321.

Baptista, F. O. (1997), 'Agricultura e capitalismo na Europa do sul', in *Agricultura, Meio Ambiente e Sustentabilidade do Cerrado Brasileiro*, Uberlândia (Brasil), pp. 45-73.

Barros, V. (1992), *Minifúndio e modernização agrícola*, INIA, Lisboa.

Blanc, M. (1997), 'La ruralité: diversité des approches', in *Economie Rurale*, nº 242, pp. 5-12.

ECC (1995), *Situação da agricultura na União Europeia. Relatório 1994*, ECC, Brussels.

ECC (1999), *Situação da agricultura na União Europeia. Relatório 1998*, ECC, Brussels.

Fragata, A. (1992), *Políticas agrícolas, diversidade e estratégias dos agricultores*, INIA, Lisboa.

Fonte, M. (this publication), 'Reconstructing rurality in the south of Italy'.

Halfacree, K. (1993), 'Locality and social representation: space, discourse and alternative definitions of the rural', in *Journal of Rural Studies*, Vol 9, Nº 1, pp. 23-37.

Hoggart, K., Buller, H. and Richard, B. (1995), *Rural Europe. Identity and change*, Arnold, London.

IEADR (1995), *A agricultura portuguesa em números*, IEADR, Lisboa.

IEADR (1995[a]), *O ano agrícola*, nº 22, Dezembro, IEADR, Lisboa.

INE (1994), *Balança alimentar portuguesa 1980-1992*, INE, Lisboa.

INE (1999), *Estatísticas agrícolas de 1998*, INE, Lisboa.

INE (1999[a]), *Inquérito à estrutura das explorações agrícolas de 1997*, INE, Lisboa.

Moreira, M. B. (1986), 'An outline on Portuguese agriculture', in *Facts and Figures on Rural Portugal*, XIII Congresso da European Society for Rural Sociology/SPER - Sociedade Portuguesa de Estudos Rurais, Braga, pp.1-24.

Moreira, M. B. (1989), 'The crisis of Portuguese agriculture in relation to the EEC Challenge', in *Agriculture and Human Values*, Vol. VI, numbers 1 and 2, Winter-Spring, pp.70-81.

Moreira, M. B. (1989), 'Portuguese agriculture and the state: an outline of the past 25 years', in W. H. Friedland, L. Busch, F. Buttel and A. Rudy (eds), *Towards a new political economy of agriculture*, Westview Press, Boulder and Oxford, pp. 289-312.

Moreira, M. B. (1996), 'Portuguese agricultural production', in Carlo Cuppo (org.) *I sistemi agro-alimentari mediterranei: confronti e prospective*, proceedings of a International conference, Ischia (Itália), 13-16 June.

Mormont, M. (1996), 'Le rural comme catégorie de lecture du social', in M. Jollivet and N. Eizner (eds), *L' Europe et ses campagnes*, Presses de la Fondation Nationale des Sciences Politiques, Paris, pp. 161-176.

Oliva, J. and Camarero (this publication), 'Itinerant rurality: the Spanish countryside after de-peasantisation and deagrarisation'.

Pereira, M. (1971), *Livre-câmbio e desenvolvimento económico. Portugal na segunda metade do século XIX*, Livraria Sá da Costa, Lisboa.

Ribeiro, M. (1997), *Estratégias de reprodução socioeconómica das unidades familiares camponesas, em regiões de montanha*, Fundação Calouste Gulbenkian, Junta Nacional de Investigação Científica e Tecnológica, Lisboa.

Rodrigo, I. (1986), 'Feminização da agricultura', in *Análise Social*, nº 92-93, pp. 643-652.

Rodrigo, I. (1992), *Da terra à profissão*, dissertação de doutoramento, Instituto Superior de Agronomia, Lisboa.

Rodrigo, I. (1998), 'Social identities and family farming in Portugal: 'the old' and 'the new' countryside', in L. Granberg and I. Kovách (eds), *Actors on the changing european countryside*, Institute for Political Science of the Hungarian Academy of Science, Budapest, pp. 215-224.

Santos, R. O. (1996), *L'installation agricole au Portugal: moment d'accumulation, de renouvellement et de modernisation*, thèse, École National Supérieure Agronomique de Montpellier, Montpellier.

Serrão, J. (dir.) (1971), *Dicionário da História de Portugal*, Iniciativas Editoriais, Lisboa.

Silva, M. C. (1998), *Resistir e adaptar-se*, Afrontamento, Porto.

Vaquinhas, I. and Neto, M. (1993), 'Agricultura e mundo rural: tradicionalismos e inovações', in J. Mattoso (dir.), *História de Portugal*, Vol.5, Círculo de Leitores, Lisboa, pp. 325-337.

Wall, K. (1998), *Famílias no campo. Passado e presente em duas freguesias do Baixo Minho*, Publicações D. Quixote, Lisboa.

11 Reconstructing Rurality in Mediterranean Italy

MARIA FONTE

Introduction

In many European Union (EU) official documents and texts Mediterranean regions are considered marginal areas, lagging behind in development. It is generally felt that, in the transition towards a developed economy with higher levels of income, these areas have to conform to the prevailing model found in Northern European countries. Nevertheless, in the 1990s, analysis of differences and perspectives that take into account differentiation in development pathways found some margin for justification. In the case of rural development these perspectives aim at highlighting the value of agro-ecological, social, economic and cultural specificities and at pursuing endogenous development models.

This objective may, however, be hindered in its pursuit of an endogenous rural development model by the deficiencies of a general, theoretical framework for understanding marginal rural economies. In a very stimulating work, Papadopoulos (1997) argues explicitly and provocatively that the dominant analysis is inadequate with regard to the rural tradition of Mediterranean countries. He illustrates in a convincing way the need to *de-construct* the concept of 'rural': " 'eliminating the rural' as it was conceived in 'Northern European' terms may give rise to refashioning the rural in order to convey meaning for southern European countries".

In Tönnies' formulation of the rural-urban dichotomy,[1] rurality was conceptualised with particular reference to the rural tradition of Northern European countries, that had its roots especially in the English model of development. Here, in the tradition of positivist sociology, the urban system was associated with industrialisation (Miller 1996), and was counter-posed against an agricultural specialisation of rural areas.

Moreover, in the case of England, in the late 19th century, the agricultural specialisation of the countryside was superseded by an international division of labour, based on importing cheap food from the colonies. Rural areas were then socially re-defined and perceived as nature reserves and recreational places for urban dwellers (Lowe and Buller 1990).

In an analogy with industrial development, this pathway to modernising the countryside is still today the paradigm against which rural development in Europe is measured and interpreted. Only recently, in the 1990s, after the crisis of both Fordist organisation in industrial production and EU agricultural policies, has the need emerged to inquire into the differentiation of national models of rural development. Hoggart et al. (1995), amongst others, proposed a distinction between four main traditions of rurality in the EU: the naturalist, the agrarian, the Mediterranean, where town and countryside seem more integrated with each other and, finally, the tradition of the margins, which is rooted in the regions on the geographical boundaries of Europe (Nordic, Transalpine and Atlantic peripheries), "where constraints on human activity arising from the character of the physical environment link the conception of rural to the sustainable use of natural resources" (Hoggart et al. 1995, 92).

In the context of this recent literature, the main aim of this chapter is to achieve a better understanding of the Mediterranean tradition of rurality in Italy and to trace the way this tradition has evolved in the recent revival of rurality and interest in de/re-peasantisation processes.

In the first part I will briefly refer to the literature on the main EU traditions of rurality, differentiating the Mediterranean from core Continental and Northern traditions. Analysis of rurality in the Mediterranean regions cannot be based on the polarity of urban/industrial versus rural/agricultural. On one side, there is a tradition of pre-industrialised cities; on the other, the urban and the rural have always been more tightly intertwined. Economic development does not entail spatial and sectoral specialisation, particularly in the form of industrial/agricultural specialisation of town and countryside.

In the second part, rural development in Central and Southern Italy is analysed to exemplify two different developmental paths of Mediterranean ruralities: the 'specific quality economy', which is notably, but not only, widespread in Central and north-eastern Italy, and a more common configuration of 'local markets economy' in the South of Italy.

Both models have common traits, but also important differences. The spatial and sectoral integration of the economy is a common feature in their histories, but the two configurations differ in the degree of formalisation and institutionalisation of the economy on one hand, and their relation to the market on the other. Specific quality economy may be seen as a local production system that inserts itself into global and segmented markets through product qualification and differentiation. In terms of 'conventional forms' economics[2] (Allaire and Boyer 1995, Valceschini 1995, Orléan 1989, Boltansky and Thévenot 1991, Thévenot 1995, Wilkinson 1997), the certification system is the 'investment in form' that has made the connection between local production and global markets possible. Rurality in the inland areas of Southern Italy remains confined within the convention of 'domestic economy', where economic transactions are based on personal, face to face relations and little formalisation.

Traditions of Rurality in the EU Countries[3]

> Hoggart and colleagues differentiate between diverse types of rural tradition in Europe (Hoggart et al. 1995, 91-109). Three of them are particularly useful... They counter-pose an agrarian tradition against a naturalist tradition. They distinguish on the one hand, the conception of rural areas as primarily productive surfaces on which the exploitation of natural resources is related to the construction of a distinct rural culture and, on the other hand, the conception of rural space as an arena for consumption, which points to a predominantly urban-centred notion of rurality. Notably, a Mediterranean tradition is referred which in fact points either to a close integration between city and countryside or to an inability to find grave distinction between the two (Papadopoulos 1997).

Rural development in Britain represents the *naturalist tradition of rurality.* Already in the mid-19th century, industrial towns had emerged and transformed Britain into the most urban country in Europe, with 80 percent of the population concentrated in towns. Rural areas underwent a process of de-industrialisation and agricultural specialisation. They lost the extra-agricultural activities, such as crafts and manufacturing, which had characterised the proto-industrialisation period, and they specialised in agricultural production, turning it into very innovative, market-oriented activity.

> Urbanisation and industrialisation thus produced a spatially polarised, as well as nationally integrated, geography in which cities functioning at the core of distinct and specialised regional economies concentrated the bulk of population and commercial and industrial activity, while rural areas became dominated by a technically progressive, market oriented agriculture (Lowe and Buller 1990, 4).

First and foremost, agriculture had a productive role, providing cheap food for the urban working class. Soon, though, and in particular after the crisis at the end of the 19th century, production lost its central role. Cheaper food was imported from the colonies and ex-colonial territories. While other European countries defended their agriculture with tariffs, Britain remained faithful to the principles of liberalism and free trade, convinced that cheap food would increase the competitiveness of its industrial sector. This attitude reinforced the agricultural crisis and, in the context of a highly concentrated system of land ownership,[4] favoured the emergence of the conservationist movement:

> Much land went out of cultivation, the rural population declined sharply and many village communities became depressed and stagnant. The cause of rural preservation was therefore given added poignancy by a sense that the traditional order of the countryside was on the verge of extinction (Lowe and Buller 1990, 9).

In this light, the symbolic representation of the countryside acquired a double meaning. On one side, the countryside became an essentially aristocratic and aesthetic model to be preserved from extinction, while, on the other side, it was linked to emerging urban needs, to new life styles and recreational models. This urban redefinition of the countryside was reinforced by the counter-urbanisation movement, which, especially from the 1960s onwards, brought into rural areas different fractions of the middle class, who were in search of the 'rural idyll' and wished to escape from the increasing inconvenience of urban life. Discontinuity of 'the rural' from agriculture and the transformation of rural areas into natural reserves and recreational areas for urban inhabitants was thus reinforced.

The Continental Northern European *agrarian tradition of rurality* is found in countries like Denmark, Ireland and France. France can be viewed as being representative of this tradition. Even in the 1950s, France was described as 'peasant' with almost half the population living in rural areas and 28 percent of the work force employed in agriculture. Abolition of feudal titles and the equal inheritance principle written into Napoleon's code, in the early 19th century, contributed to maintaining an agricultural structure that was based on the small family holding. In the 19th century agricultural production grew slowly, the surplus was re-invested in the acquisition of land and, as a result, the demand for industrial goods did not increase. "In 1871, two-thirds of the French population lived in rural areas and approximately three-quarters of all non-agricultural activity took place in the countryside, most of it in workshops rather than in factories" (Lowe and Buller 1990, 5). The French rural world continued to be marked by many small, local economies that were separated from each other, rooted in peasant agriculture and had a highly articulated social stratification. After the Second World War French rural policy was interpreted in terms of an agricultural development, whose main concern focussed on the disappearance of the peasants (Mendras 1992) or their integration into a modern and urbanised market economy (Servolin 1972).

De-industrialisation of the countryside took place in France about a century later than in the England. In contrast to England, though, agriculture is still a very important 'industry', in terms of both mass and quality production. Agricultural goods are one of the most important elements in the French balance of trade surplus. It has, by now, become a specialised agriculture, one that has experienced its rupture from the rural world, from the family farm and from nature (Hervieu 1991). After the 1970s, the model of rural development in France seems to have converged towards the English model. Contra-urbanisation has become an important phenomenon; the social composition of the countryside has changed and there is a more marked presence of a non-agricultural middle class. But rather than being tinted with naturalistic nuances, the past is perceived as nostalgia for a culture, for a specific society, and not for a landscape. New rurality is, according to Hervieu (1991), "a rurality which amounts to little more than a landscape, a physical environment, but not the matrix of a new form of rural society". Contrary to the British tradition, where peasant culture seems to have left no striking traces in the representation of 'the rural',[5] the end of the peasantry is presented as the end of the rural world in French culture, that is the end of a specific way of organising production, consumption and life styles.

In the rural sociology literature, the *Mediterranean tradition* of rurality is presented as a 'denied tradition' (Hoggart et al. 1995, 104). Here, spatial organisation is dictated by pre-industrial towns and ports (Braudel 1992, Cattaneo 1957). Analysts have observed widespread resistance to the 'rural idyll'. Conversely, the countryside is viewed as backward, underdeveloped and an obstacle to modernisation. Thus, even though agriculture in the four Mediterranean countries today accounts for a large part of the total of EU farms, farmers and cultivated land, it seems that there is neither a consolidated Mediterranean rural tradition nor a rural nostalgia. After the Second World War, rural problems were marginalised in the list of priorities for national development and, instead, the main objectives were directed at industrialisation and integration with Northern Europe. Finally, the 'Mediterranean notion of rurality remains deeply-rooted in the traditions of urban civilisation, whether this is seen through the spatial organisation of agricultural production or through the flexible interaction of urban and rural labour forces and patterns of property ownership' (Hoggart et al. 1995, 105).

A central element of the Northern-European conceptualisation of rurality seems, then, particularly unsuited for application to the Mediterranean regions. The urban – rural dichotomy, that many maintain is essential for describing the differentiation of productive and settlement models in the Fordist period, does not belong in the history and culture of Mediterranean countries.

At a territorial level, the urban and the rural seem to be connected by stronger ties, which denies a distinction of functions and roles on the basis of production, employment, exchange, land property, as well as cultural and political models. In history, this integration assumes various forms of spatial organisation, that range from the *urbanised countryside* of Central Italy to the rural towns or *agro-towns* of Southern Italy. Whether we are referring to a settlement system where small and medium towns are evenly distributed in the territory or to an agrarian system characterised by the dominance of *latifundium*, i.e. large estates or large-scale farms using day labourers, what we find is an inextricable relationship of complementarity rather than specialisation, where tracing a definite borderline between town and countryside is virtually impossible.

This characterisation is confirmed by Medieval historians. In Italy the peasant is called *contadino*, that is 'the inhabitant of the *contado*'.[6] A *contado* is the territory administered by the count, who, in Italy as well as in the South of France, used to live in town. According to Brunner (1988), the difference between Southern and Northern urban centres of Europe lies in the fact that in Northern Europe the town used to be neatly differentiated from the countryside and governed by a commercial and productive bourgeoisie, while in the South the landowning nobles living in town held on to their rural origins, giving to the Southern town a mixed character.

> While in Southern Europe the town took a noble-bourgeois character, in the North towns were neatly distinct from the countryside, where the nobles – peasants relationship persisted. In other words, in the South nobles maintained their character even when they lived in town, while in the North, they lost their 'noble' status and became 'bourgeois' (Brunner 1988, 94-95).

At the economic level, Mediterranean regions did not make a clear differentiation between economic sectors and a clear specialisation of resources (Chenery and Syrquin 1986). Agriculture kept a distinct position as an economic activity. Even when industrialisation took off under endogenous or exogenous drives, the Fordist model with its concentration of both production in factories and population in industrial towns never became dominant. The persistence, in the Centre and the South of Italy, of a rich craft tradition, the revival in some areas of small and medium enterprises, flexibility of work, informal economy and pluriactivity, constitutes a "sort of pervasive social fabric" (Papadopoulos 1997) which is hard to interpret in terms of agriculture/industry or urban/rural dichotomies. Alternatively, for entire areas we could talk of the persistence of a rural economy, where the "relationships between country and town are settled around a spatial and economic integration, rather than around separation and conflict" (Basile 1996): "a post-modern condition which was always there, as an alternative culture, and therefore is not 'post', in their case" (Leontidou 1996, 147).

In contrast to Northern European countries, where agriculture is specialised in the input-intensive field crop and livestock system – meat and cereal are the main elements of a Fordist diet (Friedmann 1982) – natural and agro-ecological features (a softer climate, the long summer droughts, the presence of hills and mountains) have led to the spread of extensive cultivation (cereals and pulses) on one side, and to the importance of trees and horticulture on the other.[7] We are dealing with an agricultural system that is labour-intensive and imposes a markedly seasonal working calendar (in comparison, for example, with the dairy farm). This phenomenon encourages pluriactivity, which prevents any sectoral specialisation of the workforce.

After the World War II, the crisis of the dominant agrarian structures in Central and Southern Italian regions, on one hand, and national and supranational agricultural policies on the other, led both to the consolidation of the small family farm and to further specialisation in tree cultivation, especially vine and olive trees mixed with extensive cereals in the dry hills, and fruit trees and horticulture in the irrigated plains (only a very small part of the territory). For the mixed agrarian systems in the hills, the dominant technological models, that have been developed for mass produced crops, are rather inadequate, especially in a very heterogeneous territory like the Mediterranean area. Agriculture thus remains backward, not very receptive to scientific techniques and unable to start the fly-wheel of development.

On the other side, however, the Fordist organisation of industrial production has not taken off. Whenever industrialisation takes place, it has either an exogenous origin, as in the heavy State industry located in the

South of Italy, or, as in Central Italy, it reflects a different model, which is widespread rather than concentrated.

The unorthodox route to economic development in Central and Southern Italy accounts to some extent for the insufficiency and lack of meaning that the category of rurality, as elaborated in the Northern European countries, seems to have here. In fact, the tradition of studies on rurality in Italy is recent and to some extent marginal, while there is a wealth of literature on regional differences, including *meridionalismo* (the purpose of which is to explain and interpret the lagging development of the South) and, more recently, studies on the "three Italies" and on "industrial districts" (Becattini 1987, 1989, 2000).

With respect to dispersed industrialisation, Saraceno (1994, 471) underlines the inadequacy of the urban-rural dichotomy:

> Northern and Central Italy's diffused industrialisation is a perfect example of the heterogeneous nature of the exchange between town and country which may involve manufactured goods, leisure activities, and different types of services occurring in both directions. In a context like this, the rural and the urban functions are no longer spatially specialised and it is meaningless to try to see it at all costs. In these cases, as argued above, a specific conceptual category should be used: the local economy, which has been proposed as an alternative to the semi-rural or the peri-urban.

Recently the concept of 'local economy' or, still better, of 'local production systems' has been widely used by economists and sociologists in order to find more significant territorial units of analysis (Becattini, 2000). While stressing the interdependence of the socio-economic system at territorial level (the term 'community' is, for a change, not very frequently used in this literature), the concept of local system leaves open for discussion the organising principles of economy and society. Rurality may be brought back to its original meaning, as a cultural rather than a spatial category.

Reconstructing Rurality in Mediterranean Italy

In relation to Italy, it seems useful to distinguish at least two socio-economic configurations of Mediterranean rurality: the 'specific quality economy', and the 'local market economy'.

The Quality Economy of the Urbanised Countryside: Local Production for Global Markets

There is now a vast amount of literature on the Marshallian district economy[8] which stems from the effort to explain the specificity of industrial development in the central and North-Eastern regions of Italy. We deal with a type of regional development based on *local production systems* (i.e. localised in a limited and defined territory) of small and medium firms, that are specialised in different phases of a production process. The spatial contiguity of firms produces a 'thickness' or intensity of inter-industrial relations, an 'industrial atmosphere', where technical knowledge and information are easily transmitted and positive externalities (external to the firms but internal to the district) generated. Rather than the firm, it is the *district* that is the production unit on which to measure economy of scale and scope, as well as other parameters of competitivity.

To discuss whether and/or when the district economy or the local systems may be considered 'rural' (Hoggart et al. 1995, 165) would go far beyond the scope and space of this essay. Here I wish to refer to rural areas, (i.e. with a relatively low population density and an extensive use of the territory) which present interesting analogies with the economic characteristics of industrial districts: specialisation on the production of a *quality specific* agricultural or agro-industrial product, externalities of localisation for enterprises, small-scale activities, cultural traditions and a high degree of institutional interrelation (Polidori, Romano, 1994). Diversification of the economy in these 'rural local systems' may come from both industrialisation of the countryside, through the diffusion of small and medium enterprises, and from the recovery of multifunctional agriculture: the development of agri-tourism and other forms of rural tourism, linked to a territory that is well-preserved, with many art-cities and full of local traditions.[9]

These are well-structured areas, where up to the 1960s the agrarian structure was based on the sharecropping family farm, firmly established on the land by means of long-term contracts. These areas have been fashioned in history as *urbanised countryside*, with widespread rural infrastructures and diversified activities. As has been shown in the quoted literature on the district economy, economic relations are driven by both co-operation and market competitive forces. Co-operation concerns especially the sharing of technical knowledge, expertise, information and research. It involves many actors, among whom local institutions stand out. Trust, this very important element of development, has an *institutional* facet and a well-defined role: professional associations, co-operatives, consortia, unions and local administration have in the course of time developed institutional devices in order to define, preserve and enhance the aesthetic and economic quality of the area – the landscape – and the local products (wine, cheese, olive oil, etc.) in a logic of collective belonging, rather than individual utility maximisation. *Institutionalisation of trust* may be considered the peculiarity of the modernisation process (i.e. rationalisation and formalisation processes) in this particular development model.

This model may be characterised as a 'specific quality economy'. The specificity may be analysed at three different levels: product, techniques, and certification systems. The product (agricultural or agro-industrial products, or rural services such as landscapes, recreational services, etc.) and the techniques derive their specificity from being linked to a territory, a tradition, a culture, a community, rather than conforming to a standard. Specificity though, as local attributes or characteristics, cannot be immediately recognised and evaluated in the global market. The certification system (designation of origin, geographical indication or traditional speciality) is the *compromise* that mediates between local and global systems as well as between different economic conventions (market, domestic, civic). Through the establishment of production regulations and procedures as well as the sensory characterisation of the product, it translates the specificity of products and techniques into *general forms*, that become meaningful not only to local but also to foreign consumers.

It may be useful to underline three points: (i) In the event of market failure with reference to quality information, the certification system leads to the formalisation of the quality characteristics of the product and, as a consequence to its valorisation through a higher price. (ii) It enables consumption of the local product to be 'de-localised'. The product can 'travel' to distant markets, and still maintain its local identity (Sylvander 1995, Letablier and Delfosse 1995). (iii) The various forms of *origin protections* can be seen in terms of *collective property rights*, which are attributed to a local community with reference to technological processes that are based on empirical, tacit and specific forms of knowledge.

Quality systems do not imply the simple re-evaluation of tradition as an economic resource. Tradition is re-interpreted, re-invented (Hobsbawm and Ranger 1992) in the light of new scientific knowledge and finally translated into the codes of 'modern' rationality. Traditions, local networks and associations become *new* economic resources (social capital) in the quality economy. Certification systems extend the rural network to new social actors, specifically to the *experts on quality*: scientists and researchers who establish the norms of production, technicians and professionals who control and protect the quality of the products or the production process, nutritionists who establish the dietary characteristics of food. Their function is one of mediation and translation between urban and rural culture; between formal/scientific and empirical/contextual forms of knowledge; between local production and global markets. In which conditions and contingencies the agrarian/rural culture will keep its status as an alternative culture or will become strictly functional to urban markets and cultures may depend on the position of these figures in relation to agriculture, the rural economy, rural culture and forms of knowledge (Tovey 1998).

As for the social characterisation of these areas, it is possible to observe a differentiation and specialisation of social figures around farming. We find various fractions of traditional productive classes such as small rural entrepreneurs (in industry and agriculture), artisans, small factory workers, employees, etc., as well as neo-peasants and neo-farmers (whose economic objective is to produce and sell quality products, including landscape itself), new rural consumers (i.e. middle class fractions that consume the rural space as the location for the first or second house and/or for tourism), new 'unproductive' or middle classes, such as technicians and quality experts. In comparison to the traditional peasant economy, we can read this diversity as a differentiation of occupations and roles relating to the production of food and agricultural or rural services in the rural local system. In contrast to the capitalist process of differentiation, though, this one does not lead to proletarisation, rather to a multiplication of professional social figures and autonomous workers.

What this new rural economy implies for the process of *de-peasantisation* has not yet been widely researched. Van der Ploeg et al. (2000, 403) speak of "a kind of *re-peasantisation* of European farming", referring to "newly emerging rural development practices" that deviate from the modernisation paradigm. If we refer to peasant economic culture in the Chayanovian terms discussed in this volume, we may notice elements of continuity as well as elements of novelties. Economic rationality in these systems may be seen as oriented towards social reproduction insofar as accumulation has well defined territorial and social limits,[10] the development model is based on independent producers, rather than a proletarisation process and, finally, the technical system aims at valorisation of the links with natural resources and biological processes. But neither the farm nor the local system are isolated anymore. They are rather part of a tighter network of global economic and social relations, consumption is de-localised, traditional, contextual forms of knowledge are translated into codified forms. Insertion into the global markets and weaker producer-consumer ties may generate tensions between constraints of the territory on one hand and growth on the other, pushing toward systems of mass production. Here is where the main threat to the specific quality economy lies.

The Difficult Evolution of the Agro-town and the Depopulated Countryside: Local Production for Local Markets

After the failure of centralised policies, inspired by a model of heavy industrialisation in the 1960s and 1970s, the district economy has been seen by many as offering an alternative path to the development policies of marginal areas in the South of Italy.[11] Indeed, profound differences in the economic systems of the two regions need to be analysed.

Starting from the land system and the agrarian structure, in contrast to the Central regions of Italy where the sharecropping family farm was solidly and firmly implanted in the countryside, in the South the *latifundium* system was based on hired day labourers, leading to a specific model of population settlement, the so called *agro-town*, i.e. a rural town, where various classes linked to the land live: peasants, small and medium land-owners, hired labour, but also artisans, employees, etc. (Jansen 1991, Bevilacqua 1997, 53).[12]

These rural towns constitute the "essential points of reference for a geography of population and a productive structure centred substantially, if not exclusively, on the *latifundium* farm structure" (Nocifora 1992, 24). The *latifundium* was the dominating productive system in Southern Italy until the middle of the 20th century and was based on one hand on extensive cultivation of cereals, pulses and of trees, like the olive, that require great availability of labour in very concentrated periods of time, and on transhumant sheep rearing (*pastorizia transumanante*) on the other. From one perspective, the *latifundium* depopulated the countryside, from another, however, it needed to dispose in some periods of the year of great masses of cheap labour. The rural towns responded to this demand: they are "the place where products may be exchanged and labour may be sold, before dawn, to the middleman of the great landowner, who comes expressly to engage the workers that are needed. People leave early in the morning and return in the evening" (Aymard 1992, 129). The functional link between town and countryside is not related to a productive specialisation between agriculture and industry, but to the social and physiological relationship between 'working time' and 'resting time': "...The town, the place of exchange, leisure and all the forms of social life is neatly opposed to the countryside, the place of work, animal life and the production of material goods..." (Aymard 1992, 131).

Only during the 20th century has social stratification in the town been articulated in significant way with new components that are completely independent from rural classes: first, with state and party bureaucrats (especially during the Fascist period), clerks and professional bourgeoisie; later, after the Second World War, with fractions of the middle classes involved in construction and trade. The rural town became the 'tertiary town', while the poor peasants and farm workers, who continued overcrowding the countryside during Fascism, emigrated abroad at first, and later, in the 60s, to the industrial towns of Northern Italy. From the perspective of social analysis, the 'agrarian block' (Gramsci 1972)[13] is replaced by the 'construction block'[14] (Nocifora 1992), while in the countryside the main changes are not those of social class, but those of demographic composition: the population grows older and human resources are impoverished after the massive emigration of men of working age.

The displacement effect that the territorial concentration of industrialisation in the Northern regions of Italy produced on Southern agriculture and rural areas, in the 1950s and 1960s, greatly outweighed the effect of the agrarian reform that had been implemented in the 1950s for re-distribution of land to the peasants. The fight for agrarian reform and against the *latifundium* carried out by masses of poor peasants in the aftermath of the Second World War, was the final struggle of the peasants as a class in Italy. Because of its limited application (only seven hundred thousand out of about 26 million hectares in total were expropriated), the agrarian reform was more a temporary regulatory mechanism of the labour market than an effective measure against the de-industrialisation of the countryside. Migration movements, in the 1950s and even more so in the 1960s, displaced from Southern Italy (first to the Americas, then to other Northern European countries and afterwards to the towns of the Italian industrial triangle – Turin, Milan, Genoa) over four million people of working age (out of a total of about 18 million in 1951), mainly peasants,[15] all of them rural and agricultural workers, thereby provoking a great loss of human, cultural and technical resources[16] as well as the disintegration of rural communities. Thus, de-peasantisation took the form of migration and depopulation, while remaining peasants lost their identity as a class "in significant relations with other groups in society" (Tovey, in this volume). Rather than products, these systems have exported manpower, human capital and resources.

In the wake of the Fordist crisis, which launched both the model of diffused industrialisation in the Centre and the north-east of Italy and the strategies of rural development in the EU, the South of Italy has reached a visible degradation of both its urban and rural territory and a very limited experience of industrialisation.[17] Rural society, deprived of the necessary human and institutional resources, could hardly trigger off an autonomous, endogenous, auto-centred process of development.

While the economic system of Central Italy is marked by integrity and autonomy of resources (human, financial, knowledge) and by its openness (to global markets), the economic system of the South stands out for its marginality, i.e. its dependency and closeness: *dependence* on external resources, in terms of remittances (money that the emigrants send back to their birthplaces), transfers of funds from central government and jobs in the factories of Northern towns; *closeness*, for not being able to insert itself in national, European or global markets.

If we consider the phenomena that are discussed when one talks of the 'new rural development', the internal areas of the South of Italy can find neither their history nor their image reflected in what is described. Starting from *counter-urbanisation*, there is neither widespread experience of expulsion from the big metropolitan areas (with some very limited exceptions) nor a permanent flux of newcomers to remote areas of the countryside. After the great exodus, rural communities lived through a process of involution and degradation, which has left them in isolation, lacking many of the services that are rightly reputed essential to modern life – security, health, education and transport services. Rather than counter-urbanisation, it is return-migration and the links that emigrants maintain with their birthplace that offer a glimpse of revitalisation to these areas. We speak here not so much of the permanent return of people of working age, but the return of retired people, and especially the seasonal return on vacation of emigrants who now live permanently in the North of Italy or in Northern Europe. Social life in Southern villages has been reshaped by this event: the main religious and family celebrations are shifted to July/August, when relatives are reunited. There is no conflict between 'newcomers' and 'old villagers', rather a reinforcing of family ties and feelings of belonging.

The *diversification* of the rural economy is neither the modern, rational evolution of the traditional forms of production nor the outcome of new processes of widespread industrialisation. It is residual, old rather than new. Depopulation of the rural towns has led to the crisis of all artisan and small industrial activities in rural areas, and no new activities have replaced them. Productive, industrial agriculture specialised in horticulture and fruit trees is concentrated in the irrigated plains close to urban centres, whereas mountain and hill agriculture, which represents over 80 percent of the territory, is the abandoned domain of ageing peasants who were unable to emigrate to industrialised Northern regions. With the exception of some areas of Campania and Puglia, the countryside in the South did not experience a general process of industrialisation. In a context of low opportunities, diversification remains a survival strategy of the family enterprise. The same activities connected to agriculture, in particular agri-tourism, are hindered not only by impoverishment of the economic and social context, but also by a model of population settlement, based, as we said, on the agro-towns. Where tourism has developed, mainly along the coasts, characteristics of mass tourism have become manifest and environmental degradation accentuated, and this constitutes today an additional obstacle to the development of the areas.

The commodification of rural activities and rural space also finds cultural obstacles in a tradition of hospitality and reciprocity, that is still alive and makes it difficult to convert them into the market convention of price. Rurality is still a way of life in the context of a domestic economy, that is only partially and locally integrated in the market. Markets remain fragmented, limited to the local dimension and unable to carry out their function as co-ordinating mechanisms of the regional economic system. The weakness of the institutional system – at both public (local administration and state representatives) and private levels (the firm and market) – contributes to the reproduction of a trust system that is closely tied to personal and kinship relationships, a system that, as is sadly known, often constitutes the substratum from which various forms of clientelism – when not of criminal association – are fed.

The Domestic Economy: Local Production for Local Markets

In this context, the rural economy remains characterised by a local network of producers and consumers, closely-knit, but isolated from the external markets. Fragmentation and heterogeneity are intrinsic characteristics of these systems, which can only be identified and classified by extensive empirical work.

At the general level, constraints of place and time derived from a domestic form of co-ordination prevail in these economies, which imply "the necessity of being personally in the presence of others" (Boltansky and Thévenot 1989). This necessity is not attenuated by a set of objective, i.e. separated from people, regulations and representation mechanisms which permit *acting at a distance* (Latour 1987). Neither social roles nor knowledge and relations are formalised. In the spatial context of local circulation, competencies are embedded in people, knowledge is tacit in form and trust remains very closely linked to personal relations.

Products are not separated from people; they are essentially crafts, which recall the skill of the artisan. Qualified by a local network of reputation they do not need a formal system of certification (Thévenot 1995). Self-consumption is important, the production-consumption circuit is short and agricultural products are mainly final products. Both physical proximity to production and customs encourage a discriminatory capacity, which is fundamental to the survival of traditional techniques and local varieties.

Insofar as depopulation and abandonment persist, these societies remain suspended between marginality and extinction. Policies tend to transform them towards economies open to mass or quality markets. Their insertion in global markets is unlikely inasmuch as they do not have the capacity to match price competition. But nor is the transition towards a more formalised quality economy simpler. Institutional weakness and certification costs[18] may constitute an obstacle as insurmountable as price competition for economies that are substantially poor and for societies where devices of translation into formal codes of communication are lacking.

While these systems may not offer great opportunities for growth, they do at any rate offer opportunities for the conservation of traditional techniques and varieties on the verge of extinction. Furthermore they provide opportunities for the valorisation of biological resources ignored by the market, the defence of a territory risking abandonment, and a decent life for people who would not find a better location (in all senses) for themselves in town.

If in the specific quality economy agricultural activities are complemented mainly by tourist activities, here the pathway to endogenous rural development could be based on the environmental and biological resources. A short-circuit agricultural development model may be conjoined with a conservationist policy for the territory and the valorisation of biodiversity.[19] But this would imply a real attention to forms of economic co-ordination specific to small, local rather than global markets, and the pursuing of agro-ecological objectives *de-coupled* from the income compensation mechanisms of the Common Agricultural Policy, that supra-national, national and local institutions do not seem ready to assume with the conviction that is needed.

Acknowledgements

This work was partly funded by MURST (Ministero dell'Università e della Ricerca Scientifica), ex-60 percent. My thanks to Ladislao Rubio for his comments. English version revised by Alexander Kirichenko. Responsibility for errors remains mine.

Notes

1 Tönnies was discussing mentality rather than spatial categories (*thanks go to Leo Granberg for this observation*).

2 Conventions are action frameworks, that coordinate economic action. Boltansky and Thévenot differentiate six main conventions: domestic, market, industrial, civic, opinion and inspiration. For a discussion of the relevance of this new economic paradigm for the analysis of agrofood system, see Wilkinson (1997); for an application to a case study see Murdoch and Miele (1999); for analytical use see Fonte (2000).

3 The first part of this paragraph, relating to the rural tradition in the UK and France, is drawn mainly from Lowe and Bodiguel (1990).

4 "Four-fifths of the land, most of it tenanted, was owned by 7000 men, and just a few hundred aristocrats controlled about a quarter of the country" (Lowe and Buller 1990, 6).

5 In the sense of what was said before: the *rural idyll* is essentially of aristocratic, aesthetic or recreational character.

6 In a figurative meaning also: "rough, vulgar, unrefined person" (see Dizionario illustrato della lingua italiana, Gabrielli).

7 Already in the Middle Ages the "Mediterranean garden" was a model not only of technological but also of aesthetic value for tree cultivation (Sereni, 1972).

8 Besides the already quoted Becattini volumes (1987 e 1989) on the agricultural and agroindustrial district, see Cecchi (1992) and Iacoponi (1996).

9 Central Italian regions, especially Tuscany, attract people from other regions and countries, so much so that a specific area of Tuscany, where the wine Chianti is produced, is nowadays known as *Chiantishire.*

10 To this end it is interesting to recall the debate on the difficulties of the industrial district economy for growth and expansion. This is, according to critics, the weakness of this specific economic system that will shift towards its transformation into the "normal" large-scale economic system. In Chayanovian terms, we can see the system as being orientated toward reproduction rather than accumulation.

11 Already in the 1970s Rossi Doria (1982), an authoritative scholar of developmental problems in the South of Italy, wrote: *"The model will not work, because the flowering of small and medium industries in those regions has happened on a ground prepared by centuries of vigorous artisan traditions and intense relations between the town and countryside. In this part of Southern Italy those centuries have been marked by agricultural misery and very modest local craft, with no commercial relations...".*

12 At the beginning of the 20th century the South of Italy (including Sicily) was inhabited by 12 million people, living mainly in towns. Towns were structurally differentiated: 12 of them had more than 50,000 inhabitants and 64 had between 20,000 and 50,000.

13 This expression was used by Gramsci (1972) to indicate the alliances for a development policy that benefited both the northern industrial bourgeoisie and the southern agrarian landowners.

14 Following the rapid expansion of the construction sector after the Second World War, the agrarian aristocracy of the South was replaced in this social block by the classes linked to this sector (Nocifora, 1992).

15 In the 1950s, in the South of Italy there were still about 3,700,000 employees in agriculture, accounting for 57 percent of the total number of workers.

16 On the contrary, in the central regions of Italy the migratory balance was slightly positive in the 1950s and 60s.

17 Industrialisation was mainly experienced by southern emigrants in Northern Europe and Northern Italy.

18 It is, for example, known that in 1998 Italy had over the 27 percent of the EU agricultural land cultivated by certified organic farms (610,000 hectares out of 2,250,000). Out of the 610,000 hectares, 70 percent was located in the South. But for the vast majority organic products are not 'valorised', i.e. certified, under labels of Southern firms.

19 In the era of biotechnology, as it is recognised, biodiversity is destined to become a fundamental resource.

References

Allaire, G. and R. Boyer (eds, 1995), *La grande transformation de l'agriculture*, INRA, Economica, Paris.

Aymard, M. (1992), 'Spazi', in F. Braudel, *Il Mediterraneo*, Bompiani, Milano, pp.123-144.

Basile, E. (1996), 'Sulla differenziazione produttiva nell'economia rurale', *Politica Internazionale*, XXIV, 5.

Becattini, G. (ed, 1987), *Mercato e forze locali: il distretto industriale*, Il Mulino, Bologna.

Becattini, G. (ed, 1989), *Modelli locali di sviluppo*, Il Mulino, Bologna.

Becattini, G. (ed, 2000), *Sviluppo locale e distretti industriali nell'economia italiana contemporanea*, Rosenberg e Sellier, Torino.

Bevilacqua, P. (1997), *Breve storia dell'Italia meridionale dall'Ottocento a oggi*, Universale Donzelli, Roma.

Boltansky, L. and L. Thévenot (1991), *De la Justification. Les économies de la grandeur* Gallimard, Paris.

Brunner, O. (1988), *Storia sociale dell'Europa nel Medioevo*, Il Mulino, Bologna (original title (1978), *Sozialgeschichte Europas in Mittelalter*, Vandenhoeck & Ruprecht, Gottingen).

Byé, P. and M. Fonte (1993), 'New Functions for Rural Space in Western Europe: A Challenge for Agricultural Techniques', *International Journal of Sociology of Food and Agriculture*, No. 3, pp. 82-96.

Cattaneo, C. (1957), 'La città considerata come principio ideale delle istorie italiane', in G. Salvemini, E. Sestan (eds), *Scritti storici e geografici*, Firenze.

Cecchi, C. (1992), 'Per una definizione di distretto agricolo e distretto agroindustriale', *La Questione Agraria*, No. 46, pp. 81-107.

Chenery, H. B. and M. Syrquin (1986), 'Typical Patterns of Transformation', in H. Chenery, S. Robinson and M. Syrquin, *Industrialization and Growth*, Oxford University Press, New York.

Esquinas-Alcazar, J. (1991), 'Un sistema globale per la difesa delle risorse genetiche vegetali', *La Questione Agraria*, No. 44, pp.49-66.

Esquinas-Alcazar, J. (1994), 'La difesa della biodiversità dopo Rio 1992', *La Questione Agraria*, No. 53, pp. 125-135.

Fonte, M. (2000), Food systems, consumption models and risk perception in the late modern society, Paper presented at the X World Congress of Rural Sociology, Rio de Janeiro, 31st July-5th August.

Friedmann, H. (1982), 'The Political Economy of Food: The Rise and Fall of the Postwar International Food Order', *American Journal of Sociology*, No. 88, pp. S248-82.

Gramsci, A. (1972), *La Questione Meridionale*, Editori Riuniti, Roma.

Hervieu, B. (1991), 'Discontinuities in the French farming world', *Sociologia Ruralis*, 31 (4) pp. 290-299.

Hobsbawm, E.J. and T. Ranger, (eds, 1992), *The invention of tradition*, Cambridge University Press.

Iacoponi, L. (1996), 'La sfida della moderna ruralità: sviluppo rurale, ambiente e territorio', pp. 51-101, in *L'agricoltura italiana tra prospettiva mediterranea e continentale*, Proceedings of the XXXIII Sidea congress, Napoli.

Jansen, A. J. (1991), 'The future of the periphery of the periphery. The case of a Sicilian Agrotown', *Sociologia Ruralis*, 31 (2/3), pp. 122-139.

Latour, B. (1987), *Science in action*, Harvard University Press.

Leontidou, L. (1996), 'Mediterranean Cities: Divergent Trends in a United Europe', in Blacksell, M. and A.M. Williams (eds) *The European Challenge: Geography and Development in the European Community*, Oxford University Press.

Letablier, M. T. and C. Delfosse (1995), 'Genèse d'une convention de qualité', Cas des appelations d'origine fromagères', in G. Allaire, R. Boyer (eds) *La grande transformation de l'agricultur*, INRA, Economica, Paris, pp. 97-118.

Lowe, P. and M. Bodiguel (eds, 1990) *Rural Studies in Britain and France*, Bellhaven Press, London.

Lowe, P. and H. Buller (1990), 'The Historical and Cultural Contexts' in Lowe, P. and M. Bodiguel (eds) *Rural Studies in Britain and France*, Bellhaven Press, London, pp. 4-20.

Miller, S. (1996), 'Class, Power and Social Construction: Issues of Theory and Application in Thirty Years of Rural Sociology', *Sociologia Ruralis*, 36 (1) pp. 93-116.

Murdoch, J. and M. Miele (1999), 'Back to Nature': Changing 'Worlds of Production' in the Food Sector', *Sociologia Ruralis*, 39 (4) pp. 465-483.

Nocifora, E. (1992), *Le città del mezzogiorno. Urbanesimo e sviluppo socio-economico*, Bulzoni, Roma.

Orléan, A. (1989), Pour une approche cognitive des conventions économiques, *Revue économique*, 40 (2), pp. 241-472.

Papadopoulos, A. G. (1997), ' "Eliminating the Rural": a Southern Response to European Integration and Globalization', paper presented at the 17th Congress of the European Society for Rural Sociology, August, 27th Chania (Crete).

Ploeg, J.D. van der et al. (2000), 'Rural Development: From Practices and Policies towards Theory', *Sociologia Ruralis* (40) 4, pp. 391-408.

Polidori, R. and D. Romano (1997), 'Dinamica economica strutturale e sviluppo rurale endogeno: il caso del Chianti Classico', *Rivista di Economia Agraria*, No. 4, pp. 395-427.

Rossi-Doria, M. (1982), *Scritti sul Mezzogiorno*, Einaudi, Torino.

Saraceno, E. (1994), 'Alternative Readings of spatial differentiation: The rural versus the local economy approach in Italy', *European Review of Agricultural Economics*, No. 21, pp. 451-474.

Sereni, E. (1972), *Storia del paesaggio agrario italiano*, Laterza, Bari.

Sylvander, B. (1995), 'Conventions de qualité, concurrence et coopération. Cas du "label rouge" dans la filière Volailles', in G. Allaire and R. Boyer (eds) *La grande transformation de l'agriculture*, INRA, Economica, Paris, pp. 73-97.

Thévenot, L. (1995), 'Des marchés aux normes', in Allaire, G. and R. Boyer (eds) *La grande transformation de l'agriculture*, INRA, Economica, Paris, pp.33-52.

Tovey, H. (1997), 'The place of food in the post-productivist countryside', paper presented at the 17th Congress of the European Society for Rural Sociology, August, 27th Chania (Crete).

Tovey, H. (1998), 'Rural Actors, Food and the Post-Modern Transition', in Leo Granberg and Imre Kovách (eds) *Actors on the Changing European Countryside*, Institute for Political Science of the Hungarian Academy of Science, Budapest, pp.20-43.

Valceschini, E. (1995), 'Entreprises et pouvoirs publics face à la qualité. Les produits agro-alimeantaires dans le marché européen', in Allaire, G. and R. Boyer (eds) *La grande transformation de l'agriculture*, INRA, Economica, Paris, pp. 53-72.

Wilkinson, J. (1997), 'A new paradigm for economic analysis?', *Economy and Society*, No. 26, pp. 26-42.

PART III:

THE GREEN RING IN THE NORTH OF EUROPE

12 Peasantisation and Beyond in Finland and Scandinavia

LEO GRANBERG AND MATTI PELTONEN

Introduction

In his celebrated study *The European Experience* Dieter Senghaas expressed the conviction that "Finland's development ... is representative of the long-term development dynamics which can be observed all over Scandinavia" (Senghaas 1985, 80). We are applying the same idea here and offer a schematic picture of agrarian development in Northern Europe by using Finland as an illustrative case and adding comparative information on the other Northern countries. It goes without saying, however, that Finland is not an average Scandinavian case, but in many ways an extreme example of the Scandinavian path. Its geographically marginal location, its harsh and damp climate, the slow urbanisation and late industrialisation impose quite demanding conditions on its development. This explains why many common phenomena are experienced in Finland in a contradictory and difficult way.

The State in Expanding the Peasantry (1750-1850)

The most important factor in the agrarian development of the Scandinavian kingdoms, Sweden and Denmark (Sweden included Finland and Denmark Norway) was the state. It could be argued that Scandinavia represented a case of the more general European phenomenon of *Bauernshüzt*, the conservation of peasantry (Jan DeVries 1976, 61). State measures to protect and promote the peasantry were several, but they were also restricted, only the tax-paying and land-owning strata of the peasantry being in favour to the detriment of the lower strata of the agricultural population. The most important of these reforms in Scandinavia was the enclosure movement of the 18th century, which acquired specific characteristics, quite clearly distinguishing it from its British and continental counterparts. Typical of the Scandinavian enclosures was the central role of the state and the methodical and quite peaceful implementation of the reform, which also received much support from the more wealthy section of the peasantry. Furthermore, Scandinavian enclosure was a compromise, and the holdings of a farm after the enclosure could remain divided in to several, parts, usually 3-4. The first enclosure acts were also implemented quite rapidly, in Denmark in about 40-50 years, and in the most important agricultural regions of Sweden and Finland in about 75-80 years (Utterström 1957, Soininen 1974, Dombernowsky 1988).

The Swede Jacob Faggot, the best known advocate of enclosure movement in the 1740s and 1750s, spoke about freeing the land-owning peasant to do what he did best, to cultivate the land and to 'free' him from other tasks which hindered his activities in connection with grain growing. He compared his agrarian policy to bee-farming – in order to get honey you must first have flower meadows. Faggot meant that the tax revenue gathered from the peasantry, the most important source of state income at the time, would be larger, when peasants were given as much freedom as possible in their work (Faggot 1746).

After the enclosure movement was initiated in Sweden in 1757 and in Denmark a year later, new reforms on behalf of the peasantry followed. In Denmark these reforms of the 1780s are called *Landboreformerne* (rural reforms). In Sweden these reforms were more widely scattered, but in 1789 all the rights of the land-owning peasantry were collected in to a unified law to secure their position. The main architect of the Danish *landboreformerne* was Count C. D. Rewentlow, himself one of the great landowners in Denmark. He was able to admit that freeing the peasantry from the yoke of the manor houses would serve the interests of those in power more than continuing the former arrangement, under which noble landowners mediated between the state and the peasantry in many ways (e.g. in collecting taxes, recruiting men for the army and in the legal system). The old Danish agrarian system was abolished in the 1780s in a series of reforms which have a special place in the history of that country. Guaranteeing the more independent status of tax-paying peasants included elimination of the remnants of serfdom (*staavnband, hoveri*), which burdened the wealthy section of the peasantry (*gaardmen*). These reforms did not, however, improve the situation of agricultural workers and servants. (Björn 1992, Kjeargaard 1994).

Important for the carrying out of this great cycle of reforms was the high price of grain. This favoured the biggest producers, who had surpluses over and above their own use and could deliver grain to the markets. It is also noticeable how most of the reforms favoured crop production and neglected dairy farming. The high price of grain even favoured Nordic agriculture; Denmark's manor-house producers flourished particularly. The period before the middle of the nineteenth century was actually called the *Kornsalg*-period (Grain-sales period). Only the most remote parts of Finland, which had become in 1809 a semi-autonomous part of the Russian empire experienced hard times every now and then.

After the enclosure movement agriculture needed a lot of new investment. New lands were cleared, drainage systems were developed and houses that were moved from older village centres were erected again or completely rebuilt. Authorities and landowners favoured new marriages among the landless population, the population started to grow vigorously and the differentiation of the agricultural population became greater (Gadd 1984, Rantanen 1997).

Finland is an exciting case, because the Swedish attitude towards reforming Finnish agriculture as well could have been negative. This possibility is underlined by the fact that in the kingdom of Denmark the enclosure act did not concern Norway, which had to wait another hundred years for its own enclosure act. Finland was very much at the centre of attention of Swedish economic politicians as a resource to be used more comprehensively. Jacob Faggot had led the reconstruction work after the Russian occupation of Finland in 1742-3 and had acquired a comprehensive knowledge of the circumstances in the eastern part of the kingdom. In Finland the role of the state was particularly important in initiating the implementation of the enclosure act (Soininen 1974, Rantanen 1997).

As a consequence of the agrarian reforms of the 18th century, the land-owning peasantry emerged and started to develop into an independent actor in the Scandinavian societies. The increasing number of independent farms in Finland is presented in Table 12.1 Of course the independent peasantry was powerful in the local community at this period, because of its resources (land, forest) and its right to employ the landless peasantry. Before the mid 19th century the political regime did not allow the peasantry the opportunity to organise itself politically, even though in Sweden the land-owning peasantry was present as an estate in the diet (*Riksdag*). Political mobilisation was impossible because the state (and the Church) effectively forbade horizontal ties between individuals. Religious movements were the only alternative means of expressing common opinions, and even this was monitored by the clergy.

Table 12.1 The Number of Farms in Finland, 1749-1998

A. The Number of Farms in Finland 1749-1901

	Independent Farms (Freeholds)	Crofter Households (Leaseholds)	Total
1749	31155	3872	35027
1775	46670	12051	58721
1805	58436	23924	82360

1815	75279	31001	106280
1840	77664	45195	122859
1860	81553	61864	143417
1882	107894	62205	170099
1901	122842	67083	189925

B. The number of farming units 1901-1998, with an agricultural area over 1 hectare

	Farm over 50 ha	Total
1901	..	211702
1910	3678	199709
1920	3773	225068
1929	3697	249184
1941	3261	245829
1950	1507	305287
1959	1572	331263
1969	1902	297252
1980	2953	224721
1990	4811	199385
1998	9082	153442*

* of which active farms (with fields under cultivation) 88070
.. no information

Sources: Rasila, Viljo (1961): Suomen torpparikysymys; Suomen talous-historia 3, 69; Farm register 1998 (www.mmm.fi/tike/tilasto/rakenne/ lukumpelto.htm)

The Emerging World Market and Agrarian Differentiation (1850-1913)

Crisis and Adaptation

In the latter part of the 19th century the agrarian world market started to expand rapidly and export prices for grain soared. This affected European agriculture, which was mostly based on small-scale farming. Because of the low prices, arable farming became unprofitable and farmers suffered all over Europe. In Scandinavia farms largely turned away from grain-growing towards specialising in dairy farming, but there were great differences between countries in how fast and efficiently they could react to changes in the world market (see Table 12.2).

The role of the agrarian sector was also important when these countries began to intensify their participation in the international division of labour. In Sweden and Finland the export of forest products grew in importance, and in Denmark the export of dairy and meat products. Only Norway was different and, instead of developing its dairy farming it specialised in fishing and shipping.

Denmark lacked forests, but other Nordic countries could benefit from exporting forest products. Until the 19th century tar had been an important peasant-made forest product. Later sawn timber and, at the end of the 19th century pulp and paper became major export products for the Nordic countries. Since in Nordic countries all landowners were also important forest owners, not only the state but also peasant landowners were important players in forest politics. Unfortunately the economically difficult situation during the period of low grain prices made peasant forest owners weak in their relationship to the emerging forest industry and other large-scale users of timber.

In Finland, the conditions for adapting to the new demands of the world market were most unfavourable, because of slow industrialisation and urbanisation and a low level of emigration compared to other Nordic countries. Finland experienced great difficulties in the countryside. Small leaseholders, who still paid most of their ground rent in labour, started to feel the insecurity of their economic position in the 1880s. At the beginning of the 20th century there were several waves of strikes by agricultural workers in the south-western part of the country and, furthermore, leaseholders began to organise themselves. The agricultural population was divided and in the civil war of 1918 they fought on both sides.

In other parts of Scandinavia the social contradiction was not as great as in Finland. The huge emigration to the United States in the latter part of the 19th century and at the beginning of the 20th century mitigated the most severe social problems, even if emigration has to be considered a result of deep social crisis. By the turn of the century one million people had left Norway for America, or half the country's population at that time (see Table 12.2). Only the emigration rate of Ireland exceeded the Norway's in the nineteenth century (Almås 1993, 52).

Table 12.2 Indicators of Agricultural Adjustment (percentage)

Indicator	Denmark	Sweden	Finland	Norway
Self-sufficiency in bread grains (1910)	60	80	41	..
Average yearly milk yield Per cow, kg (1913-1915)	2700	2200	1300	1500
Proportion of co-op. members:				..
- of cattle owners	90	17	18	..
- of milk cows	93	25	28	..
Emigration (millions) (1815-1930)	0.4	1.2	0.4	0.8

Sources: Peltonen Matti 1993, 361, Baines, Dudley 1995, 3, Gebhard, Hannes 1923, 165.

The state was particularly weak in Finland at this period. This has to do with the limited sovereignty of the semi-autonomous Finnish state as part of the Russian Empire. In Finland, as in other Nordic countries agricultural export was supported by several means. Agricultural education and extension work were initiated. The bigger issues, however, concerning import of bread grain and the position of leaseholders and the other lower strata of the peasant population were more difficult. In Sweden import duties were introduced after a long political fight. Norway and Finland practised free trade in bread grains during the whole period.

Peasant Politics and Agrarian Movements

Since the Swedish regime, the state had had an important role in expanding the peasantry. Peasants were represented in the Diet (Assembly of Estates, *Riksdag*) and land-owning peasants participated in municipal administration. Finland's becoming a part of Russia did not change the situation. Political mobilisation began late compared with Sweden and Denmark. At the end of the 19th century, several popular movements emerged in Finland. They had both educational and economic goals: education in Finnish was the mission of the Elementary School (*kansakoulu*), which were not compulsory until the 1920s and Folk High School (*kansanopistot*), which brought peasants and popular groups together. The temperance movement (*raittiusseura*) was active both in urban and rural areas. Farmers' Associations (*maamiesseurat*) were the first professional organisations for land-owning farmers and Youth Associations (*nuorisoseurat*) formed an open massorganisation for all rural youth.

All these and other similar organisations started the first wave in the construction of modern civil society in the Finnish countryside at the end of 19th century. According to Alapuro, the character of Finnish rural movements (and the modernisation process as a whole) was reactionary. National protests arose from external issues, at that time mostly Russian: such as hard-line Russian nationalist politics and Russian revolutionary movements (Alapuro 1994, 15). At the same time, the labour movement was organised locally in workers' associations and nationally in the Social Democratic Party. Tenant associations were organised in collaboration with workers' associations, the co-operative movement started from dairy co-operatives and co-operative shops and banks were established. Alapuro has analysed the social background of this activism in Huittinen, a municipality in south-western Finland, in which co-operative movement involved both land-owners and tenant farmers. Co-operative shops and banks also involved agrarian workers and craftsmen. He makes the distinction between early organisation of the peasantry and the later class-based organisation of workers and crofters. Peasant organisation contributed to bridging the distinction between earlier elite groups of the estate society and the wealthy peasantry. When workers and crofters became organised, a new distinction was created between labour and employers as well as crofters and peasantry (see Alapuro 1994, 55-82). The Civil War interrupted this development in 1918.

During the latter part of the 19th century, the ties between the state and the land-owning strata of the peasantry were weakened, but not completely severed. The quite unexpected fall in grain prices created pressure on the old political alliances in the whole Scandinavia. For some decades the peasantry became a burden to the state as an interest group demanding support and an employer group causing social unrest by stubbornly hanging on to the old customary ways of employing labour by renting agricultural land to the poorest strata of the peasantry (*torpare* and *backstugusittare*). The land-owning peasantry could now use its strong position as the owner of productive assets (land, forest) and use its employees more intensively. The rearrangement of land ownership and social relations between different groups in farming communities became unavoidable. In Finland, especially the countryside was divided into two hostile camps from the late 1890s on.

The Struggle for a Peasant State During the Long Crisis 1913-1952

Making of the Peasantry

This period of forty years – called by Hobsbawn (1994) "the age of catastrophes" – was a continuing series of crises, which started with the First World War (in which Nordic countries were mainly onlookers) and with Finland gaining independence as a lucky result of the Russian revolution. This was followed by a less fortunate result of the same revolution – the chaotic and brutal Civil War in 1918. The 40-year period included the economic depression of the 1930s and two wars against Russia, in 1939-1940 and 1941-1944. In the post-war period Finland found herself in a severe political situation under the power of the Control Commission of the Allied Powers, which continued until the peace treaty and the separate pact with the Soviet Union was signed in 1948 and a new regime of neutrality began with a cautious rapprochement with the Nordic collaboration.

During this period, farm structure was reorganised by two radical agrarian reforms. The law on abolishing tenant farming was passed after the Civil War, restructuring land-ownership to the benefit of small-holders. Minor steps in the 1920s and 1930s led to a radical sequel after the Second World War. Because of Finland's territorial losses, the Karelian immigrants had to be resettled in other parts of the country. Soldiers, their widows and some other groups were given agricultural and forest land and Lapland needed to be rebuilt. These resettlement activities continued until the 1960s and even in to 1970s to some extent.

By the beginning of the 1950s, resettlement was largely finished and the outlines for agricultural income policy were set out in the form of an agricultural income agreement. During this time, the peasantry, already undergoing the transformation process to modern family farming, had a deep impact on the cultural and economic life in Finland. The peasantry's road to political power will be described in more detail in the following.

'Abolition of Tenant Farming' and Land Reclamation Activities

The abolition of the tenancy system was carried out quite rapidly during the 1920s. By 1939, about 47.000 crofter and about 45.000 cottager households had been liberated. Redemption dues were reduced, as compensation for the former tenants' personal investment was not required. The 'liberation' affected over two fifths of all agricultural units and changed the ownership and tenancy relations in the Finnish countryside profoundly. The crofter issue had found its political expression in the crofter movement – organised action led and supported by the Social Democratic Labour Party. When the crofter question were solved, former crofters became land-owning peasants, even if with relatively small holdings. Now the peasantry was stronger than ever before. The similarity of the basic interests of the new peasantry compared with the old could be seen in the growing activity of peasant-led organisations in the countryside, not least in the activity of the agrarian political party, the Agrarian Union (*Maalaisliitto*). However, the tenant liberation left the agrarian question partly unsolved, as it was only concerned with ownership rights and not with redistribution of arable land to small holders or agricultural workers. These issues became the subjects of land settlement policies.

Land Settlement Activities

The settlement fund was created in 1920. Furthermore, the Parliament passed two laws on land acquisition (*maanhankintalait*) in 1922 and 1936. The Land Reform did not really begin, however, before the Second World War. When the first phase of the war against Russia, the 'Winter War' ended in March 1940, one tenth of Finland's population – i.e. approx. 460,000 evacuees from the regions of Eastern Karelia which had been ceded to the Soviet Union – had to be re-settled. A rapid settlement programme for rural immigrants was set up, but it was interrupted when war broke out again in summer 1941. At a victorious stage in Finland's 'Continuation War', the Karelians were able to move back home. Now it seemed that the Karelians did not need all the farms reserved for them and the national leaders were also able to promise land to the soldiers and their families as compensation for their sufferings in the war. However, the fortunes of war were reversed, and the Karelian evacuees had to be permanently resettled. The total area of arable land in the ceded territories of Karelia was 277,000 hectares with an agricultural population of approx. 220,000. Furthermore, about 100,000 Finns lost their homes in the last phase of the war against German troops in Lapland (Granberg 1992, 70-71 and 1969, 34-35).

The government set about resolving this problematic situation by means of extensive resettlement and reclamation projects. The land needed for this was primarily acquired from the state, local authorities and companies, but also from private farmers – partly by voluntary bargains and partly by expropriation. Almost 40 percent of the arable land was purchased from large farms, which resulted in a greater predominance of small farms than before the war. The Karelians were largely settled mainly before 1948, but other land reclamation activities still continued in Northern Finland until the 1960s and even in the 1970s. Clearing new arable land, building roads and constructing houses and other farm buildings was a great rural project, which was supported by the state by direct subsidies and beneficial loan arrangements.

Protection of the Peasantry

As Eric Hobsbawn has remarked, post-war capitalism was a kind of mixture between economic liberalism and social democracy (or the American new-deal policy), which in many respects followed the example of the planned economy in the Soviet Union (Hobsbawn 1994). In this mixture, agricultural policy gave priority to planning instead of free markets. The European agricultural policy of the second half of the century was an outcome of the historical lessons of the first half. The first lesson was that the free market did not guarantee such basic needs as food. The second lesson concerned the dangers for democracy if a large section of the peasantry was mobilised in support of fascist movements. After the bankruptcy of the pre-war political system and the lack of political legitimacy in Europe, getting the peasantry to support the new governments was unavoidable (Milward 1992). Thirdly, the war period was a training course for governmental planning and market intervention, not least in the food and agricultural sector. The idea of national self-sufficiency in foodstuffs was an outcome of the First World War, while the idea of income parity between farming and the industrial population was an outcome of the Second World War. A decade after the Second World War, the free market had almost entirely disappeared from the agricultural sector. Although the main features of this development were shared, national variations were considerable. In the following we will first highlight the administrative innovations in Finland and thereafter return to the role of peasantry and rural civil society in shaping the countryside.

In Finland, initiatives to support small-scale farming has been discussed in the Parliament since the 1930s, when a support scheme for small farmers buying fertilisers was suggested (Granberg 1986). The economic depression in Europe in the 1930s forced one government after another to adopt protectionism and export subsidies for major agricultural products. Some of the anti-depression measures in the Finnish countryside were channelled through co-operative banks, which also became a practice in later governmental measures concerning settlement activities and state-guaranteed loans to farmers. A similar system of market control, run by farmers' co-operatives and supported by the State, was developed in Norway (Almås 1985, 52). Farmers' organisations were thus strengthened as a consequence of anti-depression policy.

During the Second World War, various restrictions on the market economy were stipulated (*säännöstelyvaltuuslaki* 1941) and income negotiations commenced. After the war, in 1945, the government established a Committee on Prices and Salaries, which represented interest organisations (industry, labour and farmers). In the coming years, while market restrictions could only be partly dissolved, a new and more permanent system for an income and price policy was planned. Regional support and a subsidy for small farmers were introduced during the 1940s. The decision-making system concerning farmers' incomes was more or less open until 1952, when the first one-year agreement on agricultural income was struck between farmers and government. In 1956, the Agricultural Act was legislated for three years, a practise, which continued for 40 years until Finland's membership of the EU.

Government had therefore done its best to facilitate rising prosperity in rural areas in the coming decades. Many surprises were in store for the agrarian politicians, however. Let us first consider who the policy-makers were and what the origin of activities in the rural 'third sector' was.

The Peasantry as an Actor in the State Policy

After the Civil War, peasants were organised in four areas, economically in the co-operative movement, professionally both in extension organisations and in farmers' associations and politically in the agrarian party, the Agrarian Union. During the succeeding decades the peasantry became a class *für sich*, able to exert considerable influence on governmental policy. Collaboration between government and new movements in civil society is obvious in the case of co-operative banking. The Central Loan Reserve Fund of the Co-operative Banks (*Osuuskassojen Keskuslainarahasto* OKO) had to be established first (1902), before local saving banks (*säästökassat*) could be established (Alanen 1964, 354-369). The special objective of this banking activity was to support small farmers and other agrarian people in buying land and equipment for farming. From the very beginning, banks were used to channel government loans to the farmers. Banks also became the main institution for granting loans to tenants and small farmers after rented land was redeemed in the tenant liberation. During the depression, co-operative and savings banks distributed a state-guaranteed loan, the so-called franc-loan of FRF 300 million, to farmers and rural municipalities to allow them to settle their accounts.

The co-operative movement became the basis for the peasantry's economic power. What were the reasons for the success of farmers' co-operatives? First, co-operative banking capital gave the co-operatives a chance to develop into a financial and industrial grouping. Second, the public sector supported emerging co-operative business in various ways. Third, economic action was supported by political and ideological action. Fourth, as forest owners the peasantry succeeded in running co-operatives in the forest industry, thereby gaining access to the core of Finland's economy. Relatively even distribution of forest ownership gave the peasantry more resources than many other countries. Consequently, peasant activists and politicians had powerful protection in the economic sector. Later, differentiation between agrarian and industrial interests emerged. This meant that peasant interests were also divided between industrial (especially the paper and pulp industry – the former main export industry) and agrarian interests.

During the thirties, a struggle over the political orientation of the Finnish peasantry took place. During the depression, several local small-holder protests failed to develop in to a nation-wide movement. Instead, the right wing and anti-Communist Lapua Movement succeeded in organising a considerable section of the wealthier farmers in to a mass-movement. This movement culminated in the so-called Peasant March in Helsinki, which was a genuine threat to the weak democracy of the country. An attempted coup was avoided, however, and the Agrarian Union emerged as the winner from the political storms of the 1930s.

The Agrarian Union belonged to the Nordic group of farmers' parties, rare in other regions in Europe. Its ideology was based on agrarian fundamentalism, representing a third way between socialism and capitalism. It remained small before the Civil War, trying to build a profile as small-holders' party, but after the 'crofter liberation' the party changed strategy in order to represent the peasantry as a whole. In the 1920s it became the dominating party in rural municipalities across most of the country – with the exception of the Swedish speaking regions, in which the Swedish Folk Party took a similar position. The Agrarian Union participated in the Government with the social democrats in 1937 and succeeding in getting its candidate Kyösti Kallio elected as the President of Finland. This started its long period in office in Finnish politics.

A new agricultural advisory organisation continued the tradition of farmers' associations (*maamiesseurat*), but it was supplemented by small farmers' associations, the de facto successors of tenant organisations. Such organisations concentrated on promotion of agricultural modernisation.

The Farmers' Union was originally an agricultural employers' organisation but came to represent the peasantry as a whole. It took decades before former landowners and their tenants could work together in such organisations, even if the work was in quite neutral areas. In these decades, the Union of Small Farmers' Associations (*Pienviljelijäin Keskusliitto*) was still strong but, especially during and after the Second World War, the Farmers' Union gained the dominance in interest politics and became the official representative of farmers in agricultural income negotiations. The Union of Small Farmers' Associations became the left-wing alternative to the Farmers' Union until the advisory organisations were united and depoliticised at the end of the 1960s.

In the 1930s, agricultural subsidies were a question of intense political struggle because, at that time, economic liberalism still had strong supporters, not only among the right wing parties, but also among social democrats. Finland's Social Democrats also had an agrarian wing. This consisted mainly of small farmers with strong sympathy for protectionism who contributed to the party's action programme by raising issues favourable to small farmers, such as the objectives of settlement activities.

The communist movement worked underground but their legal wing tried, after the liberation of crofters, to create a union with small farmers. The Agrarian Union was rising to its post-war position as a permanent member of the government. The impact of the agrarian population in Finnish politics was strong. In the Nordic countries the situation was quite similar, in Norway, for example, the social democrats and agrarians (and fishermen) met in governmental politics in the 1930s (Almås 1985, 52). After the 1930s, the new peasantry was not only a class *an sich* but also a class *für sich*, carrying through its own interests in societal and political life.

'Regulated De-peasantisation' 1952-1990

Agricultural Sector as an Economic Burden

When the period of crises ended, Western European agriculture was to be organised in a new way. The free agrarian market was just a lost dream of economic liberalism. Most European governments now adopted the idea of income parity between agriculture and industry and this idea was institutionalised in somewhat varying ways in different countries (Tracy 1982). In the Nordic countries, agriculture was protected from the international market by import restrictions and export subsidies after the war. Attempts to deregulate agriculture after the war were less successful, with the exception of Denmark. Denmark's exports were targeted to the British market, where the price level was much lower than in Germany, Denmark's previous main export market (e.g. the price of butter was about 33 percent higher and bacon about 25 percent higher in Germany than in Britain). However, free trade did not exist, even in Denmark. Exports to Britain were based on long-term agreements and a monopoly was given to export councils (*eksportudvalg*), in which farmer organisations had a leading position (Laurinkari et al. 1984, 54-59).

The Agricultural Income Act became the cornerstone of Finland's agricultural development in the second half of the century. The average income level of the industrial worker became a standard for income levels in agriculture. The principal aim was that agricultural incomes should grow at least as rapidly as industrial worker's incomes – the political debate was mainly over whether or not to raise agricultural incomes even more rapidly, in order to bridge the gap between industry and agriculture. Complicated annual cost-income calculations tried to verify the 'objective' need for price increases. In addition, two groups of measures were applied. In the spirit of the welfare state, the legislation included subsidies both for farms in less-favoured areas and for low-income small-holders nationally. Another group of measures was the outcome of unwelcome side-effects of income policy itself. While income guarantees encouraged farmers to increase productivity and new mechanical and chemical methods contributed to this effort, overproduction started to rise. At first this was modest, temporary and concentrated on only some products, but during the 1960s overproduction turned out to be a permanent and chronic problem covering the whole range of major agricultural products. New measures were needed to cut export subsidies and to level out production.

Norway had the most radical political programme as far as income parity was concerned. In Norway, natural conditions limited production much more than in other Nordic countries and overproduction was not a major problem before the 1990s. After a series of political events (the vote on Norway's membership of the EEC and an organised tax revolt by a group of farmers in 1975) Parliament adopted a policy intended to equalise incomes between farmers and industrial workers by 1982! According to Almås, "the average income in the farm sector doubled between 1975 and 1982 and the gap between incomes of industrial workers and farmers was reduced" (ibid.

33). After 1979, however, the gap began to increase again and this objective was abandoned.

Modernisation Pressure

The Finnish countryside was full of life and hope after the war. New farms were established, new villages built, fertility reached its peak, agriculture and forestry had increasing markets at home and abroad. At the same time, production methods were undergoing revolutionary change. In the United States Fordism was adopted in the tractor industry, and after the war, Marshall Aid was used to import tractors to Scandinavia (on Denmark, see Pedersen 1988, 181-3). Finland did not receive Marshall Aid, but tractors were soon put to use in Finland, as well. Tractors and chain-saws were introduced in rural Finland in the 1950s, and these innovations became widely in use even in the small farms by the late 1960s. Together with the increasing assortment of tools to be connected to a tractor, the new technology revolution decreased the number of working hours needed on a farm – which was one of the causes of the coming structural crisis in the 1960s and 1970s.

Immediately after the war, a wave of urbanisation took place. In spite of active propaganda in favour of agrarian values, nothing could stop this migration into the towns. The amount of paid labour in agriculture started to decrease in the 1940s and dropped by half during the 1950s. The decrease continued up to the 1980s, by which time this category mostly consisted of holiday replacements funded by local authorities. Parallel to paid labour, the number of assisting family members, i.e. children, relatives and grandparents, also started to decrease in the 1940s, falling to half in the 1960s. In the 1940s, agrarian reform was able to still counterbalance emigration and the employed population remained stable. Thereafter, the migration stream from the countryside to the towns continued for half a century. The farms were thus increasingly left in the hands of the farming couple alone: in 1940, every fourth person working on a farm was a farmer or a farmer's wife, and thirty years later every second. In reality, the change was even more radical, given that a great deal of the agricultural work was seasonal. The family farmers at the turn of the 1970s had to find new methods in circumstances where help was normally not to hand when needed. This modernisation pressure only emphasised some of the more unpleasant aspects of farming: the heavy work, social isolation and a stricter gender division of labour.

Modern lifestyle was adopted in rapidly growing towns and although rural incomes were rising, the gap between incomes in industry and agriculture was great, as was the difference in the way of life. Furthermore, income differences between small and large farms were increased (Granberg 1989, 122-123). It was not a surprise that sons and above all daughters from small farms decided to move to the cities in order to start studies in the rapidly increasing number of educational institutes. The traditional bequest of a farm by partible inheritance became difficult for small farmers, who could not afford modernised agricultural methods and life style. The crisis of agricultural structure was especially dramatic in Northern, Eastern and Middle Finland, where incomes from forest work were simultaneously lost. What followed has been called the Great Migration: in the middle of 1960s the number of farms started to decrease and soon hundreds of villages became empty, as the inhabitants moved into the cities in Southern Finland or to Sweden where the standard of living was much higher at that time. Rapid migration completely changed the population structure and, in about 10 years, Finland moved from an agrarian society to a service society. What saved the Finnish population from further problems was the labour shortfall in prosperous Sweden and the welfare state project in Finland.

Regulated De-peasantisation in the Welfare State

Agricultural policy could not help reacting to changed technological and economic preconditions for farming. In Finland this reaction was delayed compared with other Nordic countries, while the resettlement policy clearly had as harmful impact on the agricultural structure from an economic point of view. In Sweden and in Norway agricultural rationalisation was accepted and began, slowly, immediately after World War II, combined with income policy and various measures to support small farmers. Denmark's liberal trade policy caused strong pressure to rationalise agriculture. Stepwise rationalisation in these countries made the change easier than in Finland. In all Nordic countries, social democrats had an important role in transmuting the new structure inside the economic sector and between economy and policy. For example in Norway, the Social Democrats won a parliamentary majority in the post-war years and their "idea ... was to solve the problem of agrarian structure by creating new jobs in the manufacturing and construction industries and in the public sector. Farming was supposed to be more profitable through enlargements, specialisation, and mechanisation" (Almås 1993, 53-54). In Norway, the number of farms first decreased slowly from 213,000 in 1949 to 198,000 in 1959 and then rapidly to 155,000 in 1969 and 115,000 in 1979 (Almås 1993, 53-54). This was parallel to the decrease in Sweden and Denmark. In Finland, similar thoughts were not aired until the early 1960s and, as far as agriculture is concerned, a new structural policy replaced settlement policy only at the end of the decade.

In the social democratic view, the question of the welfare state was related to the goals of economic growth. Social policy was a safety network for people but, much more in the economic sense. Social policy (in broad sense the welfare state) was seen as a positive contributor to economic growth. The welfare state as a whole had to be based on social thinking and ideas like equity and justice. A similar line came to be adopted even in agricultural policy, but in the years following World War II, Finland was still a hinterland in European social policy. Only minor sections of the population were protected against disease, disability or unemployment. During the period up to the end of the 1950s, there was a gradual activation in the role of the state and municipalities, e.g. in the construction of roads, schools and post offices everywhere, even in the remote rural areas.

After the middle 1960s the Comprehensive School Law (1968), National Health Law (1972) and the Day-care Law (1972) were among the most important reforms in Finland. The transition to the new comprehensive school system in Finland provided that all children of the same age group go to the same type of school. The transition to the new system was timed for 1972-77, beginning in northern areas and gradually progressing towards the South. The aim of these and other reforms in the service sector was to provide the same level of services throughout the country, a by-product of this being an increasing number of municipal jobs. Welfare services became the leading source of municipal employment. A new middle class appeared in rural municipalities, especially in education and health care – which became an argument for the establishment of new regional high schools and universities to educate the staff for the service sector. The growth of welfare services balanced the country's regional development significantly more than the industry-based regional policy. The growth of the public sector as a whole had a strong balancing effect on regional development in all the Nordic countries (Pyy and Rannikko 1995, 136-137, Brox 1984).

Beyond the Peasant State: New Romanticism, New Rurality, New Rural Policy

In Finland, the end of the post-war period of regulated de-peasantisation was marked by two major political changes. The Farmers' Party went to opposition and the Farmers' Union lost its right of veto in income negotiations. The Centre Party (previously the Agrarian Union) had been a nearly permanent government party and consequently took responsibility and power in agricultural and forestry policy, but it had finally lost this position. Because of EU membership, the Agricultural Income Act was reformed in a way, which severed the automatic connection between national input and output prices, or industrial and agricultural incomes. The future of agriculture would depend far more on the guidelines emanating from the EU than on the national policy community. The future of the countryside became increasingly affected by rural and environmental policy, coming to be recognised as the second pillar of the CAP in 1999.

There were many signs of a change in the climate. From the 1980s, the titles of numerous governmental and non-governmental organisations were changed from 'agricultural' to 'rural', and in the 1980s both the concept and the content of rural policy were discussed in the Finnish parliament. Rural policy was institutionalised by cautious steps, when a governmental delegation for rural policy was nominated to co-ordinate this policy area and a network of regionally based academic research and development institutes was established in the 1990s. Rural policy had severe difficulty, however, in attracting political support, because of the dominance of agricultural issues compared with non-agricultural. Construction of the new rural network was structured on various components, and encountered difficulty in articulating common interests of: industry, organic farmers, rural enterprises (bed and breakfasts, tourist activities) and distance-workers.

During the 1990s the decrease in the number of holdings was quickest in Finland and Denmark. Both Norway, which stayed outside EU, and Sweden, in spite of its deregulation policy before membership, changed less abruptly. At the same time the number and share of larger farms increased rapidly, especially in Finland (see Table 12.3).

Table 12.3 Number of Holdings in Four Nordic Countries in 1990, 1994 and 1998

Year	Norway		Sweden		Finland		Denmark	
	All	Over 50 ha	All	Over 50 ha	All	Over 50 ha	All	Over 50 ha
1990	84635	954	96560	15351	119502	4764	79339	16014
1994	78131	1160	90102	15571	105393	5554	69346	16426
1998	72064	1306	85307	16179	88423	8030	59761	17918

The statistics from 1996-97 show that of the 142,000 enterprises in Finnish countryside about 65,000 still concentrated on agriculture and forestry, about 23,000 combined agriculture with non-agricultural business and about 53,000 involved no agricultural activity. Agriculture provided work for 6 percent of employed and other rural enterprises 4 percent. On the other hand, the turnover of non-agricultural rural enterprises was FIM 55,5 billion compared with FIM 20,6 billion in agriculture. Even if forestry is added to agriculture the role of traditional income sources is clearly less than non-traditional rural business (MTTL 1998, 11-14 and 74). The number of summer cottages was already over 400,000 in the late 1990s, that is, five times the number of farms. An increasing number of these are used year round and have been turned from summer cottages to second homes and sometimes into the first home. A neo-romanticism seems to have been bred, renewing the national symbolism of last century, which constructed 'the beautiful Finnish landscape' and chose icons for it. However, many new symbols and signs have also been constructed under the logic of the cultural economy and as a means of attracting paying visitors. Of special importance are the festivals and events based on rural cultural heritage, which have sprung up throughout the countryside in recent decades – the tango market movement in Seinäjoki, the folk music festival in Kaustinen, the rowboat race festival in Sulkava, regional food fair in Kauhajoki etc. (for Sweden, see Ekman 1999). What is definitely new is the organised effort to sell the landscape in one way or another, which should be compared with the basic intention of the peasantry to use the landscape as a source of raw material for food and forest products.

The peasant heritage is also seen in the cinema. The wave of romantic movies located in the Finnish countryside was followed by a cultural revolution from the 1960s until the 1980s when urban values and themes were introduced to the Finnish cinema audience. Somewhat surprisingly, the 1990s then witnessed a new wave of rural romanticism in a different medium (the popular TV series such as 'Family Metsola', the radio show 'Family Kantola', which achieved its thousandth airing in 2000, movies on the themes of the Great Migration of the 1960s, the life of the loggers, the lives of popular artists with rural background and art). One cannot avoid the conclusion that during a time of increasing globalisation, people are seeking security in the common cultural identities based on the rural heritage.

The peasantry has lost its position in Finland as well as in other Nordic countries. Farmers are no longer a major class in Finnish society, not even in the most rural regions. The group of farmer organisations, the Farmers' Union among others, has lost a lot of its impact and political power. Farmers' co-operatives have lost their monopoly in food processing and a lot of economic power in society. The forest industry has lost its leading position to IT technology. Economically, farms have been left a secondary role even in the countryside. A cultural change has taken place from the old peasant culture and its remnants in family farming to constructing a new rurality, a (re)new(ed) interpretation of the landscape and of the good life. A change has taken place from peasant countryside to leisure countryside. The farmers' countryside has been lost forever – precisely as the countryside seems to be more valued and loved than ever before.

Acknowledgements

An early version of this chapter was presented at the X World Congress of the International Rural Sociology Association, August 2000 in Rio de Janeiro. Roderick McConchie (Language Centre, University of Helsinki) checked the language.

References

Alanen, Aulis J (1964), *Hannes Gebhard*, Kirjayhtymä, Helsinki.

Alapuro, Risto (1994), *Suomen synty paikallisena ilmiönä 1890-1933, Hanki ja Jää*, WSOY, Helsinki.

Alestalo, Matti (1986), *Structural Change, Classes and the State*, Research Group for Comparative Sociology No.33/1986, University of Helsinki, Helsinki.

Almås, Reidar (1993), *Norway's Gift to Europe*, Centre for Rural Research, Report No. 2, Trondheim.

Bjorn, Claus (1992), *Den gode sag. En biografi om Christian Ditlev Friedrich Reventlow*, Gyldendal.

Brox, Ottar (1984), *Nord-Norge: Fra allmenning til koloni*, Universitetsforlaget, Tromso.

DeVries, Jan (1976), *The Economy of Europe in an Age of Crisis, 1600-1750*, Cambridge University Press.

Dombernowsky, Lotte (1988), *Ca. 1720-1810. Det danske landbrugs historie II, 1536-1810.*

Ekman, Ann-Kristin (1999), The revival of cultural celebrations in regional Sweden. Aspects of tradition and transition, *Sociologia Ruralis*, vol. 39 3/1999, pp. 280-293.

Faggot, J. (1746), *Svenska landtbrukets hinder ock hjälp.*

Farm Register. www.mmm.fi/tike/tilasto/rakenne/lukumpelto.htm.

Gadd, C.-J. (1983), *Järn och potatis. Jordbruk, teknik och social omvandling i Skaraborgs län 1750-1860*, Meddelanden från ekonomisk-historiska institutionen vid Göteborgs univesitet 53.

Gebhard, H. (1923), *Småbruket i Skandinavien.*

Granberg, L. (1976), *Maatalouden rakennekriisi ja rakennepolitiikka Suomessa*, Master's Dissertation, Institute of Sociology, University of Helsinki, Unpublished.

Granberg, L. (1986), Small production and state intervention in agriculture, *Acta Sociologica* 29:3, Oslo, pp. 243-253.

Granberg, L. (1989), *Valtio maataloustulojen tasaajana ja takaajana*, Suomen Tiedeseura, Helsinki.

Hobsbawn, E. (1994), *Age of Extremes. The Short Twentieth Century 1914-1991*, Michael Joseph, London.

Jutikkala, E. (1964), *Bonden i Finland genom tiderna*, Schildt, Helsinki.

Kjaergaard, T. (1994), *The Danish Revolution 1500-1800. An Ecohistorical Interpretation.*

Laurinkari, J., S. Hiltunen, F. Just and K. Omholt (1984), 'Samspillet mellem staten, landbrugsorganisationerne og landbrugskooperationen. Landbrugskooperationen i Norden', *Arbejdsrapport* 1, Sydjysk Universitetsforlag, Esbjerg.

Milward, A. S. (1992), *The European Rescue of the Nation-State*, Routledge, London.

MTTL (1999), L. Ala-Orvola, I.P. Laurila and J. Marttila (eds.), *Suomen maatalous ja maaseutuelinkeinot 1998*, Agricultural economics research institute, Finland, Research Publications 91, Helsinki.

Pedersen, E.H. (1988), *Det danske landbrugshistorie IV*, Odense.

Peltonen, M. (1992), *Talolliset ja torpparit*, Historiallisia tutkimuksia 164. Helsinki.

Peltonen, M. (1993), *The Peasant Economy and the World Market. Finnish Peasant Farming in the Age of Agrarian Crises, 1880s-1910s*, Review, XVI, 3, 357-381.

Pyy, I. and P. Rannikko (1995), 'The rural welfare state saga', in L. Granberg and J. Nikula (eds), *The Peasant State*, University of Lapland B.20, Rovaniemi.

Rantanen, M. (1997), *Tillväxt i periferin. Befolkning och jordbruk i Södra Österbotten 1750-1890.*

Soininen, A. M. (1974), Vanha maataloutemme. Maatalous ja maatalousväestö.

Tracy, M.(1982), *Agriculture in Western Europe: Challenge and response 1880-1980*, London.

Vattula, K. (ed, 1983), *Suomen Taloushistoria, historiallinen tilasto*, Tammi, Helsinki.

13 Creating and Re-creating Modernity: Peasantisation and De-peasantisation in Ireland

HILARY TOVEY

> Taking continuing modernisation for granted is the modernist mistake par excellence, the more difficult to avoid because it is inscribed in everyday discourse (Djurfeldt 1999, 263).

Introduction

According to Djurfeldt, rural researchers need to reconceptualise the concept of 'peasants' or 'peasantry'. This is because it is not only an 'ethnocentric' term, but also one "too easily coupled with an evolutionist or modernist research programme" (1999, 268). In fact, he favours replacing it with an 'analytical model of family farming', which he argues would be a better basis for comparative and historical understandings of agrarian societies. But failing that, 'peasantry' is best treated as a Weberian 'ideal type' concept, a tool to guide empirical enquiry, but not a description of some essential reality which exists out there waiting to be discovered or whose 'exit from history' awaits documentation. 'Real peasants' are never identical with ideal-typical peasants, and it is little more than a waste of time trying to determine whether they no longer exist, or are on the point of leaving history.

Irish rural sociology would be harshly judged in the light of such comments. It has had a long history of debating whether Irish peasants have come to or are about to come to the end of their historical existence and role; and it has been particularly deeply inscribed with the assumptions of modernist and evolutionist research programmes. Yet debates within Irish rural sociology over the fate of the peasantry have generated fruitful and interesting conceptualisations, at times even anticipating some key theoretical developments in the wider European sociology of agriculture, and can be seen as contributing to our understanding of what should be included within the analytical model of family farming which Djurfeldt calls for.

This chapter starts with a brief overview of how these debates have developed, intended to contextualise my own approach to analysing Irish experiences of peasantisation and de-peasantisation. It is also my way of responding to the criticism that the concept of peasant is ethnocentric. 'Peasant' is a term rarely used in everyday speech in Ireland, because of a widespread feeling that it is denigratory which goes back to colonial representations of English-Irish relations as relations between a civilising world power and a society of backward and insular peasants. Even in Irish sociology the preferred term has been 'small farmers', although some works have contained explicit discussions of what constitutes peasant farming and peasant society. But whatever the terminology used, Irish rural sociology has consistently tried to address a quite specific problematic: the impact of socio-economic modernisation, or of the deepening integration of the Irish economy into international capitalist exchange and commodity relationships, on the family form of farming. Our way of approaching and defining this problematic has indeed been 'ethnocentric', in the sense of being deeply shaped by our understandings of local structures and local political, social and cultural processes, which makes it the more important to have a sense of how the indigenous theorisations have developed over time. But once understood, I believe, this localism does not present a barrier to more comparative analysis across the European experience.

In the rest of the chapter my emphasis is more on analysing Irish peasants as historically constituted entities, than discussing their definition as abstract types. I treat processes of peasantisation and de-peasantisation primarily as processes of the construction and the undermining of peasant projects, and look at how these occurred in particular historical circumstances and in a specific socio-geographical location. Although my discussion pays close attention to the specifics of Irish agrarian history, its intention is to contribute to producing a more general analytic framework. Irish agrarian history only appears too exceptional to offer any general parameters for understanding broad processes of European rural and agricultural change if one is captured by a narrowly teleological understanding of European history which equates 'de-peasantisation' with 'modernisation'. I argue that 'peasantisation', or the creation of an Irish peasantry, has to be seen as a relatively recent event within Irish history and indeed as one which is linked not to 'traditional' Ireland but to the beginnings of Irish 'modernity' (see also Tovey 2001). Mindful of Djurfeldt's critique, however, I do not want to treat 'de-peasantisation' as necessarily the point of exit from history of Irish peasant farmers, and their replacement by some other more 'modern' social group. I prefer to try to open up the question, in a non-deterministic way, of how the new groups, practices, policies and collective identities which are coming to exercise an influence in rural Ireland, and in Irish society as a whole, are structuring the possibilities for the continuation of small-scale family farming in Ireland, albeit in new forms.

Irish Rural Sociology and the Conceptualisation of 'Peasants'

Until around the mid-1990s, rural sociology in Ireland was rooted in the study of agriculture, and as suggested above, it has been focused (if not always in these terms) on trying to answer the question: can a non-capitalist form of production in farming survive into modern capitalist society? The terms of the discussion were largely set by the pioneering work of two American anthropologists, Arensberg and Kimball (1959, 1968). The 'peasant model' (Hannan 1979) which they had developed in order to understand rural life in a small community in the west of the country in the early 1930s was for a long time accepted by nearly everyone working in the field as an appropriate starting point from which to interpret rural social change in all areas of the country. Arensberg and Kimball identified two core constituents of Irish peasant society: individual small-farm households, organised both as families and as work units around the polarities of gender and generation; and informal institutions of mutual support, provided by the local community, which had the function of suppressing long-term reproduction of inequality between households. For some forty years afterwards, this was taken up by most researchers as simply an account of 'real peasants' in Ireland. It was not until the early 1970s, with the publication of a scathing critique (Gibbon 1973) of their findings and assumptions under the guise of a review of a contemporary publication (Brody 1973), that any explicit debate about their work began. Some focussed primarily on the quality of their ethnography, trying to assess the plausibility of Gibbon's claim that "On every score – the family, the 'mutual-aid' system, the economic and cultural stability of the system, and its politics – [this] ranges from the inaccurate to the fictive" (1973, 491). Others located the dispute more centrally as one about the theoretical models which are useful for understanding the fate of small family farming in an economically modernising society.

A key contributor to the debate was Hannan (1977, 1979, 1982, 1987). Gibbon particularly criticised Arensberg and Kimball's emphasis on peasant farming as a system which is stable, harmonious, and above all self-reproducing through time. Gibbon himself adhered to a 'classic'

Marxist position in which the expansion of capitalism and of a commodity economy inevitably leads to land accumulation, class differentiation, the destruction of family farms and the proletarianisation of farm family members. To Gibbon, this pattern of change was already clearly evident even in the 1930s (but ignored by Arensberg and Kimball) within the small-farm communities of the west of Ireland. He also assumed that their experience of change only differed from the more developed agricultural regions of the rest of the country in that they were poorer and their landholdings smaller. For Hannan, Gibbon's perspective entirely missed the specificity (or as Hannan termed it, 'deviance') of small-farming agrarian society in the west of Ireland throughout most of the 20th century. Hannan did not see peasant smallholders in the west of Ireland as just a smaller version of those elsewhere in the country, whose lack of economic resources constrained them to operate their farms at very low levels of development and eventually committed them to proletarianisation. On the contrary, they were the carriers of an authentic, alternative way of life ('peasant society') with a distinctive logic and rationality of its own. This cultural specificity was what allowed Irish peasant communities to continue to hold their land, marry, and reproduce their way of life well into the 1960s, long after orthodox economic rationality would have persuaded them to abandon the struggle. For Hannan, the task of Irish rural sociology was to chart how this 'originally demographically vibrant, subsistence economy' was changed into a 'post-peasant society' characterised by sharp differentiation between 'modern commercialised farming' and a displaced and marginalised peasant remnant (1979, 23).

Reviewing and discussing the debate at a later date, Curtin (1986) states that Hannan's identification of peasant farming with 'subsistence production' was a major mistake, which neither accorded well with the historical evidence (Arensberg and Kimball themselves recognised that the peasant economy they studied was based on production of young cattle for sale to other farmers in other parts of Ireland who fattened and finished them largely for export markets), nor allowed him to develop a coherent and distinctively different theoretical position from Gibbon. In the end, Curtin suggests, they disagree only over the timing of the process of peasant dissolution, which both agree must follow on incorporation into a capitalist market. I have argued elsewhere (Tovey 1992) that this misses a key point in Hannan's analysis: the peasant system is replaced, in Hannan's model, not by landless labourers and capitalist farmers, but by two types of family farming, one commercialised and the other marginalised.

Hannan recognised that the dominant form of agriculture emerging in 'modernising' Ireland was commercial family farming: market position was becoming more and more significant for the survival and development of these farms, as capitalism expanded throughout Irish society, but they remained nevertheless profoundly shaped by kinship and family factors (1979, 198ff). His work anticipated by several years the development within rural sociology internationally of 'commoditisation theory' as a perspective on modern farming in developed capitalist societies. Within Ireland, however, the significance of his identification of commercial family farming as a form worth studying in its own right was largely ignored. For the next twenty years (for the most recent example, see Frawley et al. 2000), researchers into what Hannan called "displacement and development" in Irish farming, or 'structural dualism' as it became more generally known, focussed almost without exception on the 'displaced', the 'bottom tier' (O'Hara 1986) of family farmers who were becoming increasingly marginalised from agricultural production, those who were identified in state policy as 'unproductive' (Leeuwis 1989) and as an economic and social problem. A great deal of energy was expended in exhaustively classifying and explaining 'marginalisation'. The emergence of successful family farm enterprises, intensifying and expanding their production, was not seen as an interesting research issue. Hannan himself, although he did identify the issue as important, had failed to investigate the economy of commercial family farming in any depth. By linking the 'peasant system model', which he created out of elements of Arensberg and Kimball's analysis, to a theory of change as 'modernisation', he encouraged researchers to focus more on transformations of 'the community' and 'community processes' in analysing change, than on class and market relationships.

In summary, Irish rural sociology in the 1970s and into the 1980s, like most other national rural sociologies across Europe, struggled to find a way of reconciling Marxist insights into the transformative impacts of the growth of capitalism which are felt as much within agrarian as other parts of society, with a Chayanov-inspired belief in the distinctive characteristics and development possibilities of the family form of farming. For Chayanov (1966), the most significant characteristic of peasant farming is cultural – their focus on social or familial reproduction rather than economic accumulation. Chayanov offered an account of re/production processes within the peasant economy which assumed the presence of societal institutions (communes, co-operatives) to support and guarantee peasant cultural orientations towards farm production; peasants living within these communal institutions and collective practices lived not just within a class relationship but a distinctive cultural world. The ethnocentrically Irish response to these different theoretical strands was to emphasise the distinctiveness of small family farming not only as a form of production but also as the basis for a distinctive (sub)society or subculture. The theoretical difficulties experienced by Irish researchers were largely a result of the fact that they applied Chayanov's analysis to the smallest and poorest farms of the western regions, which were evidently having difficulties in reproducing themselves as farming enterprises, and not to the larger (but still small-scale by international standards) family farms in other regions which reproduced themselves with considerable success, up to the past decade. The 'agrarian question' in Ireland has been predominantly a question about the fate of small farming in peripheral regions once capitalism starts to, as Kautsky put it, "take hold of agriculture"; we only began to become aware of how limited a question this is, just as sociology in Ireland lost interest in studying the agrarian at all.

The distinctively Irish approach to peasantisation and de-peasantisation has treated peasant farming as a culturally meaningful activity which requires for its survival the existence of a specific social and cultural world. This was overlaid by and confused with evolutionist and modernisationist assumptions which implied that 'de-peasantisation' is to be understood as a process of change from 'traditional culture' to 'modern rationality' in regard to farming practices. But it does not need to be understood in this way: we can avoid evolutionary assumptions and still ask questions about the cultural meaning to farmers of their farming practices, which would allow us to uncover, in contemporary farming or in previous periods, a complex of meanings that might usefully be thought of as 'peasant' and contrasted with other 'non-peasant' ways of farming. This chapter tries to develop a perspective on family farming, or simple commodity production, which locates cultural meaning, in the form of collective identity and a distinctive political project, at its centre. The concept of peasant or simple commodity production points not only towards particular structural relations between farmers and input and output markets, but also particular cultural orientations – in Bourdieu's (1989) terms, toward a 'habitus' which would guide the continuing reproduction of these structured relations. To discuss whether simple commodity production can continue to be the dominant form of production in Irish agriculture in the future, we need to ask whether the conditions which generate a collective 'family farming' habitus will exist in the future. However, it also becomes increasingly necessary to ask if there will be any productive farming at all in rural Ireland in the future: rather than agricultural production changing away from the peasant form, does 'de-peasantisation' coincide with post- or anti-productivism, the disappearance of agriculture in any form?

Peasantisation and De-peasantisation in Ireland

The Construction of an Irish Peasantry

To develop a framework for considering questions like these, I suggest that we shift our focus, to look at 'peasantisation' (and 'de-' and 're-peasantisation') as historical processes occurring in specific locations and under specific conditions. If 'de-peasantisation' is to be understood as a contemporary historical process, we can also think of 'peasantisation' as a process occurring in historical time. By this I mean in particular that we should be reluctant to describe peasants as 'traditional' social groups, which suggests that they exist prior to or outside history. The Irish case indicates that the creation of a peasant agrarian structure is something which occurs in quite specific historical circumstances, and can even be in some respects deliberately engineered or planned.

In the 17th and 18th centuries, Irish agriculture was substantially commercialised and internationalised, and food production was carried out through a social structure which was complex and hierarchical, characterised by marked class differentiation. To take as an example the case of Irish butter, this has been internationally traded for at least 300 years; even as early as the 16th century the trade was significant enough to attract severe export duties from the British colonial rulers. In the 19th century, exports of butter went not only to Britain but also to Europe, Africa and North and South America. Irish butter was highly prized for its good keeping quality on long sea voyages. Other important export products were beef, pork, barley and wheat. They were produced by an agrarian population whose relations were close to feudal: huge landowners, often absentee, at the top, with below them a complex structure of large under-tenants and 'middlemen', subtenants and sub-subtenants often lacking any legal tenure agreements, tied landholder-labourers paying for their uncertain access to land through various forms of sharecropping or payment in kind as well as in labour, and 'free' day labourers who rented small plots of land to grow potatoes for their own and their family's subsistence since their income from selling their labour was rarely enough to buy sufficient food.

In the second half of the 19th century this structure of landholding slowly gave way to the creation of a 'peasant proprietorship'. The change took place through a series of legal enactments, initially intended to provide greater security of tenure to tenant farmers, and it was linked not to intensification of feudalism but rather to the expansion of capitalism – specifically, of British industrial capitalism, in which Ireland's role was defined as that of an 'internal colony' which would provide food (and to some extent, labour) for the expanding British industrial workforce. The

British interest in creating the most appropriate political conditions for an industrial-led capitalism to flourish required weakening powerful landed interests, which might otherwise block policies favourable to industrial capital, for example around the provision of cheap food for urban workers. In Ireland, the strategy got an extra push from concerns that social order in the countryside was about to collapse, given a series of uprisings against British rule symbolised and represented by the large absentee landlords. The Land War (despite its name, this was more a political than a physical struggle) of the 1870s and 1880s (Tovey 1998, Kane 1996) led to a series of Land Acts which did not just increase tenant security but began to alter the legal basis of agricultural land ownership. The eventual outcome was the expropriation, by the early 20th century, of most large landowners in Ireland and the apportioning of their land among their tenants as the new legal owners.

Irish agriculture thus became 'peasantised' at quite a late stage of the development of industrial modernity in the larger political grouping of which Ireland was part. But this 'peasantisation' involved more than just a radical transformation in land tenure structures. It took place in two stages, distinct in terms of both the conditions giving rise to them, and their effects. Structural transformation was the first stage, involving the creation of peasant holdings out of the breakup of landed estates, as outlined above. This had been made much more possible by the radical simplification, resulting from decades of famine, evictions and voluntary emigration and bearing down hardest on the poorest groups in the rural population, of the previously very complex agrarian social structure. An immediate impact of the change in land tenure arrangements was reduction of the pressure on farmers to produce a surplus for appropriation by others. Rents paid to landlords under the old system were replaced by 'annuities' due to the state, which itself had paid compensation to the expropriated landlords, and these were both less exacting and more predictable than the previous rents had been. After 1921, the independent Irish state continued to reduce the pressure to produce a surplus for exchange, for example by removing rates and taxes from farm land, although this was presented rather as a way of increasing the competitiveness of agricultural exports. Thus if a 'peasant orientation' to farming is seen as an orientation to subsistence production, in Ireland this orientation was able to emerge only under quite specific historical circumstances, and as a result of a complex of specific causes – largely political, rather than economic or cultural – which enabled the new landholders to think about family reproduction and use values more than exchange values and the production of a surplus to pay the rent.

However, family reproduction is a much broader concept than subsistence production, and concern with family reproduction can be compatible with commercial farming. I associate the second stage of 'peasantisation' with a movement to encourage the re-commoditisation of production. At this point, I shift attention away from the state and political interventions in the management of agriculture, and towards Irish civil society. Civil society was involved in a range of ways in the creation of an Irish peasantry but I emphasise here the part played by a specific social movement, the co-operative movement in Ireland. It was largely in the mobilisation and institutionalisation of this movement that the formation of a collective identity as 'peasant proprietors' took place, giving specific cultural meaning to the new legal and economic conditions in which Irish farmers found themselves. As Turner notes, 'Collective meaning formation is necessary to the recognition of oppression, and especially to the creation of a sense of what non-oppression would be' (1996, 9).

The Irish Co-operative Movement was founded in the early 1880s. The first co-operative creamery was set up in Drumcollagher in West Limerick in 1889. From that point, the movement grew rapidly; by 1891 there were 15 co-operative societies, and 374 by the start of the 20th century, at which point the membership stood at 36.683 (Bolger 1977, 89). By 1913, 985 societies were in existence (ibid, 103). Most of these were dairy co-operatives, set up to receive milk from farmers, separate it, and process it primarily into butter, later into cheese, milk powder and other milk-based products and ingredients; and they were located particularly in the south and south-west of the country. The movement preceded Irish political independence and was to some degree oppositional to the state; and it was, like many 'modern' social movements which tend to be the creation of the intelligentsia rather than of ordinary people (Buttel 1993), instigated largely from above, with a leadership which crossed religious more than class boundaries. Its driving force was the son of a large Anglo-Irish aristocratic landowner from Co. Meath, Horace Plunkett. Plunkett realised that the collapse of rural collective organisation, in the wake of famines and the Land War, had left farmers vulnerable to exploitation by middlemen and dealers. This seemed particularly evident in the rapid growth of private commercial creameries, whose major shareholders were often butter buyers and merchants, which followed the introduction into Ireland of mechanised processes for separating cream from whey in the 1880s (Bolger 1977, 65-6).

The co-operative movement was concerned with economic relations and structures from its beginnings but it was also, particularly in its early years, a cultural and political movement. Plunkett himself explicitly associated it with other movements for independent development and autonomy which were being pursued in the same period (Plunkett 1904), particularly the literary and linguistic cultural revival, as well as movements for organisational and administrative empowerment such as Sinn Fein. Many Irish farmers were later to be mobilised through their membership in co-operatives into the general movement for national independence which led to the War of Independence of 1919-21. The co-operative movement was also a broad movement for rural development, which saw itself working for the re-valuing and re-empowerment of rural values, skills, and culture in the face of a world which was becoming increasingly urban and industrial. Irish society being at this time predominantly rural, recreating and re-valuing rural civilisation was also a way of recreating a distinctively Irish civilisation and culture.

The movement promoted co-operation as the only way to secure the conditions under which family farms could survive into the future. The removal of middlemen from the food industry was seen as essential if farmers were to receive a good price for their products. The movement sought to re-commoditise Irish peasant agriculture by drawing farmers into exchange relationships which would be neither capitalist nor socialist. Farmers were encouraged to produce a surplus for exchange, but it was to be exchanged on terms set and controlled, as far as possible, by themselves. But the movement also emphasised the importance of maintaining small producers as well as large ones, equalising access to rewards from farming, and helping to keep a vigorous, economically and socially active population in the Irish countryside. This egalitarianism shaped the organisational characteristics of the co-operative societies, for example in the emphasis on 'one man one vote' rather than 'one share one vote'. It both drew on and helped to reproduce agrarian populism as the dominant political ideology in Irish society up to probably the 1980s (Allen 1997, Curtin and Varley 1992).

A particularly significant feature of the movement was its interest in the technical improvement and modernisation of Irish farming, particularly dairy farming. The need to reform farming technically and set new quality standards for Irish food was already a core element in co-operative discourse by 1891 (Digby 1949). Plunkett described the movement as starting 'an industrial revolution' in Ireland (Bolger 1977, 75), albeit one which would, in his view inevitably in the circumstances, grow out of the industrialisation of food, rather than the importation of industries with few indigenous linkages. In these ways, the co-operative movement in Ireland, which I am claiming as an integral part of the process of Irish 'peasantisation', must be recognised as a movement for agrarian modernity. The new social arrangements in rural Irish life which it aspired to bring about were guided by its conception of what a 'modern' agrarian society should be like. The need to reform Irish butter production standards, for example, was seen as essential not only for economic reasons, to recapture the British market which had been lost to Scandinavian and French producers, but also for cultural ones, to demonstrate the modernity of the Irish nation. In its emphasis on technical development as the way forward for farmers, the co-operative movement helped to promote the conditions in which, years later, modernisationist understandings of rural development as the increasing specialisation of farms in agricultural production could flourish, leading to the destruction of many of the farms which the movement initially set out to support and sustain.

The co-operative movement never managed to collectivise all Irish farmers, but it did provide Irish dairy farmers, particularly in the south and south west of the country, with an identity and a historical 'project'. It contributed to the creation of a cultural ideal of family farming, which over time became 'naturalised' in the sense that it was represented as the only natural way to organise agricultural production, or as a natural expression of Irish history and national identity. The 'peasant project' to create and maintain a technically advanced, uniform, small-scale producer society remained a feature of the ideological discourse of the main farming organisations in Ireland until well into the 1980s (Peillon 1982). It shared with Chayanov the belief that the modernisation of farming did not necessarily require accumulation of land, but that the availability of familial motivations and family-based skills could lead to a form of highly technologically developed, high value-added farming on relatively small holdings which, in conjunction with a co-operatively organised food processing and agribusiness sector, could guarantee reproduction for a large number of farm households. Other historical conditions supported an on-farm logic of peasant reproduction; but the co-operative movement provided Irish farmers with identity, collective institutions, and a project for a peasant society understood not as just one element within a broader Irish society but as the core and basis of that society.

Institutionalising the Peasant Project

Irish peasant studies, as we have seen, have generally agreed that a peasant society existed, if in increasingly anomic and marginalised form, up to the 1960s (Brody 1973, Hannan 1979). But they understood as 'peasant' only the small-farm communities of the Western seaboard, dependent largely on dry cattle production for a cash income and on mixed farming for subsistence; the term does not include larger, more commercialised and individualised farms in the east and south. My use of the term clearly does include these other farms, which have been the backbone of the co-operative milk industry in Ireland; yet even with this enlarged usage I would still agree that 'de-peasantisation' in Ireland began to become marked really only from the 1960s on. What were the relations with other groups in Irish society which helped to institutionalise the peasant form of farming and to maintain it up to that point? Four seem particularly important.

Relations with the state I have suggested already that the colonial state in Ireland had a strategic interest in supporting the peasant or family form of farming. After Independence, while we might debate whether the term 'peasant state' (Granberg and Kovách 1998) could usefully be applied to the new Irish state, farmers remained a strategically important constituency for it. Many voters were peasant farmers or the family members of such farmers. During the decades (1930s and 1940s in particular) when it was state policy to develop economic self-sufficiency and to reduce external (largely British) market dependency, the peasant form of farming was regarded as essential; it was called on to ensure that much of the food needs of a large proportion of the population would be met by their own efforts, and to absorb a surplus labour force in the absence of much industrial and service employment (which it critically failed to do). Irish peasant farmers were well integrated into the political system during the first half of the 20th century, either through representation by others (particularly the then populist party Fianna Fail) or, more briefly, through direct participation in a farmers' party.

Relations with agribusiness, or *industrial capital* as this operated within the food and farm input industries: as early as the 1930s and 1940s, the co-operative movement had begun to weaken as a social and cultural force. It concentrated increasingly narrowly on running the creamery societies, neglecting other existing or possible forms of co-operation, and while the central leadership grew more autocratic, the participation of ordinary members in running their organisations fell off. Increasingly, 'Economic and technical considerations superceded social concerns. Co-operation was now less a social creed and more a business device' (Bolger 1977, 121).

Nevertheless, private capital continued to be relentlessly pushed out of the dairy agribusiness sector (the pattern of agribusiness development in relation to meat and grains remained rather different), and by the early 1970s, when the processing and distribution of liquid milk also began to come into co-operative hands, co-operative agribusiness had virtually a 100 percent monopoly control over the milk produced by Irish dairy farmers. Share capital and turnover in the societies greatly increased. Farmer control over agribusiness meant that many societies experienced great difficulties in acquiring enough development capital to expand in the ways their increasingly professionalised managers would have liked; most of the surplus generated from the business was returned to producers in the form of higher prices paid for their milk. Smaller producers were not always protected, but nevertheless it seems likely that many more of them remained in dairying than would have been the case if they had been dealing with privately owned industrial capital.

Relations with finance capital Grant and MacNamara (1996) note that relations between banks and farmers are of central importance to development in modern agriculture. The way in which banking capital polices access to and use of credit by farmers can push farm families towards more technical methods of production and more formal levels of rationality in relation to the farm business – or indeed, away from them. Up to the 1970s, Irish farmers were generally regarded by Irish banks as a source of savings to invest outside agriculture, and often overseas, rather than as suitable recipients of loans and credit. Only very large farmers were encouraged to borrow from the banks (Smith and Healy 1996, 125). The main area within the food industry which was seen in this period as a suitable location for investment was the (largely privately owned) beef industry – exporters of live cattle, cattle slaughtering and bulk packaging companies. Hostility to the co-operative movement since its inception, from Catholic priests and from small business owners particularly in the grocery trade, had encouraged the banks to find legal and other reasons why they could not lend money to the co-operative societies. It was not until the 1960s, when the state became active in encouraging the expansion of dairy farming into new regions of the country in anticipation of EU membership and access to CAP market supports, that banks became involved in financing the development of creameries, and then this was almost entirely via the state-owned Agricultural Credit Corporation. Thus throughout the first half of this century, both farmers themselves and the institutions in the food industry which they controlled largely escaped penetration by external capital.

Relations with other classes in society, particularly industrial classes whether labour or business. The populist ideology articulated in farmers' organisations, as well as in the main political party Fianna Fail, led farmers generally to regard 'business' with some suspicion. Farming organisations, for their part, tended to see industry as competing with agriculture for the scarce resources of state support and funding. Ideologically, the Irish labour movement and the farming organisations shared a vision of 'a workers' and small farmers' Republic'; in practice, of course, the status gulf between even small property owners and landless workers, urban or rural, was very difficult to bridge (Ruane 1979). Cross-class relations thus tended to confirm the collective identity of Irish farmers as a self-sufficient and self-contained peasantry. Moreover, while indigenous industry did exist at this time in Ireland, and in fact expanded during the decades of Protectionism, up to the 1960s no industrial class existed in sufficient strength to counterpose an alternative vision to the agrarian model of modernity.

De-peasantisation – Process and Relationships

De-peasantisation was not just a structural transformation, but also a huge cultural shift, in which the agrarian model of modernity was not only eclipsed by industrial and global alternatives but actually became transformed into its opposite, a symbol for all that was backward, insular and traditional in Irish society. Agrarian Ireland experienced a process of urbanisation which manifested itself in at least two different ways: in the anomic collapse of confidence among farmers in the value of their skills, knowledge and way of life, which Brody (1973) traces to increasing contact with urban ideas and orientations; and in the shift of material and ideological power from the countryside to the city which accompanied capitalist economic development in Ireland as elsewhere. The process of cultural de-peasantisation was accompanied by significant structural changes in each of the four relationships sketched above, with the result that Irish farmers found themselves in an increasingly new relational situation from the 1960s on.

We might date the first step in the process of de-peasantisation to the late 1950s, when a radical reinterpretation of the notion of 'development', and 'development policy' began to take place within the state itself, among certain civil servants and a small circle of urban intelligentsia, particularly economists (Daly 1998, O'Hearn 1998). Economic nationalism and protectionism were abandoned and Ireland's situation was reinterpreted as that of a 'small open economy' which must engage in global economic transactions to survive. Multinational-led industrial development was seen as crucial for the provision of employment and of opportunities for profit for native Irish capital; agriculture (and later other extractive industries such as mining, forestry and tourism) became drawn into the new development vision as an arena of 'natural advantage' to be made as competitive as possible in global market exchanges.

By the mid-1960s, the state-devised Second Programme for Economic Expansion was explicitly accepting that the solution to the problem of very low incomes in farming was to transfer labour from agriculture into other economic sectors – presented as a normal part of economic modernisation. However this remained "a somewhat unique policy position" (Frawley et al. 2000, 16) in Ireland. If the change in development strategy began to gain acceptance in the 1950s, its effect on farmers was delayed or moderated, for largely political reasons, for another two decades: in the 1960s by the political needs of the Irish state, although farming organisations were already beginning to express alarm – and to organise civil unrest – at what they saw as a shift of state support from agricultural to industrial development; and in the 1970s by those of the EU. EU Accession in 1973 enabled the Irish state to transfer to the EU the problem of finding funding for the mounting array of farm producer subsidies which threatened it with financial disaster (Matthews 1982). It was not until the 1980s that the effects of the emphasis on competitiveness began to be directly experienced on a large scale. They were carried into agriculture in two ways. The first was the increased commitment to implementation of that part of the Common Agricultural Policy which in Ireland was called the Farm Modernisation Scheme. State financial and technical support began to be tightly targeted on selected farmers, the rest being expected to move into off-farm employment such as in the new factories which multinational capital was being encouraged to establish in rural areas, and to make their land available to the 'modern' group. The Farm Modernisation Scheme was explicitly a policy of 'supporting the strongest', that is those who already possessed adequate resources of land and capital and who were willing to adopt productivist transformations of their farm operations. It was accompanied by a policy decision – only beginning to be reconsidered in the late 1990s (Kinsella et al. 2000) – that part-time farming be treated as a transitional step to moving out of farming altogether.

The pressure for global export competitiveness was translated into Irish agriculture, secondly, as a shift in state support for the Irish food industry from farmers to the industrial food processors. A striking re-interpretation of the nature of relations between farmers and food processors appeared in policy and promotional documents during the 1980s (Tovey 1991). Farmers were referred to as 'suppliers' to the food industry, where previously they were represented as 'members' of 'their' co-operative processing societies. The problems experienced by those with small-sized holdings and little access to capital, which previously legitimated state intervention and support on behalf of farmers, began to be presented as problems facing the food industry (in the form, for example, of seasonality in the supply of milk to processors) which could only be solved by removing out of farming those who could not become the sort of supplier the industry needed. Larger dairy farmers, and intensive producers in livestock and horticulture, began to find that banks were interested in making credit available to them, on fairly favourable terms. The smallest farmers still found banks unsympathetic, and preferred to find ways of self-financing any improvements they made to their farms (Leeuwis 1989). But for many farmers in the middle, debt-interest repayments began to escalate, and although foreclosures by loan companies were limited in number because of their impact on a public which still tended to see winning 'the land for the people' as one of the few real achievements of the Irish fight for independence, many farmers in this category lost effective control of their business to financial capital for years. By the end of the

1980s, finally, the largest dairy co-operatives began to privatise and to launch themselves on the Irish stock exchange. This coincided with an ongoing process of mergers and takeovers within the food industry generally which is tending towards ever-increasing concentration in structure of agribusiness. The collapse of the peasant project for modernity has occurred alongside the universalisation of capitalism within Irish society.

Structural and Cultural Change in Contemporary Irish Agriculture

De-peasantisation (and also peasantisation) involves both structural and cultural processes of change. Structural patterns of 'de-peasantisation' in Ireland are not clearly following 'the classic thesis of the transformation of agriculture under capitalism' (Newby 1978) – concentration of land, expropriation of the landholders, and their transformation into labourers for rural or urban capitalists. Both movement from family farming into more capitalist forms of management, and land concentration, have been limited in extent. Far more striking has been the growing concentration of production, output and income on a steadily contracting group of farms. Commins and Keane estimated in 1994 that 20 percent of Irish farms held about 40 percent of all agricultural land, and accounted for some 60 percent of marketed output; today they are probably closer to accounting for 80 percent. Instead of the increased uniformity among Irish farmers aspired to by the co-operative project, we are finding increasing heterogenity between farm households, within and between classes, regions, and commodity groups (Commins and Keane 1994, Commins 1996).

The broad changes in economic and political relations which underlay de-peasantisation affected different commodity producers differently. Commins and Keane identify the greatest contrast in contemporary Irish farming as that 'between systems which are based on, or incorporate, dairying and those systems based on cattle production' (1994, 93). Irish dairy farmers have an average farm size of around 33 hectares, which is nearly twice the average for cattle farmers (19 hectares). Measured by gross margins produced, dairy farming is six times more productive than dry cattle farming (Census of Agriculture 1991). Income differentiation is correspondingly marked. In 1996, large specialist dairy farms had an average income of IR£ 58,000 while small farms in dry cattle production had an average income of IR£ 1,800 (National Farm Survey). Dairy farmers tend to be younger than other farmers (nearly two-thirds were under 54 years of age in 1991, compared to less than half of dry cattle farmers), they are much less likely to have any other occupation besides farming, and they also make more use of family labour than other farm types do. In these ways, contemporary 'specialist' dairy farmers still approximate reasonably closely the ideal of family farming associated with peasant modernity in the early years of the co-operative movement. But specialisation in different commodity lines is once again very spatially patterned. Since the early 1980s, the small-farm counties of the north-west have been losing their national share of production for all farm enterprises, but particularly of dairying. Cattle and sheep production are now the mainstay of farming in the west and north west, tillage and dairying are concentrated in the south and south-east; but farms in the south and south-east still have larger cattle and sheep enterprises than in the west and north west (Frawley and Commins 1996, 10). Agricultural production itself is gradually becoming concentrated in the south and south-east of Ireland, while the rest of the country becomes rural space.

Between farm households, dependence on income from farming is also increasingly variable. It was estimated in 1999 that income from off-farm employment, on average, made up about 40 percent of the total income of farm households (Kinsella et al. 2000, 485). Frawley and Commins described the extent to which farm households have reduced their dependency on farming as a means of livelihood, through off-farm work, on-farm diversification, and take-up of state welfare payments, as "one of the remarkable features of Irish farming in recent years" (1996, 20). Less than one quarter of farm households now depend solely on their farms as a source of income. Those least likely to do so are the smallest farm enterprises, located in more western or northwestern areas, but the trend is increasingly wide-spread. Increasingly, it is the access which household members have to non-farm income sources, and the type of those income sources (welfare transfers versus employment, level and sector of employment) which determines the socioeconomic status of the farm household, rather than the size of the holding or the scale or type of the farm business. More and more of these are now rural rather than agricultural households, even if they continue to hold some land; it becomes increasingly difficult to establish even how many 'farmers' or 'farm households' there are in Ireland today.

As a pattern of cultural change, 'de-peasantisation' can be linked to the disappearance of the logic of peasant reproduction emphasised in the Chayanovian peasant model. While there is little research on contemporary Irish farming which addresses this issue directly, some evidence can be assembled which seems to suggest that relationships between enterprise and household are changing in significant ways. On both successful dairy farms, and on part-time or 'non-viable' small dry stock farms, the meaning of 'farming', and the meaning of 'family farming', are undergoing interesting revisions. On smaller farms, as we have already noted, the retreat of productivism is transforming these into 'rural households' in which the farm ceases to have much economic or cultural centrality; in more western areas in particular, men who own – and to some degree, work – some farmland are increasingly likely to record themselves on censuses and other surveys not as farmers, but as factory workers/builders/whatever their best other occupation is, as the social stigma attached to being a small farmer increases. At the other end of the scale, intensive dairy farmers increasingly think of the income generated by their farm as family (rather than farm) capital, to be invested inside or outside farming, wherever it will bring the best return (Tovey 1995, 2000), and are encouraged by the farming media and farm advisory services to think of themselves less as part of a rural world than as members of the rising class of independent Irish entrepreneurs. The growth of an entrepreneurial discourse around farming reveals not only changing subjective understandings inside farming, but also of a shift in the wider status system of Irish society, in which the business person and entrepreneur has enjoyed remarkable upward mobility in recent years, while the status of the farmer has heavily declined. In Irish society generally, agriculture has lost much significance other than economic. It is becoming just another, decliningly profitable and attractive, 'sector' within the economy. De-peasantisation includes not just the de-centralisation of farming within rural households but also the de-centralisation within society of a rural productivist orientation to nature.

While the above presents 'de-peasantisation' as the emergence of heterogeneity in Irish agriculture, both between farmers, and between regions, an alternative account might present it as the *re*-emergence of heterogeneity. Accounts of agrarian society in Ireland before the Land Wars depict a society which was much more complex, varied and stratified than that which emerged in the late 19th and early 20th century. More broadly, we might argue that heterogeneity has historically been the norm for agrarian societies in most parts of Europe, and that the construction of a homogeneous 'farming sector' – whether peasant, or perhaps, socialised agriculture – has always required quite specific political and ideological interventions to achieve it. Much 20th century agricultural development policy consisted of attempts to overcome any bases for variability which derived either from nature or from culture – pushing forward the adoption of technological changes in production which made output less dependent on specific qualities of land and local climate, and which replaced local by scientific knowledges about production practices. At the same time, economic development strategies for agriculture insisted on the specialisation and professionalisation of the farmer as agricultural worker alone. However part-time farming, or household pluriactivity, has probably been historically the more 'normal' way in which agrarian households secured a living, and the attempt to create 'full-time family farms' which became a feature of both American and much European state policy in the early modern period can be seen as an attempt to wrench aside normal historical trends through a conscious act of will. What effect is the re-emergence of heterogeneity likely to have on the future – or futures – of Irish small scale family farming, in the period after de-peasantisation?

Rural Change and Peasant Futures

If certain features of early modernity appear to have favoured the emergence and consolidation of peasant agriculture, while the middle, 'Fordist' or 'organised capitalist' period led to de-peasantisation, it is tempting to argue that some aspects of the conditions of late modernity, such as the emphasis on consumption styles and practices as a source of identity and status, on diversification of products and the close interrelatedness of what we consume with who we are, could encourage the re-emergence and re-consolidation of a peasant farming world. To assume that this could happen in an evolutionary way, however, is dangerous. Such 're-peasantisation' could not happen without a struggle: it seems to require the development of some broader collective projects for reorganising social relations and rethinking society which would justify maintaining a food system centred on small-scale, localised, ecologically sensitive and perhaps humanitarian production practices, projects which seem likely to run into some powerful opposition. Any capacity which farmers themselves might have had to articulate a non-urban, or non-consumer-space vision of the countryside, seems to be steadily eroded by the ongoing transfer of control over the 'rural environment' and its uses from rural to urban authorities – environmental management specialists, tourism and heritage experts – in which the specific knowledges and skills associated with livelihoods gained from productive work on nature lose status and credibility. The achievement of 'peasantisation', in the Irish case, was possible only in quite specific historical circumstances, which included the existence of a social movement in support of peasant reproduction but which was itself part of and integrated into broader social and cultural processes of change; and the existence of a complex of political and power relations with other key social groups which tended to insulate peasant farmers from the more destructive effects of penetration by industrial and financial capital. The possibility of similar conditions recurring again seems remote (Tovey 1998).

However a recent, much more optimistic, analysis of the future for peasant agriculture across the EU (Sociologia Ruralis 2000, 4) does foresee 'a kind of *re-peasantisation* of European farming' (van der Ploeg et al. 2000, 403, italics in original). This seems to mean primarily a movement against the scientisation of farming which was central to modernisation models of agricultural development. Family farmers are starting to take back control over their on-farm production processes, the way they source inputs and use their resources and the diversification of outputs, from external scientific, technical and administrative authorities. "The farmers' *determination* to maintain control and their willingness to develop new capacities to ensure they do so" is linked by the authors to "the old and well-known *resistance paysanne*" (2000, 403, italics in original).

The emergence of this anti-modernisationist movement comes from farm families' recognition that the modernisationist model, with its emphases on enlargement of scale, intensification of production and commoditisation of inputs and outputs, offers little hope of escape from the 'agricultural squeeze' which either draws farmers ever deeper into a 'regulationary treadmill' or expels them from farming. The new paradigm, on the other hand, offers the hope that farm families can increase their income and employment levels, on and off the farm, without simultaneously losing control over their resources and working lives. It is still largely confined to the level of rural practices, as a self-driven process which articulates "the well-understood self-interest of increasing sections of the European farming population" (2000, 395); but the practices are increasingly being facilitated by rural development policy within the EU and may success in establishing themselves as the new policy orthodoxy. If that happens, the European 'peasantry' seem likely to be accorded a renewed, central place within rural development: "Although coalitions with new rural dwellers, urban consumers, and environmentalists, for example, are certainly necessary, farmers will continue to be the focus of such rural coalitions and arrangements" (2000, 404), and will be particularly strategic actors in the development of future European rurality (van der Ploeg and Renting 2000, 531-2).

One important effect of this analysis is to direct our attention to the very different ways in which a 'post-modernisationist' set of policies for countryside development may be understood and implemented. Most discussions of what follows on from the heyday and decline of modernisation models for agricultural development, to date, have understood this to be 'post-productivism' – where the productive work of food production is subordinated to or even replaced by roles for farmers as landscape managers, heritage conservers and tourism hosts. Van der Ploeg et al. offer us instead a scenario in which a highly productive small-scale agriculture, working through revised sets of market and consumer relations and with revised criteria for product 'quality', is the lynchpin of a multi-layered, cross-sectoral process of revitalisation of the rural economy. The important implication is that an anti-modernisation form of rural development need not be either post-productivist or post-peasant.

How we assess the possibilities of this sort of re-peasantisation depends rather critically on how pluriactivity is understood, by researchers, by policy-makers and state agents, and by farm families themselves. In the Irish case, we have little information on the latter (although work by O'Hara 1998 on why many farm wives work off-farm suggests that the cross-generational reproduction of pluriactive farm households may be less secure than at first appears), but we do have more information on the two former issues. Arguably, both researchers and state agents still regard pluriactivity more as a way of keeping families living on their holdings in the countryside, than as a strategy for developing small-scale, high quality, high income farming. Thus, the recent study of low income farm households by Frawley et al. concludes that 'over the longer term the main option for realistically addressing the problem of low incomes in many farm households is to expand opportunities for other employment' (2000, 6). They recognise that "a proportion of low income farmers could be helped to maximise the market-based portion of their farm income" (2000, 5) but there is little suggestion in their study (other than improving farm efficiency) how this might be done. Organic farming, for example, is not addressed as a potentially significant alternative line of production. Kinsella et al. appear to place more emphasis on the centrality to rural development of the continuance of farm work, even in areas increasingly peripheral to agriculture (2000, 484) since they research pluriactivity on Irish farms in terms of its effects in stabilising the farm population. Yet their concluding discussion of "Enabling policy measures" ignores questions of agricultural development: what is important is that rural development policies must "go beyond sectional approaches' to address 'issues such as local employment generation, infrastructure and environmental improvement" (2000, 493). Infrastructural improvements, they suggest, should be targeted on shortening travelling time to off-farm work; strategies should be developed for improving 'labour efficiency' on the farm in order to increase labour availability off it; agricultural research and advisory agencies should develop new packages for pluriactive farms which "address the need for alternative technological trajectories and farming systems that meet the specific requirements of the pluriactive households" (2000, 494).

Thus both studies prioritise the development of the *off-farm* activities of farm households over the development of their farm enterprises. Additional income accruing to farm households from their extra-agricultural activities is undeniably of great importance to their continued survival as households, but it is hard not to suspect that the dominant view of the state and the research establishment in Ireland remains little changed from the Farm Modernisation Scheme of the 1970s: that 'part-time farming' is not 'real farming'; that the collapse in productive agriculture on Irish farms will continue to increase and spread; that only very few very modernised farms will ultimately be worth trying to save; and that the real question of rural development is how, under these conditions, to maintain countryside demography, rather than any version of 'peasant agriculture'.

Re- or neo-peasantisation depends not only on supportive social relations with powerful groups in society but also on the formation of peasant collective identities and movements. Van der Ploeg et al. emphasise that the rural development movement which they associate with re-peasantisation is still based primarily at the level of new practices. It is not clear whether these practices will lead on to more discursively articulated projects and policies. Whether they do or not, it seems certain that they will not simply try to reproduce the 'peasant society' of the past. Notions such as an orientation to 'farm family reproduction' – securing family reproduction on the land across generations – lose much of their appeal in what Giddens (1994) and others call the 'reflexive' world of late modernity. Can 'real peasants' survive the disappearance of such ideal-type characteristics as these?

References

Allen, K. (1997), *Fianna Fail and Irish Labour, 1926 to the Present*, Pluto Press, London.

Arensberg, C.M. (1959), *The Irish Countryman*, Mass: Peter Smith, Gloucester (first edition 1937).

Arensberg, C. M. and S.T. Kimball (1968), *Family and Community in Ireland*, Harvard University Press, Gloucester (First edition 1940).

Bolger, P. (1977), *The Irish Co-operative Movement - its History and Development*, Institute of Public Administration, Dublin.

Bourdieu, P. (1989), 'Social space and symbolic power', *Sociological Theory* 7:1, pp. 18-26.

Brody, H. (1973), *Inishkillane*, Allen Lane/Penguin Press, London.

Buttel, F. (1993), 'The production of agricultural sustainability – observations from the sociology of science and technology', in Patricia Allen (ed), *Food for the Future – Conditions and Contradictions of Sustainability*, John Wiley, New York.

Chayanov, A. (1966), *The Theory of Peasant Economy*, Irwin, London.

Commins, P. (1996), 'Agricultural production and the future of small-scale farming', in C. Curtin, T. Haase, H. Tovey (eds), *Poverty in Rural Ireland. A Political Economy Perspective*, OakTree Press/Combat Poverty Agency, Dublin.

Commins, P. and M.J. Keane (1994), *Developing the Rural Economy – Problems, Programmes and Prospects*, National Social and Economic Council, Report no. 97, NESC, Dublin.

Curtin, C. (1986), 'The peasant family farm and commoditisation in the West of Ireland', in N. Long et al., *The Commoditisation Debate*, Agricultural University, Wageningen.

Curtin, C. and T. Varley (1991), 'Populism and petit capitalism in Ireland', in S. Whatmore et al. (eds), *Rural Enterprise*, David Fulton, London.

Daly, M. (1998), 'The Statistical and Social Inquiry Society of Ireland', in K. A. Kennedy (ed), *From Famine to Feast: economic and social change in Ireland 1847-1997*, Institute of Public Administration, Dublin.

Digby, M. (1949), *Horace Plunkett – an Anglo-American Irishman*, Blackwell, Oxford.

Djurfeldt, G. (1999), 'Essentially non-peasant? Some critical comments on post-modernist discourse on the peasantry', *Sociologia Ruralis* 39, 2 (April), pp. 262-70.

Frawley, J. and P. Commins (1996), *The Changing Structure of Irish Farming – Trends and Prospects*, Rural Economy Research Series no. 1, Teagasc, Dublin.

Frawley, J.P., P. Commins, S. Scott, F. Trace (2000), *Low Income Farm Households – Incidence, Characteristics and Policies*, Oak Tree Press in association with Combat Poverty Agency, Dublin.

Gibbon, P. (1973), 'Arensberg and Kimball revisited', *Economy and Society* 2, 4, pp. 479-98.

Giddens, A. (1994), *Beyond Left and Right: The Future of Radical Politics*, Polity, Cambridge.

Granberg, L. and I. Kovach (1998), 'Beyond socialism, beyond the peasant state', in Granberg and Kovach (eds), *Actors on the Changing European Countryside*, Institute for Political science, Budapest.

Grant, W. and A. MacNamara (1996), 'The relationship between bankers and farmers – an analysis of Britain and Ireland', *Journal of Rural Studies* 12, 4, pp. 427-37.

Hannan, D.F. (1977), 'Patterns of intergenerational replacement in traditional Irish agriculture', paper read at the Annual Conference of the Sociological Association of Ireland, April.

Hannan, D.F (1979), *Displacement and Development – Class, Kinship and Social Change in Irish Rural Communities*, Economic and Social research Institute, Dublin.

Hannan, D.F. (1982), 'Peasant models and the understanding of social and cultural change in rural Ireland' in P.J. Drudy (ed), *Ireland - Land, Politics and People*, Irish Studies no. 2, Cambridge University Press, Cambridge.

Hannan, D.F. and R. Breen (1987), 'Family farming in Ireland', in B. Galeski and E. Wilkening (eds), *Family Farming in Europe and America*, Westview Press, Boulder, Oxford.

Kane, A. (1996), 'The centrality of culture in social theory; fundamental clues from Weber and Durkheim', in Stephen P. Turner (ed), *Social Theory and Sociology*, Blackwell.

Kinsella, J.S., Wilson, F. de Jong, H. Renting (2000), 'Pluriactivity as a livelihood strategy in Irish farm households and its role in rural development', *Sociologia Ruralis* 40, 4 (October), pp. 481-96.

Leeuwis, C. (1989), *Marginalisation Misunderstood – Different Patterns of Farm Development in the West of Ireland*, Wageningen University Press, Wageningen.

Matthews, Alan (1982), 'The state and Irish agriculture, 1950-1980', in P.J.Drudy (ed), *Ireland – Land, Politics and People*, Cambridge University Press, Cambridge.

Newby, H. (1978), 'The rural sociology of advanced capitalist societies', in Newby (ed), *International Perspectives in Rural Sociology*, Wiley, Cambridge.

O'Hara, P. (1986), 'CAP socio-structural policy - a new approach to an old problem?', in *The Changing CAP and Its Implications* (Conference Report), Dublin, An Foras Taluntais/Agricultural Institute.

O'Hara, P. (1998), *Partners in Production? Women, Farm and Family in Ireland*, Berghahn, Oxford.

O'Hearn, D. (1998), *Inside the Celtic Tiger*, Pluto, London.

Peillon, M. (1982), *Contemporary Irish Society – an Introduction*, Gill and Macmillan, Dublin.

Ploeg, J.D. van der and H. Renting (2000), 'Impact and potential: a comparative review of European rural development practices', *Sociologia Ruralis* 40, 4 (October), pp. 529-43.

Ploeg, J.D. van der, H. Renting, G. Brunori et al. (2000), 'Rural development: from practices and policies towards theory', *Sociologia Ruralis* 40, 4 (October), pp. 391-408.

Plunkett, Horace (1904), *Ireland in the New Century*, Irish Academic Press, Dublin.

Ruane, J. (1979), 'Rural industrialisation and the problems of the new rural proletariat: a case study'. Paper presented to VI Annual Conference of the Sociological Association of Ireland.

Smith, L.P.F. and S. Healy (1996), *Farm Organisations in Ireland (A Century of Progress)*, Four Courts Press, Dublin.

Sociologia Ruralis (2000), special issue on 'The Socio-Economic Impact of Rural Development: Realities and Potentials' (eds) J.D. van der Ploeg, H. Renting, M. Minderhoud-Jones, vol 40, no. 4, October.

Tovey, H. (1991), '"Of cabbages and kings": restructuring in the Irish food industry', *Economic and Social Review* 22, 4, pp. 333-50.

Tovey, H. (1992), 'Rural sociology in Ireland – a review', *Irish Journal of Sociology*, 2, 96-121.

Tovey, H. (1995), 'Farming, development and the environment in Co. Meath', Report on the Irish stage of the research project *Land, Culture and Crisis: from Productionist Success to Fiscal and Environmental Impasse on European Farms 1940-1990*, funded by EU DGXII under the SEER Programme (unpublished).

Tovey, H. (1998), 'Rural actors, food, and the post-modern transition', in L. Granberg, I. Kovach (eds), *Actors on the Changing European Countryside*, Budapest, Institute for Political Science.

Tovey, H. (2000), 'Milk and modernity: Dairying in Contemporary Ireland', in H.K. Schwarzweller and A.P. Davison (eds), *Dairy Industry Restructuring*, Research in Rural Sociology and Development, vol 8, JAI Press, New York/Oxford.

Tovey, H. (2001, forthcoming), 'From agrarian to ruralised: modernity and the construction of "traditional" Ireland', in H.Donnan, D. Meagher, J. Ruane (eds), *Alternative Modernities*, Pluto Press, London.

Turner, Stephen P. (1996), 'Introduction', in Turner (ed), *Social Theory and Sociology*, Blackwell, Oxford.

14 Societal Paradigms and Rural Development – A Theoretical Framework for Comparative Studies

IVAR JONSSON

> Life in town appears to be poisonous for all creative talents at their start. Thus, there are rotten patches and deserts in the life of those generations that now live (Jonas Jonsson Minister of Justice 1927).

Introduction

The concept 'rural' has many connotations. Everyday discourse about it is often based on prejudice referring to backwardness in terms of culture, education and brutish human relations. Rural areas are increasingly criticised for being an economic burden due to subsidies for agriculture and in some cases for aboriginal peoples' communities. These common views of the countryside reflect negative prejudices just as romantic views of rurality often reflect positive prejudices. According to these, the countryside is a place where people live in harmony with nature and each other, and lead a democratic life that can only truly be realised in small communities. These opposing themes are frequent in films, novels and popular music. Both reflect a very old distinction between urban and rural life: urban life is based on manufacturing, services and cultural institutions, rural life is based on agriculture and mining. This simple dichotomy is obviously unrealistic and does not tell us much about how rural life has become integrated into industrial societies. It is also unable to explain the apparently chronic problem of depopulation of rural areas, or why policies are implemented that work against their development.

In this chapter we discuss rural development as embedded in the wider society. This embedding is the outcome of struggles between socio-economic forces that attempt to realise different 'societal paradigms', i.e. trajectories of social and economic planning. Any reduction of de-peasantisation to a necessary development from one mode of production to another, such as from feudal to capitalist modes of production, is criticised here as reified theorising. As will be argued below, farmers'[1] culture in Iceland and their political organisation and struggle led to economic development that resulted in an economy based on small scale production rather than large scale manufacturing.

We start by presenting a framework for analysing the inter-relationship between changing societal paradigms and rural development. Then we illustrate the theoretical discussion by analysing the unusual history of rural development in Iceland during the 1920s and 1930s. This period is chosen because of the unusually strong position of Georgeists and social anarchists attempting to realise an alternative path of modernisation.

Embedding the Rural through 'Societal Paradigms'

Rural areas are embedded in the institutional structure of the wider society. This means that rural development results from conflicts between socio-economic forces and is therefore 'structured' in an active way (Giddens 1984, 34-6). The structuration of the countryside is a consequence of the balance of power between socio-economic forces. This refers to organised interest groups and their different degrees of control over power resources such as organisations, political skills, money, information, ideological production, access to political elites and to institutions of the state (Jonsson 1989, 1995).

Struggle between the different socio-economic forces concerns the fundamental principles upon which societal development is to be based. These principles refer to fundamental ideas of how society is best organised, and when they become dominant over long periods we may call them 'societal paradigms'. These paradigms are mediated through hegemonic politics (Jonsson 1995, 32-44) in which organised socio-economic forces fight for their interests within the framework of alternatives that structural conditions allow.[2]

There are four types of societal paradigms. 'Techno-economic paradigms' refer to ideas about what constitutes the best practices and means of organising production, services and consumption in general. Secondly, 'power-political paradigms' refer to ideas about how collective decision making is best organised and its enforcement institutionalised. Thirdly, 'reproduction-social paradigms' refer to ideas about the roles different social groups and institutions should play in regard to sexuality, biological reproduction, care of children, old persons and the disabled, and socialisation; Fourthly, 'ethical-prescriptive paradigms' refer to general ideas about right and wrong conduct in different situations, the aims and quality of life. We now take a closer look at these different categories.

Ethical-prescriptive Paradigms

Prescriptive practices may take long time to become institutionalised, as Weber showed for the Protestant Ethic (1958). It took Protestant asceticism over two centuries to become institutionalised practice in the form of the 'spirit of capitalism' which has been the dominant ethical paradigm in the industrialised countries of North America and Europe since the eighteenth century. While this ethic is fundamentally *individualistic*, others have been the basis for *collectivist* kinds of 'spirits of capitalism', such as the Catholic-dominated corporatist capitalism of South-Europe and the Buddhist-dominated statist capitalism of SE-Asia, Japan, S-Korea and Taiwan (Chan and Clark 1992).[3] In recent decades environmental ethics have become increasingly important in western capitalist countries, becoming institutionalised in national political parties and governmental policies and in the policies of international organisations. Following the rise of gene technology, biological ethics has recently become an important object of political debates.

Techno-economic Paradigms

The history of western capitalism is characterised by shifting techno-economic paradigms. A fundamental element of the capitalist techno-economic paradigm is the principle of private property. It took European societies centuries to introduce private property as the way to regulate wealth and resource use. This involved the gradual institutionalisation of more free exchange of privately owned resources. The process however was never complete, and in the period from the second world war to the 1980s market exchange was increasingly re-regulated as can be seen from the growth of the welfare state (Esping-Andersen 1990), environmental regulation etc. Interestingly enough, agriculture has always been much more regulated than other sectors in European economies.

As Marx argued in his very early articles *Wood*[4] and *Mosel*[5] in Rheinische Zeitung, (1842-3), private property is not natural but a political product. The principle of private property is still being questioned, increasingly so in recent times. The struggle of aboriginal peoples for their traditional rights to territorially based resources has been particularly important. United Nations Convention 169 (1989) guaranteed aboriginal people rights to co-determination regarding exploitation of natural resources. The ongoing struggle for implementation of this convention in different countries shows that the principle of private property is constantly being actively embedded in historical contexts.

A second element of techno-economic paradigms is the spatial distribution of production of goods and services (Storper and Walker 1989). Where one invests and locates production geographically depends on many different factors, such as local skills, local cultures, infrastructure, distance from markets, governmental regulation and support (financial and military). Ideology is also an important factor, in the forms of localism (Putnam et al. 1994), nationalism, colonialism and globalism.

A third element is infrastructures, that is, ideas about best practice in techniques and in organisation of labour, and how to organise the transport of inputs and outputs, which involves connecting producers and markets by means of physical (road, air and sea traffic) and symbolic transport systems (information on production and market trends). Recent history has seen the rise and fall of the Fordist paradigm of mass production (Aglietta 1979, Freeman 1987); a post-Fordist techno-economic paradigm has become increasingly important in recent years, based on economies of scope, small batch production, information technology, flexible labour markets and flexible firms organised in local and global alliances and networks (Dunford 1990, Sayer and Walker 1994).

A fourth element is ideas about best practice in the organisation of training and skills, including what kind of skills are needed and how and where skills training should take place (Lash and Urry 1994). The final element is ideas about how best to organise consumption. These refer not only to controlling the *magnitude* of consumption through state monetary and fiscal policies, but also *what kind* of goods and services should or should not be provided. Ideas about how to organise consumption also include *where* to consume and with whom. The dominant locus of consumption in western capitalist societies is the nuclear family which fetches its usables from shopping centres (Hirsch and Roth 1986). But there have also been others, such as the collective forms of consumption of very early Christian communities around the first century DC, the Catholic tradition of monasteries (Gunnarson 1973, 107-30), or the urban communes of the 1960s and 1970s with their collective households (Kanter 1979).

Power-political Paradigms

These express ideas about how collective decision making is best organised and its enforcement institutionalised. They include ideas about the desirable level of concentration of power, and ideas concerning who are the legitimate actors in the political life of a given community. They range from the totalitarian to the minimalist (Guerin 1970), and involve definitions of insiders and outsiders, or who has the right to participate in collective decision making.

Political power has become more decentralised in European societies in recent centuries. The sovereignty of European monarchs was 'handed down' to consultative assemblies drawn from the three fundamental classes, aristocracy, bourgeoisie and peasants, in the 19th century and by the turn of the 20th century representative democracy was becoming general. Suffrage was gradually extended to men of no property, and after the First World War, to women. Younger people were gradually allowed to participate in the political community (Mosesdottir 1998, Hettne et al. 1998). After the Second World War, class divisions were incorporated into the structures of many European states. Neo-corporatism is based on tripartite collaboration between the state, trade unions and employer associations, through which these actors attempt to adjust wages to productivity development (Schmitter and Lehmbruch 1979, Jonsson 1989).

The inter-mediation of regional interests is organised in different ways in different countries. Before the implementation of the nation-state system in Europe, regional interests were represented through the hierarchical system of feudal lords with their subordinate counts etc. In many contemporary western parliamentary systems the rule of proportional representation is counteracted by regional disproportional representation, such as the separation in Germany between Bundestag and Bundesrat to secure regional interests (Mosesdottir 1998). In most countries the nation state is organised in terms of three levels of jurisdiction, the state, counties, and municipalities, where the state is predominant and municipal officers and politicians have relatively limited power determined by parliament. In the case of federations, regions are more autonomous and possess legislative power to differing extents in areas such as taxation, education and culture, health and social services, but their autonomy never extends to foreign or monetary policy. Autonomous areas are often organised on the basis of ethnicity and there appear to be increasing numbers of such areas.

Since the late 1970s, different forms of home rule have been established in Europe, for example in Greenland, Spain and UK (Jonsson 1999, Keating 1995). With the shift from nation states to supra-states like the EU one should expect greater pressure for increased regional autonomy in the future.

Reproduction-social Paradigms

The central ideas here concern ethnicity, gender and generation in the context of the family or household. In earlier European agricultural society, social and biological reproduction was located in the extended family, with co-ordinating regulation by the church. Industrialisation and urbanisation, helped the diffusion of the nuclear family, with increasing co-ordination by the state (Hirsch and Roth 1986). With the growth of the welfare state after the Second World War and the expansion of social services, schools and care institutions, the role of the family as locus of socialisation and care was greatly reduced. The role of extra-household institutions and actors varies between both countries and periods. Different constellations of extra-household institutions and their interrelations may be called 'societal regimes', i.e. structured constellations of class, gender and ethnic relations. One may distinguish three main types, liberal, ecclesiastic and egalitarian (Mosesdottir 1995, 1998) centred respectively around the market, religious organisations, and socialist parties. Concrete examples might be the USA as liberal, Germany as ecclesiastic and Sweden as egalitarian.[6]

All western systems appear to be highly class, gender and ethnically segregated in terms of biological reproduction, care and socialisation. Married couples of mixed race and/or ethnicity are relatively rare. The same goes for class. Socialisation and care of children in the home is predominantly based on women's homework, and nursery staff and primary school teachers tend to be predominantly women. The staff of health institutions and institutions for the elderly are predominantly women. Germany is a predominantly male breadwinner society with the husband being the main source of household income. Sweden and the USA are characterised to different degrees by dual breadwinners. There is also a big difference between the three countries in terms of the extent of collectivisation of socialisation and care functions. Of the three, public expenditure as a proportion of GNP is largest in Sweden and smallest in USA (Mosesdottir 1998).

The Ontology and Historiography of Societal Paradigms

The ontological status of societal paradigms can not simply be reduced to the level of pure ideas (see Figure 14.1). As Kuhn claimed (1970), paradigms are as much practical experience as ideas and theories. Nor are they simply abstractions of behavioural motives, nor rationalist abstractions in the sense of Weberian 'ideal types'. Societal paradigms are practices within horizons which are influenced by reflexive experiences of action. Reflexive experiences are 'existential' in the sense that they are actively 'interiorised' (Sartre 1972). This means that one's ideas about past and present do not result from one's social context in a simple determinist way. Our reflections are not simply products of our 'habitus' (Bourdieu 1977) or social background, but result from our horizons which we know are changing and can be changed. Paradigmatic practices also have causal consequences to the extent that they are shared by many people and endure over time, through active struggles in which actors have unequal access to power resources. Access to power resources depends on an individual's or organisation's relation to economic wealth and to ideological and cultural symbols and their means of production and reproduction.

Societal paradigms are norms of expected 'best practice'. They are institutionalised in rules, and organise behaviour in different social fields. Institutionalisation of a paradigm creates relations of power and domination between individuals and groups, and is itself the outcome of actions that are grounded in existing structures of power and domination. However, power relations between groups change over time and need constantly to be actively reproduced by means of changing power resources, so the paradigmatic ideas and norms can not be reduced to power structures as such. The effectiveness of structures is always mediated in practice.

Societal paradigms and their institutionalisation depend on the balance of power between different groups in society. It is in this context that rural development should be analysed. Feudal Europe was characterised by decentralised political power and economic activity, in the form of geographically based hierarchies of counts and lords, and self-sufficient farming. The rise of industrial capitalism made centralisation the main principle of political and economic organisation. Economies of scale were basic to the organisation of the state and production. Essentially this presumed standardisation of everything from inputs and outputs of factory production to regulation and laws concerning inputs and outputs of the state. The standardised inputs of the state were the national citizen and his/her claims, while the output was standardised by formulating laws and regulations within the framework of constitutions (see Weber 1958 and Lukács 1971 on the phenomena of rationalisation and reification).

Modernity is the result of this process of standardisation and scale economies. However, there have been different pathways in modernisation as there have been different societal paradigms. Urbanisation and ruralisation have followed different processes in different European countries in accordance with differences in the balance of power between groups. An analysis of rural development in Iceland is interesting in this respect.

Rural Development in Iceland and Societal Paradigms

There has long been a strong tendency for the Icelandic population to concentrate in two districts, the capital Reykjavik and Reykjanes, which lie side by side on the south-east corner of Iceland, the Reykjanes peninsula. It takes less than one hour to drive across the total area of the two districts.

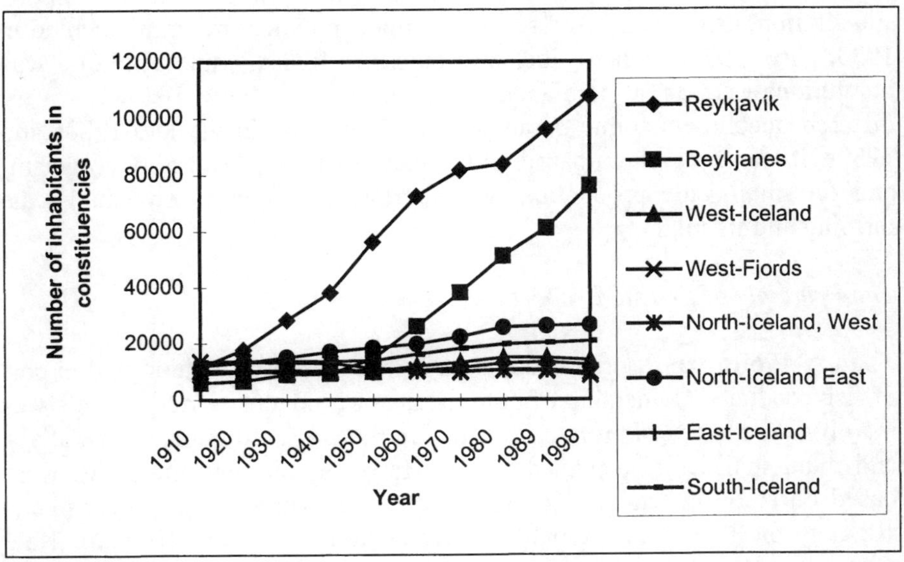

Sources: Statistical Bureau of Iceland 1984 and Central Bank of Iceland 1999.

Figure 14.1 Distribution of the Population in Iceland between Constituencies, 1910-1998

Historically there has not been a consensus about this pattern of development. Let us have a closer look at how societal paradigms behind this demographic pattern have changed over time.

The Development of Techno-economic Paradigms

The cornerstone of the techno-economic paradigm in Iceland is the neo-classical and Ricardian idea of comparative advantage (I. Jonsson 1991, 1995). This is evident in the emphasis on investment in fishing and fish-processing which was the dominant export-orientated strategy until the late 1960s. Export of energy to multinational corporations in the aluminium and ferro-silicon industries became important in the 1970s. These are both capital intensive, involving investment in trawlers, advanced assembly line technology in freezing plants, and hydro-electric power plants. This strategy of investing in capital intensive export industries to exploit what is considered to be a comparative advantage is not peculiar to Iceland as a peripheral country. It is characteristic of peripheral areas of other countries as well, for example the Nordic countries (Granberg 1998).

Until the 1960s, Icelandic industrial policy was based on import-substitution, using tariffs and selective import policies first implemented in 1933. In 1973, when Iceland entered EFTA, this policy was institutionalised as a permanent feature. As we show below, a very different techno-economic paradigm was influential in the late 1920s and 1930s. It was a paradigm based on the ideas of Henry George (Georgeism) and on small-scale production, with particular emphasis on small-scale farming and fishing.

The Principle of Private Property

The pre-1960s export-orientated strategy was based on fishing and export of fish products. Ownership of both means of production and products was mainly in the hands of private firms and individuals, although co-operatives and municipalities also played a part. However, the new industries were based on state ownership. The state owns the hydroelectric power plants although the largest municipalities have stake as well. The Icelandic state owns at least 51 percent of the new firms in the industries established in the 1960s and 1970s, e.g. the aluminium, ferro-silicon and diatomite multinationals.

Banks in Iceland have in principle been owned by the state. The first state owned bank was established in 1881, Landsbanki Islands. In 1904 a private bank, Islandsbanki, was established with Danish capital, but went bankrupt in 1930. State-owned banks were established in 1929 and 1930, for the agricultural sector and for fishing. In 1953 a bank was established to serve the manufacturing sector with the state as a minor shareholder. Verslunarbanki Islands, a private bank owned by firms and individuals in the trading sector, was established in 1961 (Nordal and Kristinsson 1987, 216-17). These last two banks and a small bank owned by the trade unions merged in 1990 with a privatised bank to become Islandsbanki. There are also some smaller banks owned by trade unions and municipalities. Finally, there have been regional development funds and industrial development funds set up by the state to steer loans towards regional and industrial innovation (I. Jonsson 1991). Strong pressures for privatisation of the banks emerged in the 1980s and 1990s. In 1998 the Bunadarbanki was 'handed out' to the population: every citizen received shares in it which they are free to sell on the stock market. There are also plans to sell the largest bank, the state-owned Landsbanki Islands, but this has proved difficult due to large sums of state-secured loans, mostly in the fisheries sector. The state has thus played a major role in the financial sector in Iceland. In 1998, for example, 55 percent of all loans and bonds were given by the two state-owned banks (Central Bank of Iceland 1999).

Property relations in industrial production and services have been a matter of constant political struggle in 20th century Iceland. But whether there should be private, co-operative or public ownership has been more a matter of pragmatics than dogma. The right wing Independence Party accepted state ownership when private capital was not able to invest and the centre party, the Progressive Party, which has had strong ties to the co-operative movement, accepted private ownership when political pragmatism required it. The first decades of the century are particularly interesting in terms of the struggle: these established the ground for the balance between private and collective forms of ownership in the decades that followed. At this time the co-operative movement was becoming a major economic force, as we discuss below. In 1935 the hereditary tenure of public land by farmers became law. In short, the growth of private property and raw capitalism was severely limited. This was done through active struggles over hegemonic politics, in which the government of 1927-32 was particularly important.

Iceland got home rule in 1904, and a political system consisting of the continuation of the political groups and parties established during the independence struggle in the 19th century. After 1910, class struggle became increasingly important politically. The Progressive Party (PrP) was established in 1916 to represent the interests of farmers and agriculture. The Trade Union Conference and its political organisation, the People's

Party (PP), were also established in 1916. Both parties were opposed to large-scale industry in the 1920s and early 1930s.

Spatial Distribution of Production of Goods and Services

The Progressive Party built a minority government in 1927 with the support of the People's Party. PrP leaders were influenced by the writings of John Ruskin and social-anarchists such as Kropotkin and were opposed to large-scale industry and urbanisation (Asgeirsson 1988, 28-30, Fridriksson 1991, 34-5). They advocated that the nation should live in the countryside. Land held by large farmers should be split up so that peasants could start new small farms. The state was also to provide new areas for new peasants – access to new land that had not been cultivated before. Land around fishing villages and towns should be split up so that workers and fishermen and their families could keep animals and cultivate vegetables and in principle be 'semi-peasants' with one foot in fishing and the other in farming. This ideology, that based modernisation on small peasant farming and the countryside, dominated the PrP until 1944, when the party congress accepted that modernisation should also be based on small fishing communities (Asgeirsson 1988, 140).

The PrP had from the 1920s to the 1940s a strong position in Icelandic society, engaging in either one-party minority government (1927-32) or coalition governments in which it was the dominant party. I analyse its power base in more detail below. Here I mention briefly only three factors. *First*, its over-representation in the parliament, Althingi, due to a mismatch between the constituency system and rapid urbanisation. *Second*, its close ties with the co-operative movement whose leaders were among the founders of the party. The co-operative movement had been established in the late 19th century by farmers who organised imports of goods and exports of agricultural products as an alternative to Danish domination of Icelandic trade. It later expanded to other sectors, such as fishing and manufacturing, and played a leading role in industrialisation and modernisation outside Reykjavik. However, it experienced major crisis in the 1980s and its central organisation, SIS, was dissolved while individual co-ops increasingly collaborated with private firms (I. Jonsson 1991, 1995). Figure 14.2 highlights the growth of the co-ops in terms of the numbers of members. *Third*, the PrP had a close relationship with the main popular movement in Iceland, the 'youth societies' (*Ungmennafelag*), established around the turn of the century and still the main movement in the countryside. It organised everything from social events such as debates on political issues to voluntary road building. It emphasised sport and opposed alcohol consumption. As a nationalist movement it particularly opposed the 'Danishness' which was thought to characterise Reykjavik. As a result, its ideology was pro-countryside and anti-urbanisation and it supported small industries as against big industry.

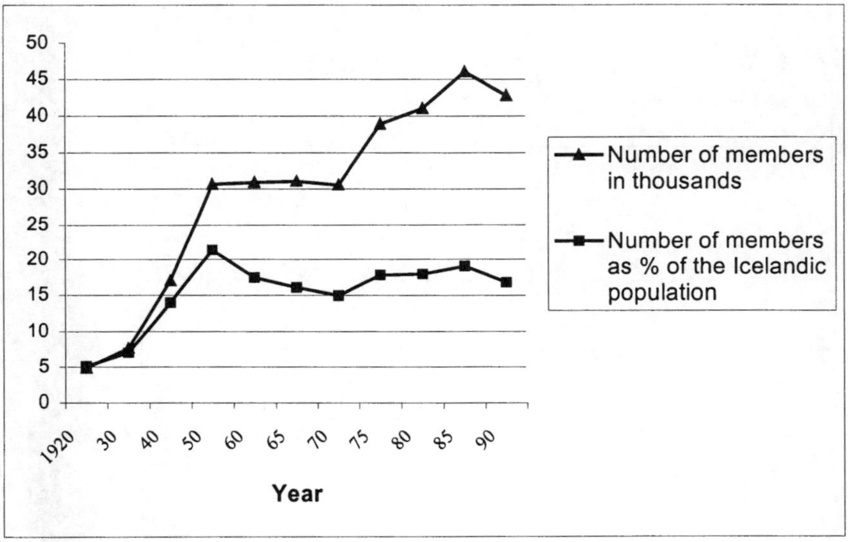

Source: Statistical Bureau of Iceland 1997 and 1998.

Figure 14.2 Members of Coops in Iceland, 1920-1990 (in thousands and as percent of population)

The minority government formed in 1927, in which the PrP was supported by the PP, was in power until 1932. It was extremely effective in developing an infrastructure for the countryside which would work against rural de-population while strengthening small scale production. It established a bank for the agricultural sector, Bunadarbanki Islands in 1929, and supported investments in freezing plants in harbour villages along the coast for the export of lamb meat (Johannesson 1948, 267-8). As Figure 14.3 shows, state expenditure on agriculture expanded fast in these years. Production of hay, milk and meat increased tremendously, made possible by state loans and subsidies for improvements in agriculture. Existing homefields were levelled and new homefields created. Land drains were dug, haybarns, manurestores and fences built like never before (Sigurdsson 1937). All this helped to modernise agriculture and expand its ability to produce for the fast-increasing domestic market as well to export meat.

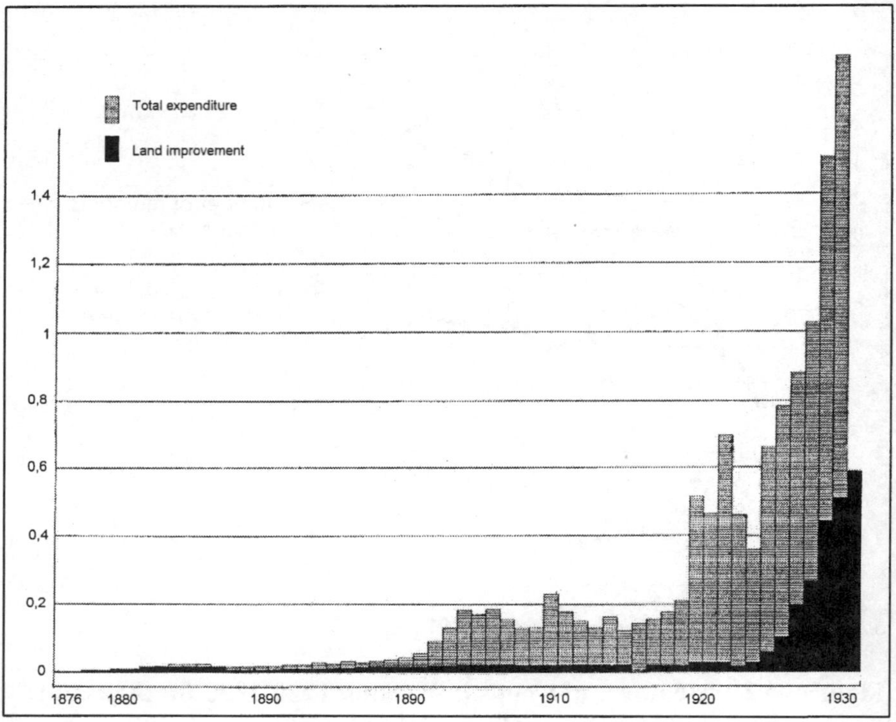

Source: Atvinnumalaraduneytid 1931, 193 (copy of original graph)

Figure 14.3 State Expenditure on Agriculture in Iceland, 1876-1931 (millions of kronas)

Although agricultural restructuring led to a substantial increase in arable land per farm, it was not based only on the modernist idea of economies of scale. State subsidies for farm improvements were differentiated by size and large farms did not get as much as small farms. It was government policy that farms should be split up among all the children instead of letting the oldest son inherit the whole farm when the old farmer died. The improvements in agriculture would increase the amount of cultivated land and thus generate enough land for all the children to establish their own farms (Asgeirsson 1988, 86-7).

This was a conscious attempt to develop an alternative to implementing capitalist large scale farming (Asgeirsson 1988, 87). In debates in the parliament and media from 1920 to 1927 different transport and energy production policies were put forward. The right advocated concentrating agriculture in the more productive south of Iceland and wanted to build railroads to bring agricultural products to the capital. Those favouring small-scale farming argued for investing in freight ships that would serve all parts of the country. The 1927 government chose the second strategy and established a state-owned coastal liner company. Railroads were never again seriously discussed (Asgeirsson 1988, 85). It also rejected right wing politicians plans to build a hydro-electric power station in the South that would primarily serve Reykjavik.

Technology and Infrastructures

The emphasis on developing the agricultural economy led the government to concentrate on building infrastructure that would serve the countryside and a de-centralised economy. It built new roads and bridges (see Figure 14.4), new telephone lines and stations that linked distant farms to the telephone network. The number of stations increased from 22 in 1926 to 64 in 1930 and the total length of telephone lines increased from 8800 km to 12750 km (Atvinnumálaráðuneytið 1931, 57-58). Transport of passengers and cargo by coastal liners also increased: by 1931 there were two liners which sailed 32 times to the harbours around Iceland (Atvinnumálaráðuneytið 1931, 77). State expenditure on building harbours increased from 111,685 kronas in 1927 to 210,786 kronas in 1929, although in 1930 it was cut to 90,000 kronas. Safety at sea was also improved. By 1930 six new lighthouses had been built and many of the 49 old ones greatly improved (Atvinnumálaráðuneytið 1931, 60-65).

The fundamental aim was to build infrastructure that would reduce the trend of population concentration in Reykjavik and other larger towns. However, the government had other aims as well. To undermine the ideological hegemony of the ruling capitalist class and of Reykjavik as a power centre, it embarked on revolutionising the education system. The aim was to make the countryside the cultural base of Icelandic society. It started building schools in the countryside and 'froze' the capital, Reykjavik, in this respect.

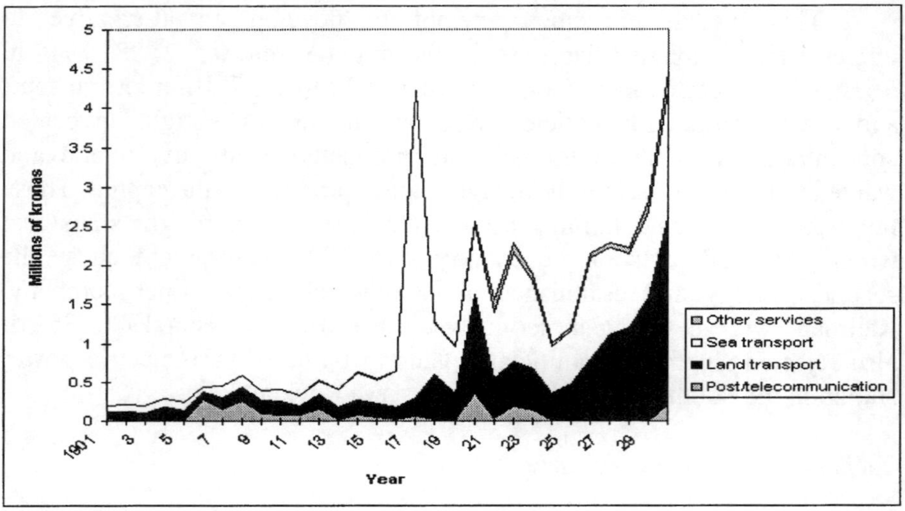

Source: G. Jonsson 1991

Figure 14.4 Government Expenditure on Infrastructure, 1901-30

Schooling and Skills

As part of its longterm policy of developing agriculture and related production, the government started intensively to build up the school system in the countryside. State subsidies were available for building primary schools in the countryside but not for schools in villages and towns (Fridriksson 1992, 34). Three large new 'regional schools' for secondary and college education, described as 'intellectual power-stations of the countryside' were built in South-, West- and North-West Iceland. The state paid half the cost of building secondary schools in the countryside, but only 40 percent in towns (Fridriksson 1992, 35). Yet the schools were not controlled by the state, but by local people in the regions. Three new 'housemother' schools were also established in the countryside, operating in close relationship to the regional schools (Fridriksson 1992, 42).

The only school that educated students for university was located in Reykjavik. The government froze expenditure to this school and limited recruitment to 25 students. Assuming that the school's basic function was to reproduce the ruling class in Reykjavik, it even put a radical socialist in as its headmaster (Fridriksson 1992, 29). And to undermine the monopoly of the ruling class in Reykjavik, it allowed the main high school in North Iceland, in Akureyri, to graduate students for university studies. The government's ideology was that primary and secondary education should be as practically orientated as possible and should reflect Icelandic culture and working life. Its policy was therefore to decentralise the education system, develop regionally based educational systems and minimise copying of foreign formal education.

Consumption

Until the third quarter of the 19th century, trade in Iceland had been in the hands of Danish merchants. Icelanders were allowed in 1853 to trade with other nations than Danes, but it was not until the 1870s that they gradually took over trade in Iceland. The main force behind this was the farmers' co-ops. The first farmers' coop was established in 1844, to get better deals from Danish merchants; it distributed profits according to the amount each member purchased. Farmers' consumer co-ops, and later, workers' co-ops in the fishing villages, increased gradually in the following decades. In 1902, they established a common national organisation, *Samband íslenskra samvinnufélaga* (SÍS). The co-operative movement became a major force in wholesale and retail trading and gradually expanded into production as well, in fishing plants, garments, skins and shoe factories as well as agriculture-related production such as dairies and slaughterhouses. In 1942, 50 co-ops were members of SIS, spread all around Iceland (Gudmundsson 1943, 106). They were one cornerstone of the network of power of Icelandic ruralists in the first half of the century.

In regard to who consumes and how, forms of consumption have always been private rather than collective, despite the strength of the co-operative movement particularly in the countryside and fishing villages. It organised collective forms of supply of goods and services, but the family was and is the basic unit of consumption. Even the 1935 law on co-operative farming (Johannesson 1948, 78) referred only to developing agricultural villages in which farmers could share machinery and reduce input costs, but still cultivate their individually-owned land. It is only in social services such as health and education that consumption has been collective in form, organised by the state in state institutions.

Power-political Paradigms

Employer and labour movements were late to develop in Iceland compared to some other European countries. The first village trade unions were established around the turn of the century and a national organisation of trade unions was established as late as 1916. The interests of farmers had already been well secured by the establishment in 1899 of the Agricultural Society of Iceland, which is still the main body of agricultural policy formation and has always employed more staff than the Ministry of Agriculture.

The first employers' unions were established almost simultaneously with the trade unions. The first union of deck-hands was established in November 1894 in Reykjavik (Báran) while the first union of fishing vessel owners was established in September the same year. Merchants in Reykjavik established their association in 1899. The Icelandic Chamber of Commerce was established in 1917 (Gislason 1990, 133) and the Association of Icelandic Trawler Owners (FIB) in 1916 as a response to wage demands by a new union of trawler deck-hands established in 1915 (Gislason 1990, 154). Other associations, of importers and wholesalers, of industrial employers, and of retailers, were established in 1928, 1934 and 1939.

The national organisation of trade unions grew slowly, and represented only wage-earners in the private sector. Public sector employees established their own confederation in 1942 and employees with university degrees founded theirs in 1958 (Thorleifsson 1977, 316). The labour movement in Iceland has always been relatively weak due to this separation into three main confederations (I. Jonsson 1991, 1995). Up to 1936, the national organisation of trade unions (ASI) did not have the formal right to negotiate wages. It was not until the PP and PrP coalition government of 1934-7 that the PP finally secured by law the negotiating rights of trade unions and built the foundations of the Icelandic welfare state. The rate of unionisation remained low until after the Second World War.

In this context of a relatively weak labour movement and a strong employers' movement, on one hand, and relatively underdeveloped state apparatuses on the other, the role of the state in the economy and society was particularly important. The 1927 Progressive Party government did not carry through decentralisation of the state by moving ministries and institutions to the countryside, nor cut down state activities. It regarded the state as an instrument for improving the economy and intellectual life in the countryside. No party in Iceland has ever fought for regional autonomy, probably because the population is very small, widely dispersed in small villages, and culturally homogenous. In short, the predominant power-political paradigm in Iceland in this century has emphasised centralisation of power rather than decentralisation.

Reproduction-social Paradigms

Being unusually ethnically homogeneous, Icelanders have at least in this century been keen on reproducing the population as 'Icelanders'. Following the nationalist struggle for independence, they have been particularly suspicious of Danes, but also of foreigners in general. For example during the Second World War the Prime Minister of Iceland, a member of the Progressive Party, warned the US government that it would be very unwise to have black sailors from the American navy based in Iceland as that could threaten national reproduction.

To the government of 1927, it was crystal clear that true Icelandic culture was to be found in the countryside. A strong tradition of folk cultural activity existed, based on reading the old sagas and other texts in the evenings on the farms. Rural culture also preserved old traditions of rhyming and storytelling. The government not only built new schools in the countryside to strengthen its cultural autonomy, it also established a Cultural Fund to buy paintings and works of art thought to be particularly Icelandic as exemplars for young artists. It published books thought to be exemplary for Icelandic writers, which had roots in the culture of the countryside, and supported geological research and natural scientists who studied Icelandic nature (Atvinnumalaraduneytid 1931, 234-6). It was assumed that this would undermine the decadent culture of the towns, particularly Reykjavik, regarded as polluted by Danish and foreign culture.

Concerning gender relations and sexuality, the government of 1927-32 emphasised the traditional roles of women as wives and men as breadwinners, reflected, for example, in the building of 'housemother schools' in the countryside (see above). Social services developed later in Iceland than in the other Nordic countries; they were the responsibility of municipalities until 1936, when an Act on Social Security was introduced and the state became in principle responsible for this sphere. Care of children, old persons and the disabled was basically the responsibility of their families. Recipients of social benefits did not have right to vote until 1934.

Central government expenditure on social services increased fast between 1927 and 1930, from 3,998 million kronas (1930 prices) to 5,292 million kronas (1930 prices) or 32 percent in real value. The bulk of the expenditure was on health and education. Figure 14.5 shows the growth of expenditure 1901-1930. Together central state and municipalities' expenditures may be roughly estimated to have been around 4.4 percent of GNP in 1927[7] (Ministry of Social Affairs 1942, 26-7, G. Jonsson 1991, 176, Statistical Bureau of Iceland 1997).

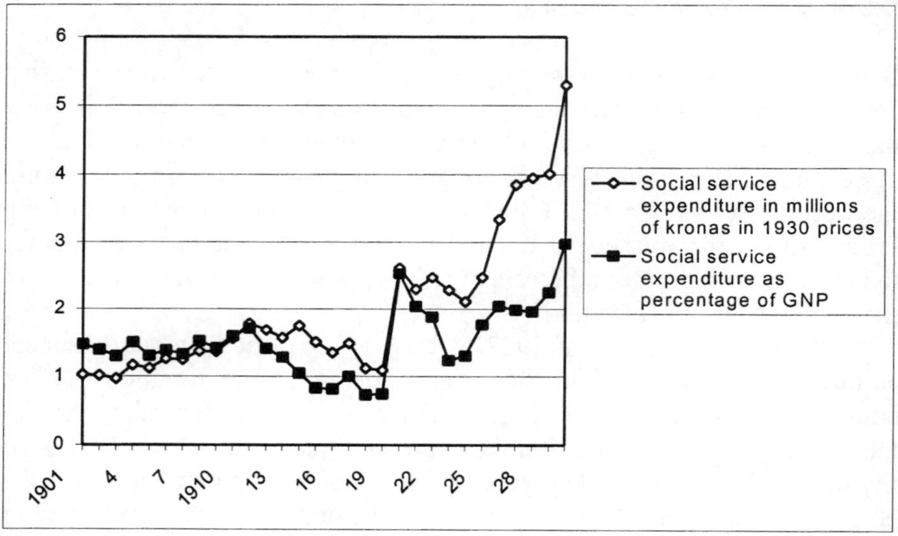

Sources: G. Jonsson 1991. (Price index: 2/3 prices of consumption goods and 1/3 prices of construction costs; Thjodhaghsstofnun 1992).

Figure 14.5 Social Service Expenditure by Central Government in Iceland, 1901-1930

Before 1930 there was only one state-owned hospital in Iceland, a hospital for mentally disabled persons, Kleppur, established in 1907 (Jonsson 1950, 80). There were also some tiny private hospitals owned by doctors and catholic nuns. A large hospital that would serve the whole country was constructed under pressure from the MP of the Progressive Party, Jonas Jonsson, who became Minister of Justice in 1927 and was opened in 1930 (Jonsson 1950, 57-8). However in a country where around 40 percent of the population was living in rural areas, it was difficult to provide a health service for everyone. District doctors and nurses played a central role in the health system and the 1927 government emphasised increasing expenditure in this field, especially in the most remote areas.

Ethical-prescriptive Paradigms

The ideology of the Progressive Party was deeply rooted in asceticism. The farmers' culture was seen as ethically ideal while town culture was thought to be degenerate. When people moved from the countryside to the towns, it was claimed that they moved into communities that were sick from severe alcoholism and other degenerate activities. The basis for a healthy physical and intellectual life was only to be found in the countryside. The new Minister for Justice said at a meeting of the Progressive Party in autumn 1927:

> It is noteworthy to highlight our experience that children have on the average learnt more from staying two to three months in the countryside than attending school the whole winter. It is also a noteworthy fact that no artist and no writer has been born or grown up in town in this country, except for a few worthless examples. Life in town appears to be poisonous for all creative talents at their starting phase. Therefore, there are rotten patches and deserts in the life of those generations that now live (Jonas Jonsson fra Hriflu, quoted in G. Fridriksson 1992, 33).

Such ideas were popular in Iceland in the first half of the century. They presumed that farmers were the only clean race in nations that were under constant threat due to the growth of the urban working class (Asgeirsson 1988, 45). Jonas Jonsson wrote in 1915 in the journal of the Icelandic Youth Association that the main aim of the nation was to "empower the Icelandic race". The year before he had claimed that idleness in towns caused degeneration and that the upper-classes degenerated because of their effortless life and the lower classes due to their crummy houses, lack of sunshine and foul air (Asgeirsson 1988, 47).

Social Forces and Political Strength

Why did 'ruralism' become so strong in Iceland and manage to determine social and economic development until late in the twentieth century?

Firstly, there are structural reasons. Iceland is a latecomer in terms of modernisation and industrialisation. In 1910, 54 percent of the population was living from agriculture; this fell to 47 percent in 1920 and 38 percent in 1930 (Thjodhagsstofnun 1992, 147). The bulk of the population either lived in or had grown up in the countryside. The population was very small (78.470 in 1901 and 108.861 in 1930), so there was no large domestic market for large-scale industry. Agricultural technology was at a fairly low level of development. Unemployment and misery was great among the poor in the towns and fishing villages, and many faced the real threat of being forced to move there.

Secondly, both working and capitalist classes in the towns and villages were in the early stages of being organised in the first three decades of the century. Neither was strongly institutionalised ideologically;

there existed neither working class nor bourgeois culture as had developed for over a century in England, for example. However there was a strong tradition of social movements and ideological activity in the countryside. There was a long tradition of social activity related to *farmers' economic affairs*, a strong tradition of *intellectual activity*, and finally there was the Icelandic Youth Association (*Ungmennafelag Islands*) which was already organised in most municipalities around the turn of the century, with more active members than any political party would have for the next 50 years. Two main organisations dealt with farmers' economic affairs: the Agricultural Society, established in 1899 as a general association of existing local agricultural societies (a number of regional associations were established around the same time (Johannesson 1937)), and the co-operative movement. The agricultural societies concerned themselves with improvements in agriculture and agriculture-related education. However, since agricultural improvement and the welfare of farmers was strongly felt to depend on relations with Denmark, the agricultural societies became increasingly politicised over the 19th century and became a forum for political discussion as well as technical. This network of farmers became a power resource of the Progressive Party; it was so strong that, as mentioned earlier, the Agricultural Society of Iceland has consistently had the status of a quasi-ministry of agriculture.

The co-operative movement grew fast in the last decades of the 19th century. By the turn of the century it dominated local markets in many rural areas and fishing villages. After establishment of the SIS in 1902 its growth leapt again as it now had enough turnover to invest in new activities and exploit economies of scale in import, export and production of goods (Gudmundsson 1943, 190-1). In 1920, turnover of imported and exported goods by SIS was 5.6 percent of GNP (Fridriksson 1991, 125, Thjodhagsstofnun 1992, 31). Assuming the turnover of the local coops was at least the same, we estimate that co-operative turnover was near 15 percent of GNP in the 1920s as it grew constantly in those years. By the late 1920s most coops had joined SIS.

However, SIS was more than an economic force. It was a social movement that had large number of members; and it was an organisation that invested in intellectual and ideological production. In 1919, the General Meeting of SIS decided to invest 20 percent of profits every year in a Cultural Fund. This covered the costs of the movement's college, established in 1917 as a business school with the important ideological function of introducing the ideology of co-operation to the students. Many of its students became organic intellectuals and members of the political and economic elite in Iceland. The fund also covered the costs of SIS's journal (*Samvinnan*) and other publications, and the cost of public lectures by important ideological leaders which were held in the countryside in 1921-2, 1924-7, 1929-30 and 1934.

The co-operative movement and its party, the Progressive Party, had intellectual roots in the farmers' study-societies, particularly in the Thingeyja-counties in North-East Iceland, which had started in the mid-19th century. Collections of books owned by study-societies became the basis for the Thingeyja County Library (Hoskuldsson 1993, 452-70). Its 1910 catalogue shows that it contained 635 books printed in foreign languages, mainly Danish, and 25 foreign journals, again mainly Danish. Six of these journals concentrated on the co-operative movement and its ideology. In addition to literature by classical authors – Dickens, Dostojevski, Goethe, Gorki, Hugo, Ibsen, Heine, Zola etc. – there were many social-anarchist books that appear to have laid the ground for the ideology and theoretical base of the co-operative movement and the Progressive Party, including Kropotkin's *Memoirs of a Revolutionist* and *Fields, Factories and Workshops*, Jaeger's *The Bible of Anarchy*, William Morris's *News from Nowhere*, 10 works by Tolstoy, and Henry George's *Progress and Poverty* and *Social Problems and Protection or Free Trade* (Syslubokasafn Thingeyinga 1910).

In 1915 the journal *Rettur*[8] was started by the director of the library, Benedikt Jonsson, Jonas Jonsson, later Minister for Justice, and other social anarchists and Georgeists. It became the main forum for ideological discourse among Icelandic farmers. In its first ten years the journal was very much preoccupied with the writings of Henry George. Georgeism in Iceland became much more radical than the original ideas of Henry George himself and more radical than the Georgeist movements in Norway and Denmark. Icelandic Georgeists supported public ownership of land along with hereditary tenure (G. Jonsson 1991, 106-7) and fought against legalisation of the sale of state and church land, as it was bought by richer farmers while the situation of tenants got worse.

This tradition of ideological production and intellectual leadership was given state backing by the 1927 government, as discussed above; the Cultural Fund which it established strengthened the theoretical and ideological base of the ruralist movement. For example between 1929 and 1932 it paid for publication of a three-volume translation of Charles Gide's book *Social Economy* (Économie sociale), a Georgist alternative to both marxist and neo-classical economics (Atvinnumalaraduneytid 1931, 235). The Progressive Party also had its own daily newspaper, Timinn, established in 1917, which presented its immediate political ideology and legitimated its daily practical politics. Here too radical Georgeists had a strong position. Jonas Jonsson, was from the start one of three members of the editorial board, along with the director of SIS, Hallgrimur Kristinsson, and remained a board member for decades.

Leaders of the ruralist movement were always conscious of producing 'organic intellectuals' (Gramsci 1978, 5-6), that is people who organise society in general on behalf of a social group or class. The aim was to undermine the role of the 'organic intellectuals' of capital, i.e. journalists, engineers, academics, state bureaucrats, party politicians, private 'think tank' specialists, or trade union leaders. These people are often incorporated into the ruling class. The ruralists and the co-operative movement created the means to generate organic intellectuals by establishing study societies among farmers, libraries and journals. SIS, as we have seen, established the Co-operative College in 1917 in order to produce leaders for the movement (J. Sigurdsson 1996). The ruralists were also very active in youth societies and the Icelandic Youth Association, organising meetings on political issues of the day, giving speeches and writing in its journal.

The Icelandic Youth Association (IYA) was the third element of the popular power basis of the ruralist movement. The first societies for young people started in the late 1880s and early 1890s in rural municipalities. In 1907, the Icelandic Youth Association was established by 8 Youth societies (Juliusson et al., 172-3), spurred on by implementation of home rule and nationalist optimism to develop what was and probably still is the largest ideological movement in Iceland. By 1912, IYA had 44 member societies and 16.080 members, i.e. almost 20 percent of the population (Magnusdottir 1997, 10).[9] During its first decades it was ruralist-orientated, although many of the societies were established in towns and villages.

Transcended Business, Political and Social Networks

The analysis above suggests that the ruralist and co-operative movements developed into a very unusual web of networks that transcended the societal spheres of business firms (coops), politics and intellectual production, interrelating networks in the different societal spheres. The web had historical roots in the organisation and institutionalisation of farmers' economic interests, but it was also a result of conscious strategic struggle by organised politicians and organic intellectuals.

The main strength of the network was that it covered the whole of Iceland at a local level, as well as being an integrated economic, intellectual and political network. This is illustrated in the following scheme (Figure 14.6).

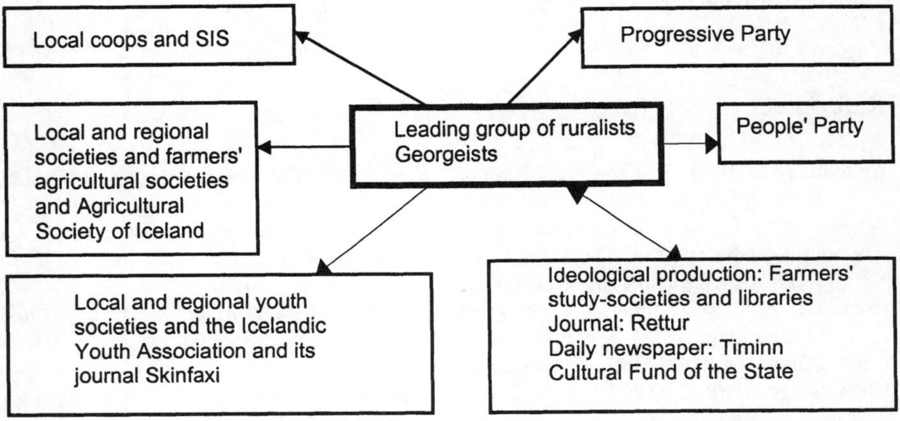

Figure 14.6 The Transcended Power Network of the Ruralists

Acknowledgements

I would like to thank Lilja Mosesdottir for valuable comments on this chapter while writing it and Jon Sigurdsson, former Rector of the university of the cooperative movement in Iceland (Samvinnuhaskolinn a Bifröst), for his valuable comments and assistance while collecting material for this chapter in Iceland.

Notes

1 The term *farmer* is *bóndi* in Icelandic. Bóndi is a general term that refers both to *great farmer*, i.e. rich farmer, and *peasant*, i.e. a farmer who owns his own small farm. *Crofter* or *hjáleigubóndi* in Icelandic is a farmer that rents land owned by another farmer. The state has owned land since the reformation in 1550 that used to belong to the Catholic Church. Such state farms were and still are rented out cheaply so the farmers on the state owned farms could become relatively rich.

2 See D.V. Porpora 1998.

3 As S. Chan and C. Clark (1992) claim, the statism of Taiwan is based on the ethics of Sunism, which emphasises the obligations of the state to secure the well-being of its citizens.

4 'Proceedings of the Sixth Rhenish Diet. Debates on the Law of Thefts of Wood'.

5 'Vindication of Correspondent writing from the Mosel'.

6 Note that these labels reflect only tendencies and that the point here is not to reproduce the myth of Sweden as an 'egalitarian' country as exploitation based on

class and ethnicity exists there as in other countries. But, these social relations are regulated differently according to the dominant societal regime in Sweden.

7 Expenditure on education included.

8 *Rettur* resembled a Danish journal *Ret*, the journal of the Danish Georgeist movement (G. Jonsson 1991, 107).

9 In 1910 the population of Iceland was 85183 (Statistics Iceland 1998).

References

Aglietta, M. (1979), *A Theory of Capitalist Regulation*; *The US Experience*, New Left Books, London.

Asgeirsson, O. (1988), *Idnbylting hugarfarsins*, Sogufelagid, Reykjavik.

Atvinnumálaráduneytid (1931), *Skýrslur um nokkrar framkvamaemdir rikisins 1927-1930*, Atvinnumálaráduneytid, Reykjavik.

Bourdieu, P. (1977), *Outline of a Theory of Practice*, Cambridge University Press, Cambridge.

Central Bank of Iceland (1999), *Hagtölur manadarins.*

Chan, S. and Clark, C. (1992), *Flexibility, Foresight and Fortuna in Taiwan's Development; Navigating between Scylla and Charybdis*, Routledge, London.

Dunford, M. (1990), 'Theories of Regulation', *Environment and Planning D: Society and Space*, vol. 8, pp. 297-321.

Esping-Andersen, G. (1990), *The Three Worlds of Welfare Capitalism*, Polity, Cambridge.

Fridriksson, G. (1991), *Med sverdid i annarri hendi og plóginn i hinni; Saga Jonasar Jonssonar fra Hriflu*, Idunn, Reykjavik.

Fridriksson, G. (1992), *Dómsmálaráðherrann; Saga Jónasar Jónssonar frá Hriflu*, Idunn, Reykjavik.

Giddens, A. (1984), *The Constitution of Society*, Polity, Cambridge.

Gramsci, A. (1978), *Selections from Prison Notebooks*, Lawrence and Wishart, London.

Granberg, L. (1998), 'Constructing the Barents Region: The Play and the Audience', in Granberg, L. (ed), *The Snowbelt*, Kikimora Publications, Helsinki.

Gudmundsson, G. (1943), *Samband islenzkra samvinnufelaga 1902-1942* (Samband islenzkra samvinnufelaga).

Guerin, D. (1970), *Anarchism: From Theory to Practice*, Monthly Review Press, New York.

Gunnarson, G. (1973), *De stora utopisterna 1; Från Platon till levellerna*, Tidens förlag, Stockholm.

Habermas, J. (1987), *The Theory of Communicative Action*, Vol. 2, Polity, Cambridge.

Hettne, B. et.al. (1998), *Den globala nationalismen*, SNS Förlag.

Hirsch, J. and Roth, R. (1986), *Das Neue Gesicht des Kapitalismus; Vom Fordismus zum Post-Fordismus*, VSA-Verlag, Hamburg.

Hoskuldsson, S.S. (1993), *Benedikt a Audnum: islenskur endurreisnarmadur*, Mal og Menning, Reykjavik.

Johannesson, Th. (1948), *Althingi og atvinnumalin; Landbunadur og utvegsmal – hofudthaettir*, Althingissogunefnd, Reykjavik.

Jonsson, G. (1991), *The State and the Icelandic Economy, 1970-1930*, Dr. Phil.-thesis, London School of Economics and Political Science.

Jonsson, I. (1989), 'Hegemonic Politics and Capitalist Restructuring', in *Thjódmál, Yearbook of Political Economy and Social Sciences 1989*, National Institute of Social and Economic Research, Reykjavik.

Jonsson, I. (1991), *Hegemonic Politics and Accumulation Strategies in Iceland 1944-1990; Long Waves in the World Economy, Regimes of Accumulation and Uneven Development. Small States, Microstates and Problems of World Market Adjustment*, D.Phil.-dissertation, University of Sussex.

Jonsson, I. (1995), *West Nordic Countries in Crisis: Neo-Structuralism, Collective Entrepreneurship and Microsocieties Facing Global Systems of Innovation*, New Social Science Monographs, Copenhagen Business School, Copenhagen.

Jonsson, I. (1999), 'Development, learning-processes and institutionalised racism', in *Ethnic and Racial Studies*, Vol. 22. January 1999, Routledge, London.

Jonsson, S. (1950), *Althingi og heilbrigdismalin*, Althingissogunefnd, Reykjavik.

Jonsson, S. (1984), *Sjavarutvegur Islendinga a tuttugustu old*, Hid islenska bokmenntafelag, Reykjavik.

Juliusson, A.D. et al. (1992), *Islenskur soguatlas, 2. bindi. Fra 18. old til fullveldis*, Idunn, Reykjavik.

Kanter, R.M. (1979), 'Communes in cities', in Case, J. and Taylor, R.C.R., *Co-ops, Communes & Collectivities. Experiments in Social Change in the 1960s and 1970s*, Pantheon Books, New York.

Keating, M. (1995), 'Spain, peripheral nationalism and state response' in McGarry, J. and O'Leary, B., *The Politics of Ethnic Conflict Regulation*, Routledge, London.

Kuhn, T. S. (1970), *The Structure of Scientific Revolutions*, The University of Chicago Press, Chicago.

Lash, S. and Urry, J. (1994), *Economies of Signs and Space*, SAGE, London.

Lukács, G. (1971), *History and Class Consciousness*, Merlin Press, London.

Marx, K. (1842-3), 'Proceedings of the Sixth Rhenish Diet. Debates on the Law of Thefts of Wood and Vindication of Correspondent writing from the Mosel', *Rheinische Zeitung* 1842-3.

Mosesdottir, L. (1995), 'The State and the Egalitarian, Ecclesiastic and Liberal Regimes of Gender Relation', *British Journal of Sociology*, December 1995.

Mosesdottir, L. (1998), 'Breaking the Boundaries. Women's Encounter with the State in Sweden, Germany and the United States', in Christiansen, J. et al., *Working Europe; Reshaping European Employment Systems*, Ashgate, Aldershot.

Nordal, J. and Kristinsson, V. (eds, 1987), *Iceland 1986*, Handbook Published by the Central Bank of Iceland, Reykjavik.

Putnam, R.D., Leonardi, R. and Nanetti, R.Y. (1994), *Making Democracy Work: Civic Traditions in Modern Italy*, Princeton University Press, Princeton.

Sartre, J.P. (1972), *Being and Nothingness*, Methuen & Co Ltd.

Sayer, A. and Walker, R.(1994), *The New Social Economy*, Blackwell, London.

Schmitter, P.C. and Lehmbruch, J. (1979), *Trends Towards Corporatist Interest-Mediation* SAGE, Beverly Hills, California.

Sigurdsson, J. (1996), *Bifrastarævintýrið og Jónasarskólinn: skerfur og saga Samvinnuskólans og Samvinnuháskólans frá upphafi til 1995*, Samvinnuháskólinn á Bifröst.

Sigurdsson, S. (1937), *Búnadarfélag Íslands, aldarminning*, Bundarfelag Islands, Reykjavik.

Statistical Bureau of Iceland (1974), *Tölfraedihandbók 1974*, Reykjavik.

Statistical Bureau of Iceland (1984), *Tölfraedihandbók 1984*, Reykjavik.

Statistical Bureau of Iceland (1988), *Kosningaskyrslur 1874-1946 and Kosningaskyrslur 1949-1987*, Reykjavik.

Statistical Bureau of Iceland (1997), *Hagskinna*, Reykjavik.

Statistical Bureau of Iceland (1998), *Landshagir*, Reykjavik.

Statistics Iceland (1998), *Landshagir*, Reykjavik.

Storper, M. and Walker, R. (1989), *The Capitalist Imperative; Territory, Technology, and Industrial Growth*, Basil Blackwell, Oxford.
Syslubokasafn Thingeyinga (1910), *Bokaskra Syslubokasafns Thingeyinga i Husavik*, Syslubokasafn Thingeyinga, Husavik.
Thjodhagsstofnun (1992), *Thjodhagsreikningar 1901-1945*, Thjodhagsstofnun, Reykjavik.
Thorleifsson, H. (1977), *Fra einveldi til lydveldis*, Almenna bokafelagid, Reykjavik.
Weber, M. (1958), *The Protestant Ethic and the Spirit of Capitalism*, Scribner's, New York.

PART IV:

THE FUTURE OF PEASANTS IN EUROPE'S GREEN RING

15 Afterword

LEO GRANBERG, IMRE KOVÁCH AND HILARY TOVEY

Begg and Meurs, in their chapter on agrarian Bulgaria, discuss how our interpretations of the present may be shaped by the different ways in which we approach the past. Historical accounts may emphasise discontinuities with the present, or may emphasise continuities, or may try in some way to address both. The various chapters of this book offer us different judgements on the current situation of the rural inhabitants of 'green ring' countries: most see them as undergoing forms of de-peasantisation but some note possibilities for re-peasantisation. These differences probably reflect not only real differences in social patterning and process between countries, but also differences in the historical perspectives of the authors.

As editors, however, our approach in this book has been primarily to emphasise continuities. In our framework, peasantisation took place before de-peasantisation, producing agricultural structures which later became the objects of erosion, adaptation, changes. Peasantisation is not a universal process. It was in some historical circumstances interrupted, in others prevented, and in still others delayed. In the countries in what we are here calling the green ring, which form a rim around the early industrialised core of Europe, late and (often) institutionalised peasantisation meant that rural questions have had great importance in political life, and rurality has had a strong impact on mentality and cultural life in such countries, even in the cities.

Thus we have tried to avoid ahistorical uses of the term 'peasantry', such as its common use simply to describe people in farming communities, however their structural and relational circumstances may vary. We have also not felt it necessary to restrict our authors to one single definition of peasantry. Our interest is in peasantry as a social construction, including self-constructions – traces of constructions of the peasantry can be followed back to 19th century debates between romantic nationalists and other cultural groups, for example. The meanings given to a peasantry are interconnected with societal and political processes during the period of the making of modern nation-states, when, in some green ring countries at least, peasants themselves became activated in political life. In many cases, the ethnic roots of new nation-states were sought in their past agrarian culture.

Peasantisation

Peasantisation refers to processes of construction and institutionalisation of a peasantry, both as a concept and as a class or social stratum for itself. An ideal-typical account – which may be more typical of the countries in the north of the green ring than of the rest – would roughly divide these processes into the following: early structural arrangements by the state; a developing market orientation and social movements by the peasantry; and the social and political institutionalisation of these movements. The first phase started with legislative acts, when some basic obstacles to free farming were identified (serfdom, lack of ownership rights, village order, land tenure arrangements). State intervention opened up a path for the individuation of peasants, often leading first to subsistance farming before re-marketisation and commoditisation processes reached the small peasantry. The development of a market orientation was facilitated by social and cultural changes. This is clearly to be seen in the intertwined cultural and economic processes underpinning the turn from crop farming to milk production in Northern Europe from the late 19th century on, which was catalysed by a co-operative movement among small and middle-sized peasantry. The final phase of peasantisation was its institutionalisation in economic, political and cultural life. The depth of institutionalisation, and the forms it took in different cases, are critical to understanding the differences between different countries in the green ring. Here we are not only referring to differences in the state systems and societies of these countries, we are suggesting that differences in the peasantisation process are a key to the different development paths chosen by their states. This differentiation, in turn, is the starting point for de-peasantisation processes.

Serfdom disappeared from Eastern Europe, through several reforms, generally much later than from Central and Western Europe. A strong landed aristocracy also dominated Denmark in the Middle Ages, although later structural reforms moved the country onto a similar path to other Nordic countries. These reforms released peasants from the village order as well as from much of the economic dominance of the landed aristocracy or other large estates. Particularly in the Russian Empire, however (where serfdom was abolished in 1861), this phase was not completed before the socialist regime interrupted the peasantisation projects, and Stalinist collectivisation led to the well-known catastrophe for market-oriented peasants. In Mediterranean countries too, peasantisation was one competing project among others. In Greece peasant revolts caused an agrarian reform, which established a rural structure dominated by small peasantry; in southern Italy the partial agrarian reform after the Second World War did not succeed in completing a thorough-going structural change. In Spain and Portugal, reforms were totally hindered by the political dictatorships, which froze the agrarian structure in favour of large-scale, mostly non-capitalist farming units.

As far as pre-socialist Central and Eastern Europe is concerned, it is apparent that a wide variety of structural changes took place: one can find both a Nordic-like development of an increasingly market-oriented free peasantry along with their interest organisations, and large feudal estates preserving remnants of serfdom, as well as all the variations in between. In their first phases, new political regimes after the World War II in some countries reflected an attempt to extend the peasantisation project. However, land redistribution and other reforms, followed by intervention, collectivisation, and then permanent control by the emerging socialist state, could not facilitate successful peasantisation. Only in Poland and Yugoslavia did the small peasantry manage to safeguard themselves against collectivisation projects, and survive as a social class. Their opportunities to articulate their own interests were very limited, however, under the political or ideological dominance of the communists.

It is no surprise, given this experiment, that the concept 'peasant' has varying meanings in different countries and regions. Peasant may refer to a wealthy and trustworthy landowner-citizen (in Scandinavia) or to a pure rural inhabitant with a very low status (in Poland). In Portugal, ordinary speech treats peasant (*camponês*) as a synonym for a traditional rural lifestyle and it is nowadays banished in favour of another term *lavrador*, referring to a farmer or agricultural producer. In fact a similar process has taken place in many countries, but sometimes (e.g. in Finland) the older term has made a partial comeback with positive connotations – reflecting support for the cultural aspects of re-peasantisation. In Central Europe, on the other hand, re-peasantisation has strong negative connotations.

In Europe, institutionalisation of the peasant 'project' seems to have developed furthest in Iceland (see Jonsson). Here the peasantry not only established their own political party, economic interest organisations, and a strong co-operative movement, but they even succeeded in blocking urban development focussed on the capital city, Reykjavik. The difference between Finland or Denmark and Ireland lies in fact in the degree of economic and political institutionalisation. In the Nordic countries, peasant-owned co-operatives managed to build their own economic empire not only in dairy products but also in capital markets (co-operative banks) and in the main areas of national industry (in Finland this was forestry - they met their first limit when their attempts to move into the metal industry were hindered by the Finnish capitalist elite). This level of institutionalisation was not achieved in Ireland nor in any other Western European countries outside the North. In all the Nordic countries the peasantry were also directly represented in Parliament by their own political party. In Ireland the political parties had a broader background and orientation, and over the period of transformation towards an urban capitalist society, rural interests were left to the domain of an increasingly minority fraction inside the existing parties.

All this suggests the decisive impact which the state has had on the peasantry. Peasantries have developed or disappeared according to the direction of state policy. In some cases the state reformed legislation long before the peasantry began to demand such reforms (Sweden), in other cases the state determinedly resisted any reforms, even if they would have been very rational from the point of view of a capitalist market economy (Portugal and Spain) and/or the general welfare of the society (socialist countries). A 'free' peasantry is not, as it is often presented to be, an outcome just of developing capitalist markets; it is much more an outcome of an interplay between state, markets and social classes in society.

De-peasantisation and After

In a limited sense de-peasantisation deals with modernisation of farming. In a broader sense the concept also refers to the decreasing economic, social, cultural and political importance of farmers and food production workers in society. De-agrarianisation, or de-agriculturalisation, is an important part of the process; these concepts highlight the ways in which rural residence becomes dissociated from farm activity, which is also seen as connected with other social and economic changes such as the development of pluriactivity, changing gender and generation roles in farm families, and rural exodus (see Oliva and Camarero, Kasimis and Papadopoulos). In Western Europe, de-agrarianisation has been largely a consequence of agricultural modernisation. Failures in such social rearrangements led – under conditions of technical revolution in agriculture – to the heavy out-migration of women and young people from rural areas. There are also, however, signs of a new trend since the 1980s, in which the migratory balance has turned to urban-to-rural. For example, in Spain retired workers, middle class people and neo-rurals have moved to the countryside. Their lifestyles are very diverse, and generally they do not migrate in search of a job but for lifestyle reasons, "rather in search of a paradise out of the congested cities" as Oliva and Camarero put it. In some cases immigrants from the South or from the East of Europe are searching for a temporary job in regions of intensified agriculture, but they do not perceive themselves as future peasants in these regions since for them the job is only a temporary strategy, and it is doubtful – if not impossible – that many would be able to use their extra income to establish their own peasant farm at home in Eastern Europe.

In some regions a revitalisation of rural communities is evident. The development of a specific 'quality economy' is one sign of such a trend, for example in Portugal's Northern Coastal area or in the urbanised countryside of Tuscany in Italy. In these regions agricultural and tourist activities have been organised to complement each other, forming what can be called a 'quality' or a 'culture economy'. In these cases, the presence of large numbers of non-farming residents does not imply the marginalisation of agriculture. Regions striving towards this type of economy can be found throughout the rural EU. Another model for possible re-peasantisation is where peasants are left on their own, after extensive out-migration, living in a 'local markets economy'. In these regions, the weakness of the institutional system can contribute to the reproduction of a trust system that is closely tied to personal and kinship relationships. This model is typical in Southern Italy (see Fonte). In these two types of regions, spatial organisation is not based on a polarity of urban/industrial versus rural/agricultural, but on an urbanised countryside and on rural towns. Whether regional revitalisation based on a quality economy indicates a process of re-peasantisation remains to be further studied; as Fonte notes, much may depend on whether economic relations operate to encourage the reproduction of local knowledges and production practices, or to dissipate them. A third variant is Polish re-peasantisation, which took place during the socialist period on both the farm and the community levels. Re-peasantisation was strengthened on one hand by peasant resistance against the state, and on the other because of the underdeveloped market conditions in the country. Both Gorlach and Starosta writing about Poland, and Bruckmeier and Kopytina in reference to Russia, draw attention to the way in which attachment to the rural community, fostered under these conditions, can constitute a valuable 'social capital' for rural development. In these ways, rural community processes as well as more strictly agricultural processes enter into our assessments of the possibilities for re-peasantisation. Yet even in this most peasant country of the post-socialist world, social processes seem to be leading to a complete de-peasantisation of Polish society in the near future.

In a sense, collective farming was the vehicle which brought modernisation into the countryside in the socialist countries. The subsequent de-coupling of collectives and state-owned farms shows that dependence on a specific development path had an impact on the structural solutions found in post-socialist agriculture. Existing material entities, class structure (not least the role of the middle class and new elites), and the ideological heritage from socialism all contribute to steering the development path after socialism (Alanen).

In post-socialist countries, the thesis of re-peasantisation has been presented in several different connections. Specifically, some hundred thousands of new family farms have in fact come into being after socialism. Our chapters do not, however, offer much evidence to verify the thesis. It is mostly in Russia that some elements can be identified which might lead to a later re-peasantisation. Individual market-oriented family 'farmers' have not made any wide breakthrough in Russia. However the modes of life of subsistence farmers show some peasant features, but also some features which originated in socialism. If anything like a peasantisation process is taking place, it is an unconscious process rather than a conscious 'project', because people's survival strategies have been defensive and reactive (Varis). On the other hand, Russian agrarian reform lost momentum in the 1990s, which only means that the history of inconsequent and incomplete agrarian reforms in Russia continues.

The contemporary meaning of rurality or the countryside in post-socialist countries is quite ambiguous. Its value as cultural heritage has been underlined in official speeches, yet this seems to coincide with a transformation of its image from positive to negative in popular life. In the post-soviet Baltic States, the loss of a moral foundation for society is experienced as a real problem in the countryside, and similar remarks have been made about other post-socialist countries (see Alanen and Majerova). The presence of such anomie might be claimed to be the complete opposite to peasant societies which are supposed to be founded on a deep-rooted normative order. On the other hand, the concept of 'peasantisation' reminds us that peasant societies have always been constructed, in specific historical circumstances; the stable normative order which we now attribute to them in retrospect may not have been very evident to those living through their early decades.

Overall, chapters in this book generally suggest that de-peasantisation has been the dominant tendency in the European green ring during the last 30-50 years, depending on the country in question. During that time, the economic role of the agricultural population has diminished radically. The peasantry are still capable of political contributions, but only from minority positions without even a dream of reaching political dominance in any country. Some social and cultural features, typical of peasant community, still exist in restricted regions. But in the same period, globalisation has transformed local cultures in such a way that a specific peasant culture is hard to distinguish from commercial (and tourist) activities in the countryside. Peasant culture has been adopted as an element in neo-rural programmes and projects for cultural heritage, while at the same time this culture has experienced a huge loss of energy to reproduce and develop itself. Re-peasantisation seems to be possible only in combination with other efforts, when a peasant project is connected to some other more 'globalist' project, such as preserving the cultural heritage of humanity, or the sustainability of biodiversity.

The Green Ring Revisited

How useful has the concept of a 'green ring' turned out to be, as a framework for interpreting the experiences of change and transformation of fourteen diverse countries in Europe? It is worth noting, first, that in developing this concept we are to some extent questioning the usefulness of other frameworks which have been offered in recent discussions of European rurality. One influential argument has been the need to distinguish a north-south axis in rural Europe and to recognise that rural development models pursued for Northern Europe are not necessarily appropriate or applicable to Mediterranean countries (see Fonte). For others, a North Europe-South Europe (or a West Europe-East Europe) distinction is still insufficient, and we need to pay close attention to the different national traditions in thinking about 'the rural' which have been historically central to the culture of different nation-states in Europe, and to the different models or understandings of how the rural should be managed and developed to which these have given rise.

Our concept of a European green ring questions both the suggestion that identifying national differences in the understanding of the rural is more important than trying to find commonalities, and the argument that there is a common Northern European rural paradigm which is inappropriately imposed on the countries of the European South. It is part of our intention to open up and deconstruct the idea of a Northern European paradigm, by examining the agrarian histories and experiences of different Northern European countries. In some cases (for example Ireland) we find that these reveal as much or more commonality with Mediterranean rurality (in terms for example of land reforms, farm structures, or the dispersion of industry into the countryside) as with that of the more 'core' countries of Northern Europe. What the concept of a green ring suggests is that the major division within Europe in relation to rurality is a division between those countries in which urban industrialisation developed early and produced a fairly rapid reorganisation of their class structures and patterns of social and political relationships around industrial development; and those countries, largely found on the geographic periphery, in which agrarian interests, groups, culture and concerns retained considerable societal significance until late into the 20th century. We have tried to suggest in this book that taking this common perspective on the European rural rim can open up possibilities for enhanced understandings of historical patterns of rural change, in particular of the different ways in which rural and agrarian social groups have been implicated in state-led projects for societal modernisation.

It is nevertheless clear that national differences in rural traditions, and in particular, differences in historical political experiments, remain highly significant and give rise to a diverse set of agrarian patterns and processes even across the green ring countries themselves. This book also supports the argument that, in terms of rural development, the flexibility and scope to identify endogenous models which can take account of the specific social, cultural, economic and indeed ecological characteristics of particular societies should be as far as possible devolved down to national and even to local levels. To that extent, we share the concern of Mediterranean rural scholars who have emphasised the problems which follow from trying to squeeze a South European rurality into the conceptual confines of Northern European assumptions. As Central and Eastern European countries move closer to membership of the EU and start to anticipate in their policy regimes the norms and imperatives of EU agricultural and rural policy, this issue becomes increasingly pressing for them too.